I0044405

Aerospace Technology: Systems, Models and Designs

Aerospace Technology: Systems, Models and Designs

Editor: Stephen Baggins

New York

Published by NY Research Press
118-35 Queens Blvd., Suite 400,
Forest Hills, NY 11375, USA
www.nyresearchpress.com

Aerospace Technology: Systems, Models and Designs
Edited by Stephen Baggins

International Standard Book Number: 978-1-63238-663-2 (Hardback)

Cataloging-in-Publication Data

Aerospace technology : systems, models and designs / edited by Stephen Baggins.
p. cm.
Includes bibliographical references and index.
ISBN 978-1-63238-663-2
1. Aeronautics. 2. Aerospace engineering. I. Baggins, Stephen.
TL545 .A47 2019
629.13--dc23

Contents

Preface

Aerospace refers to the human endeavor to fly in space or the earth's atmosphere using science and engineering. Aerospace technology covers the design and operation of aircraft and spacecrafts. Aerospace can be further divided into the two main categories- aeronautics and astronautics. The study of aerospace technology falls under the domain of aerospace engineering. Functional safety is a crucial aspect of aerospace. Some of the crucial areas of study in this field are radar cross-section, fluid mechanics, astrodynamics, propulsion, aeroacoustics, etc. The ever-growing need of advanced technology is the reason that has fueled the research in the field of aerospace technology in recent times. This book is a compilation of topics that discuss the most vital concepts and emerging trends in this field. It attempts to understand the multiple branches that fall under this discipline and how such concepts have practical applications. It will serve as a reference to a broad spectrum of readers.

Significant researches are present in this book. Intensive efforts have been employed by authors to make this book an outstanding discourse. This book contains the enlightening chapters which have been written on the basis of significant researches done by the experts.

Finally, I would also like to thank all the members involved in this book for being a team and meeting all the deadlines for the submission of their respective works. I would also like to thank my friends and family for being supportive in my efforts.

Editor

1

Knowledge Management Patterns Model for a Flight Test Environment

Roberto da Cunha Follador[1,2], Luís Gonzaga Trabasso[2]

ABSTRACT: This paper investigates how Knowledge Management patterns in a Brazilian Air Force flight test environment can be simulated using a System Dynamics approach. The research has been conducted initially by a literature review on the main Knowledge Management and System Dynamics theories. Data for this research has been collected in a previous study consisted of documental research regarding the flight test environment Knowledge Management and a questionnaire-based survey which identified both a low Knowledge Management maturity level and the flight test core competence as the capability of performing flight test campaigns. The issued problem was the tradeoff between actions focused on performing flight test campaigns *versus* Knowledge Management to transfer the core competence inside organization in order to keep it in a high level. A system dynamics quantitative model has been developed as a result of this research. Fluxes and stokes were identified within the model and the relation between them emerged by identifying systemic feedback loops that may compromise the Knowledge Management and the core competence transferring. These features enable a holistic visualization and better understanding of the problem as well as the possibilities of identifying ways of improvement.

KEYWORDS: Systems dynamics, Core competence, Knowledge transferring, Quantitative model.

INTRODUCTION

The Flight Test (FT) activity is based on knowledge attained by experience or research, and its core competence, for the Brazilian Air Force, may be expressed as "the capability of performing flight test campaigns" (Follador and Trabasso 2015) and should be kept and propagated inside the organization.

The mission of the Flight Test Organization (FTO) reads: "It is a Brazilian Air Force Organization specialized in the field of Science and Technology and its mission is to deliver specialized technological services regarding flight test, aircraft instrumentation and data telemetry to support research, development and certification of aeronautical products and to train specialized personal in flight test" (Brasil 2011, p. 7).

The FTO executes great part of the Brazilian Air Force (BAF) activities regarding flight test and offers training courses for BAF test pilots, test engineers, instrumentation engineers and flight test technical personnel. Knowledge inside the FTO environment, mainly that related to its core competence, must be preserved and the activities regarding Knowledge Management (KM) must be developed such as genesis, maturation, use, preservation and dissemination.

Follador and Trabasso (2015) presented a documental research regarding the flight test environment and also submitted a questionnaire for KM maturity level assessment, identifying a low KM maturity level inside the FTO in BAF and the flight test core competence as the capability of performing flight test campaigns. These findings, in addition to a high demanding activity in performing Flight Test Campaigns, indicated that may exist a commitment between actions focused on performing flight test campaigns *versus* KM activities regarding the transference

1.Departamento de Ciência e Tecnologia Aeroespacial – Instituto de Estudos Avançados – São José dos Campos/SP – Brazil. **2.**Departamento de Ciência e Tecnologia Aeroespacial – Instituto Tecnológico de Aeronáutica – Divisão de Engenharia Mecânica – São José dos Campos/SP – Brazil.

Author for correspondence: Roberto da Cunha Follador | Departamento de Ciência e Tecnologia Aeroespacial – Instituto de Estudos Avançados | Praça Marechal Eduardo Gomes, 50 – Vila das Acácias | CEP: 12.228-900 – São José dos Campos /SP – Brazil | Email: follador@ieav.cta.br

of the core competence inside the organization preventing an ideal sharing of important knowledge.

The objective of this paper is to simulate and analyze the KM patterns, in a BAF FTO, in order to apprehend the relationship between the activities of performing flight test campaigns and those of KM within the organization. This problem has been addressed using a system dynamics approach, where a quantitative model is proposed to identify the main stakeholders of the FTO KM system, pointing out fluxes and stocks as well as the relationships between them.

This article is organized as follows: the second section presents the literature review and describes the contribution of this paper. The third section describes the methodology used to construct the model. The following section presents the results and discussions about FT KM model, regarding the commitment between performing flight test campaigns and KM activities necessary to maintain the core competence in the organization. The final section summarizes the contribution of this research and discusses future studies.

LITERATURE REVIEW

Sterman (2002) suggested that, in a world of accelerating complexity and change, thoughtful leaders increasingly recognize that the management tools that had been used have not only failed to solve the persistent problems encountered, but may, in fact, be causing them. Examples of organizational policy resistance are: well-intentioned efforts to solve pressing problems create unanticipated side effects. "Learning about complex systems when you also live in them is difficult. We are all passengers on an aircraft, we must not only fly but redesign in flight" (Sterman 2002, p. 4).

To understand the system a literature review was performed on the topics of KM, System Dynamics (SD) and FT.

KNOWLEDGE MANAGEMENT

KM has been researched in the last decades (Davenport and Prusak 1998; Nonaka *et al.* 2001; Nonaka and Takeuchi 2004; Grundstein 2008; Lloria 2008; Albers 2009). An organization that does not understand the importance of knowledge nor manages or improves it, based on the environment demands in which it is inserted, may be faded to a condition of capabilities loss or stagnation.

Nonaka and Takeuchi (2004) explore the concepts of explicit and tacit knowledge and propose 4 modes of knowledge conversion in a process called "The Spiral of Knowledge", namely: (1) Socialization; (2) Externalization; (3) Internalization; and (4) Combination.

The organizational environment must provide conditions and means for allowing knowledge to flow. Bock (1999) says that it is important to consider KM in terms of 4 integrated dimensions: content, culture, processes, and infrastructure. Content is related to the following question: what kind of knowledge is important to my organization? Prahalad and Hamel (1990) state that some knowledge, known as core competence, is the source for organizations' technological success.

"Culture sets both the limits (constraints) and the direction of movement of behavior within the organization: culture dictates the acceptance of all organization change" (McNabb 2007, p. 113). Therefore, the organization's cultural dimension is a major factor to KM success, so people and their beliefs may be responsible for the cycle of knowledge acceptance inside an organization.

Albers (2009) proposed a practical approach to implement KM based on 5 basics steps:

1. Select KM team.
2. Establish KM strategy and business case.
3. Perform knowledge assessment and audit.
4. Perform information technology (IT) assessment.
5. Develop project plan and measurement systems.

Fonseca (2011), in a study about KM in an aeronautical company, pointed out that the KM system in that environment had low effectiveness. His findings have been confirmed by institutional characteristics like lack of cultural organization and low adherence to KM initiatives. These characteristics may be also present within the FT organization, where the high workload of FT workers may contribute to reinforce the tendency of low KM effectiveness.

FLIGHT TEST KNOWLEDGE MANAGEMENT

Kimberlin (2003) states that the FT activity is dependent on efficient KM. The BAF Basic Doctrine Manual (DCA 1-1) defines the FT activity as "the action that consists in use air force resources to identify the flying qualities and the performance of aircrafts and systems" (Brasil 2012, p. 57). This statement indicates the dynamic characteristic of the FT activities, because FT deals with new systems that accompany the airspace technology continuous evolution. The correct evaluation, via flight tests, of the new systems guarantees their proper work and ensures flight safety.

Dealing with the state-of-the-art technologies available, an FT organization must have a KM concern in order to keep its operational capabilities and to be able to process the market technology push.

There are research examples relating FT and KM. Gray (2005) proposes a mathematical model to investigate and understand a FT technique called boundary-avoidance tracking, identifying the causes of dangerous flight conditions known as Pilot Induced Oscillation (PIO). Follador *et al.* (2009) analyse FT methods to evaluate aeroelastic structural vibrations using Operational Modal Analysis (OMA).

An example of KM applied to FT techniques is the aeronautical certification manuals, such as the Advisory Circulars (AC), from the Federal Aviation Administration (FAA). They suggest a great number of techniques that are recognized as a guide to conduct FT activities. The AC 25-7 — Flight Test Guide for Certification of Transport Category Airplanes (United States 2012) — indicates its purpose as: "This AC provides updated guidance for the flight test evaluation of transport category airplanes. These guidelines provide an acceptable means of demonstrating compliance with the pertinent regulations [...]" (United States 2012, p. 1).

In the AC 25-7, it is worth noticing that FT knowledge is constantly evolving in line with new technology capabilities, in order to provide compliance with regulations. The evolution of the FT techniques demand constant training of Human Resources to provide them with the required ability and understanding to deal with the new techniques. Another source for FT knowledge is the reports produced after each FT campaign. It can be learned from the reports the issues regarding the whole campaign, namely, motivation, applied resources FT techniques, and final results.

Although there is a huge amount of written material regarding FT, great part of the knowledge attained during FT campaigns is related to tacit knowledge, acquired during on-the-job learning, and must be shared with the organization through the process of "externalization" (Nonaka and Takeuchi 2004). This kind of knowledge is part of the FTO core competence, identified as the capability of performing FT campaigns (Follador and Trabasso 2015).

Another important issue regarding FT reports is related to its main part, the 7-part paragraph: "There are key features likely to be required in a complete flight test report: conditions, results, analysis, role relation, conclusions, recommendations and standards compliance" (Gratton, 2010, p. 1).

In the researcher's judgment, the core part of the 7-part paragraph is the role relation or Mission Impact. This part deals with the impact of the tested system characteristics on the mission it is supposed to perform. The test team must joint the operational knowledge with the FT information obtained and predict if the mission is still possible to be accomplished or how it will be affected by the issues revealed by the FT. A lack on this operational knowledge may lead to a misinterpretation and a false FT result that may produce serious problems in the future system FT or even in the aircraft operational performance.

Follador and Trabasso (2015) also identified the following characteristics in the KM environment analyzed in their study:

- Lack of infrastructure dedicated to KM.
- Lack of KM specialist.
- High turnover of trained staff.
- Flying Test Course (CEV) and library are the main stocks of registered knowledge.
- CEV and library not linked.
- High workload of the FT teams, impairing them to conduct knowledge sharing.
- Frequent occurrence of rework.

These findings, especially the 6th characteristic, emphasize that the operational knowledge may be lost because of the FTO high workload on performing FT campaigns; consequently, there is little or no time to maintain the FT pilots updated with the evolution of operational missions.

SYSTEMS DYNAMICS

SD is a methodology largely used for analyzing systemic behavior in organizational or social systems, by means of representing causal relations between its elements and by analyzing its evolution in time (Forrester 1971, 1994, 1997; Sterman 2002; Villela 2005; Morecroft 2007; Figueiredo 2009; Ford 1999; Amaral 2012). The awareness for SD is growing because of its unique capacity for representing the real world. It deals with complexity, non-linearity and feedback structures, inherent to social and physical systems. It is a methodology that provides better understanding of complex systems evolution (Forrester 1994; Sterman 2002; Meadows and Wright 2008).

SD response is usually represented by particular behavior and pattern like exponential growth and oscillations and by its direct relation with feedback loops characterized as reinforced, balanced, delayed balanced and their combination (Ford 1999; Gonçalves 2009).

Sterman (2002) proposes the iterative approach for SD modeling. Results of any step can yield insights that lead to revisions of the previous step, as depicted in Fig. 1. The first step of the SD modeling process is to understand the problem (Ford 1999; Sterman 2002; Morecroft 2007).

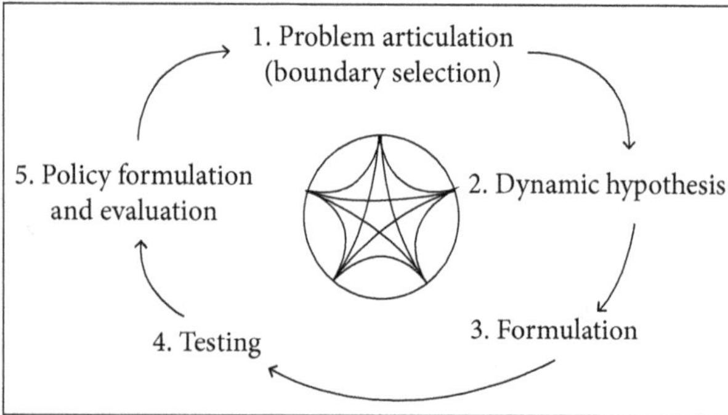

Figure 1. Iterative SD modeling process (Sterman 2002).

Ford (1999) proposes 8 steps for SD modeling, namely:

1. Get acquainted with the system.
2. Be specific about the dynamic problem.
3. Construct the stock and flow diagram.
4. Draw the Causal Loop Diagram (CLD).
5. Estimate the parameters values.
6. Run the model to get the reference mode.
7. Conduct sensitivity analysis.
8. Test the impact of policies.

Several authors studied the relation between qualitative and quantitative approaches to SD modeling (Ford 1999; Wolstenholme 1999; Coyle 2000). These 2 approaches are complementary as seen in the 8-step method proposed by Ford (1999): the first 4 can be classified as qualitative whereas the remained 4 steps belong to quantitative approach to SD modeling.

One important aspect of Step 5 — Estimate the parameters values — is to choose the correct time horizon. Sterman (2002) says that it should extend far enough in history to show how the problem emerges and describes its symptoms. A common mistake at this step is to estimate a time horizon too short that does not allow for correctly estimating the length of delays in the system. "[...] A good rule of thumb is to set the time horizon several times as long as the longest time delays in the system..." (Sterman 2002, p. 119).

The Step 8 — Test the impact of policies — may be applied both in qualitative and quantitative SD modeling (Coyle 1996). This may result in new policies to be applied to the model in order to achieve improvements in the system performance and to minimize the problem addressed (Morecroft 2007).

Meadows and Wright (2008) identify 12 leverage points, as places to intervene in a system with new policies, in order to change the structure of systems to produce more of what is needed and less of that which is undesirable. Examples of the leverage points are:

- Numbers: constants and parameters such as subsidies, taxes, and standards.
- Buffers: the sizes of stabilizing stocks relative to their flows.
- Delays: the lengths of time relative to the rates of system changes.

The literature presents several studies using SD as a tool for modeling subjects as: technology innovation (Swinerd and McNaught 2014), improvement of anemia control (McCarthy *et al.* 2014), adult obesity trends (Fallah-Fini *et al.* 2014), infection screening among young women (Teng *et al.* 2015), information sharing, psychological safety performance effects (Bendoly 2014), and diffusion of technological innovation (Swinerd and McNaught 2014).

Regarding specifically researches on KM, the following topics are found: KM performance (Chen and Fong 2015), KM process in airlines (Zaim *et al.* 2013), organizational IT investment strategy and market performance (Liao *et al.* 2015), analysis of technology districts' evolution in a knowledge-based perspective (Dangelico *et al.* 2010), and capacity planning (Špicar 2014).

Gonçalves (2011) analyzed the balance between the provision of relief and recovery and the capacity building in humanitarian operations. This is carried out by modeling the performance of 2 polar resources allocation strategies: one focusing in relief and recovery efforts and the other on capacity building. The author reports a counterintuitive behavior from relief/recovery and capacity tradeoff that is characterized as capability trap.

Repenning and Sterman (2002) define capability trap as the phenomenon that arises when people's efforts to achieve performance targets come at the expense of maintenance and learning, thereby eroding the health of the system. The SD model developed by Gonçalves (2011) is presented in Fig. 2, where a causal loop diagram depicting the feedback processes for an organization allocating resources between providing aid and building capacity can build on the capability trap phenomenon.

A similar condition reported by Gonçalves (2011) was identified in the KM system for the FT environment addressed in the present paper. Follador and Trabasso (2015) also report the high workload as an obstacle for knowledge sharing between flight test teams, which may be characterized as a capability trap.

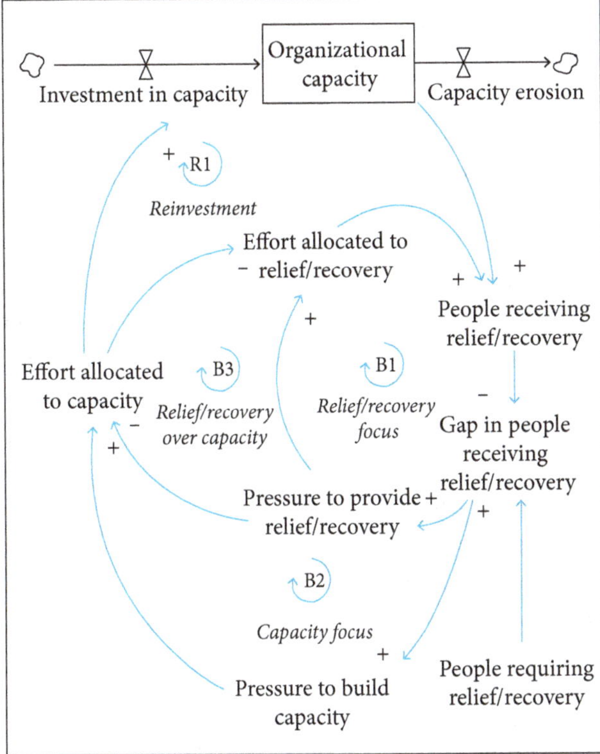

Figure 2. Counterintuitive behavior from relief/recovery and capacity tradeoff (Gonçalves 2011).

The searched literature showed the increased relevance of SD in the modeling of complex systems, but it did not address specifically the KM regarding the knowledge transferring problem in an FT environment. Thus the present study aims to identify a counterintuitive relation between the dedication of performing FT campaigns and the risk of losing the organization core competence using a SD modeling process. Then a course of actions is proposed as means to mitigate the risk of losing the FT core competence.

MODEL DESCRIPTION

The SD modeling process used is based on that proposed by Sterman (2002) and Morecroft (2007). For this development it was chosen the methodology proposed by Sterman (2002), following the 2nd path proposed by Morecroft (2007):

- Identification of the problem under study.
- Identification of the dynamic response.
- Conception of a map of the main sectors involved.
- Model the system checking for flows and stocks.

The acquaintance with the KM system was achieved based on the results presented by Follador and Trabasso (2015) and Fonseca (2011) about KM systems. The KM acquaintance in a flight test environment provided enough information to formulate the dynamic problem: there is a poor knowledge transferring inside FT knowledge management environment. This, in turn, implies rework and the loss of important FT knowledge achieved mostly during FT campaigns.

It was understood that a KM system may have a good initial knowledge transferring, because, to became an FT pilot or engineer, one must go through a capacitating course where the basic knowledge is provided. After this initial stage, the knowledge transferring opportunities tend to stabilize at a lower level along the time. This assumption of dynamic response is presented in Fig. 3. This curve corresponds to one of the fundamental modes of dynamic behavior, corresponding to a reinforced loop followed by a balanced loop plus another reinforced loop (Sterman 2002).

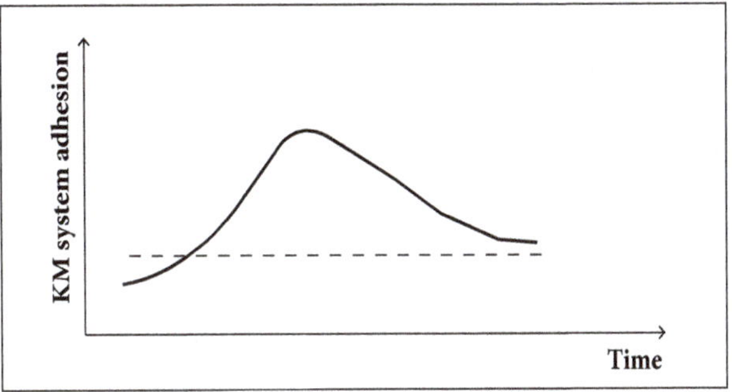

Figure 3. Dynamic response of knowledge transferring in a KM system.

The dynamic mode reflects the behavior of a crescent adhesion in the initial moments of a KM system responding to a reinforced positive loop, due to the value it may have inside the organization. Then, a temporal progression occurs and the adhesion behavior reaches the maximum value, influenced by a balanced loop. Finally, the dynamic mode directs the KM system behavior to a lower level, represented by the dashed line, caused by a reinforced negative loop, due to systemic issues in the FT organization that correspond to the problem addressed in this paper.

After the KM dynamic response identification, a map of the main sectors involved in the system was conceived based on the research on KM maturity level performed by Follador and Trabasso (2015). This map involves 3 main areas that affect

KM: culture/people, processes and technology/infrastructure, as it is presented in Fig. 4.

For example, the CLD for process sector was constructed based on the research performed by Folador and Trabasso (2015) and Fonseca (2011). The CLD of KM process sector is presented in Fig. 5.

Once the CLDs for all KM sectors are constructed, they are grouped to construct the stock and flow diagram for the KM system model, depicted in Fig. 6.

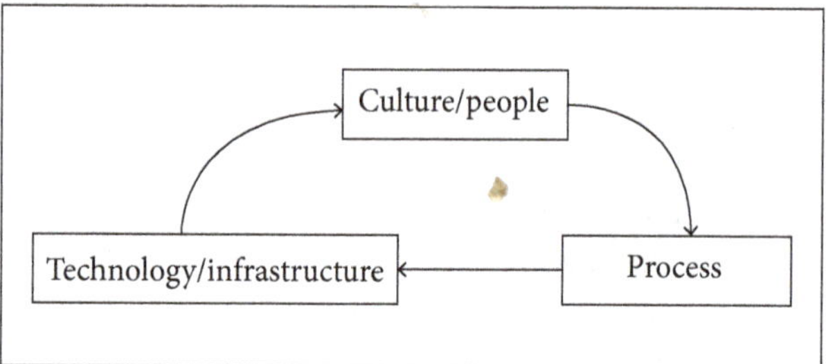

Figure 4. Map of the main sectors involved in the KM system.

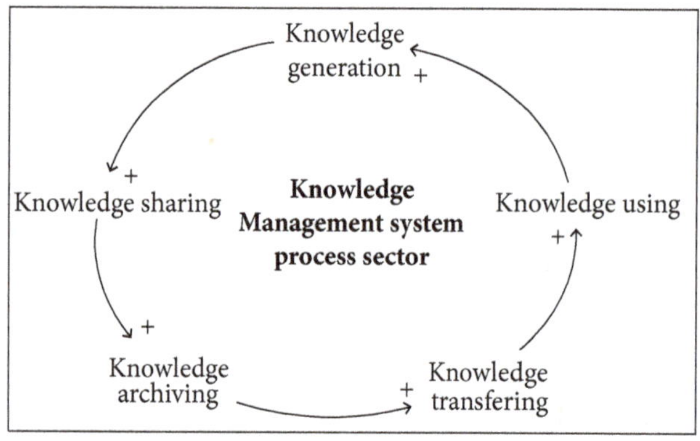

Figure 5. CLD of the KM process sector.

To construct the KM model it was used the software Vensim® version 6.3 provided by Ventana Systems in an academic version. This software is an integrated framework for conceptualizing, building, simulating, analyzing, optimizing, and deploying models of complex dynamic systems.

The model was constructed by the identification of "stocks" (or levels) as {*Organizational Knowledge*} and "flows" (or rates), like [*Knowledge use*] and [*Knowledge erase*]. Influence variables where also identified and added to the model, such as <*researching*>, <*contracting*>, <*capacitating*>, and <*on the job training*>.

Considering the interactions among flows, stocks, and variables, it was possible to identify reinforced and balanced feedback loops represented by the symbols such as (R) and (B), respectively. These loops provide the possibility of studying the causal relations in the KM system that are traduced into equations and into curves of behavior over time. Once the main stock-and-flow structures (or levels and rates of change) and feedback processes characterizing the system are captured, it is possible to translate them into a mathematical simulation model (Gonçalves 2009).

RESULTS AND DISCUSSION

In Fig. 6, the stock {*Organizational Knowledge*} participates in one loop involving [*Knowledge transferring*], [*Knowledge use*], and [*Knowledge generation*]. This loop deals with the knowledge generation effort of the organization. Several variables are relevant in this loop, such as <*researching*>, <*contracting*>,

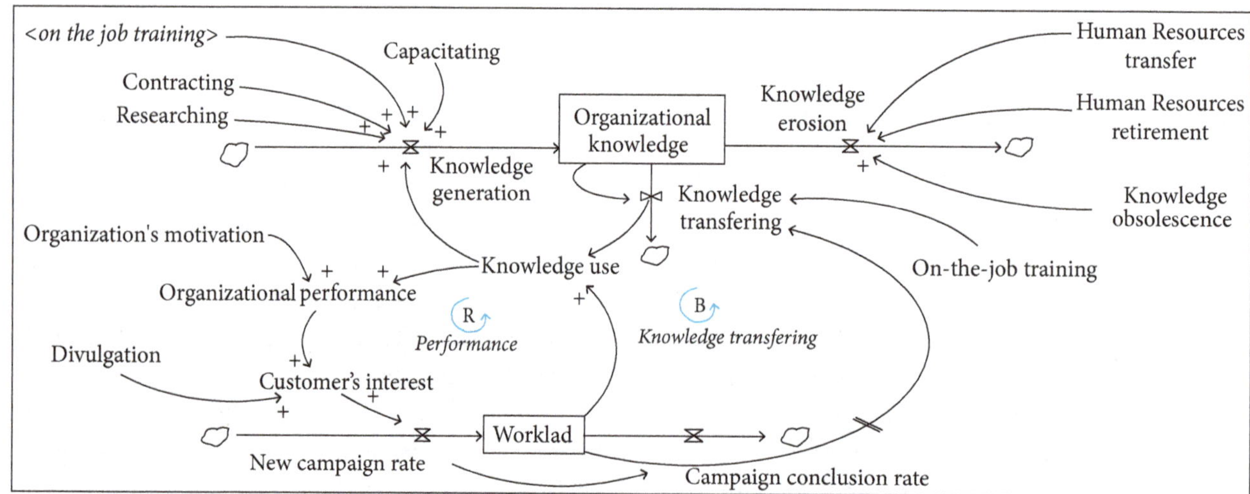

Figure 6. Proposed stock and flow diagram for the KM system model.

<*capacitating*>, and <*on the job training*>. In terms of FT, the variable <*on the job training*> bears singular importance, as it carries the experience exchanged during execution of FT campaigns or flight tests. Even the operational knowledge, as a basis to understand the mission impact, is transferred from pilot to pilot during flights or participates in operational campaigns.

Another important variable in the model is <*knowledge use*>. It participates in 3 feedback loops, namely, Knowledge generation (reinforced loop), Performance (reinforced loop), and Knowledge transferring (balanced loop).

These 3 loops interact among themselves and are influenced by a high Human Resources attrition rate, denoting 2 conflicting concerns to the FTO: (1) performing FT campaigns; and (2) transferring knowledge among the teams involved in FT campaigns execution. This conflict shows that, with a low number of FT teams in this organization and the stress to perform flight test campaigns continually required by the main customer (BAF), the KM activity and the organization knowledge may be jeopardized. This can be seen from Figs. 7 and 8, where the simulations response for <*organizational knowledge*> and <*knowledge transferring*> are presented.

The results represent a counterintuitive observation: the greater the effort to maintain the FT organization totally focused on performing FT campaigns, the greater the risk of losing its core competence and ability to continue fulfilling the customer's needs in the future. Despite the effort to generate and transfer knowledge by the FTO, the additional workload influences the system and it may be prevalent over the other activities. Furthermore, in a medium to long term, it may threaten the knowledge transferring, characterizing a capability trap that may prevent the organization from sharing and continuously learning its core competence.

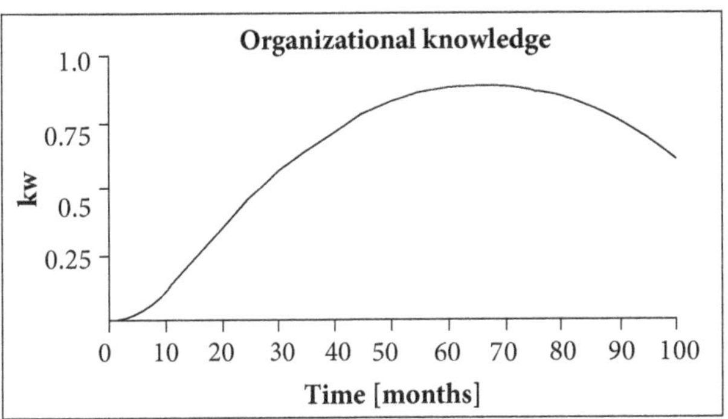

Figure 7. <*organizational knowledge*> response simulation.

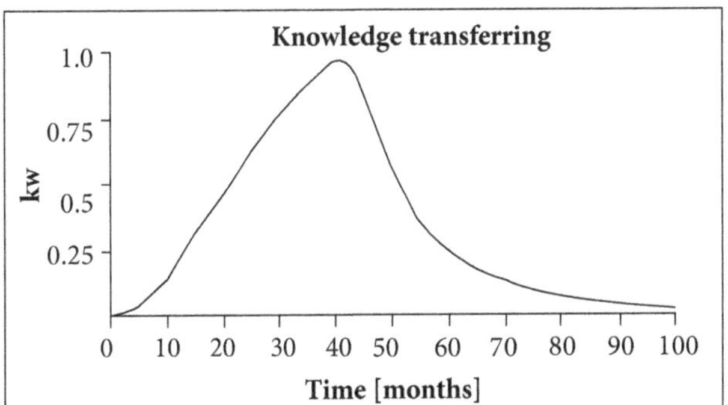

Figure 8. <*knowledge transferring*> response simulation.

It is important to notice the great similarity with the curve presented and the dynamic hypothesis for knowledge transferring inside the FTO. This finding is important because it brings awareness of the importance to look for leverage points in the model to introduce new policies for providing a more stable knowledge transferring and a preservation of the organizational knowledge while attending the requirements of FT campaigns.

CONCLUSION

This paper has shown that it is feasible to construct the SD model to represent a KM system inside an FT environment, providing means for observing the existing feedback loops in the KM system and identifying a simulated response for knowledge transferring very similar to the dynamic hypothesis formulated.

The SD model has provided means to understand the main variables in the FT environment regarding KM and to observe how they influence the KM system. Also, it was possible to understand a counterintuitive aspect, regarding the necessity of performing FT campaigns and to preserve FT knowledge, where efforts to maintain the FT organization focused on performing FT campaigns may offer a risk of losing its core competence and the ability to continue fulfilling the customer's needs in the future. This pattern characterizes the existence of a capability trap within the FT organization.

Further research will be addressed to improve the model where new policies will be proposed in order to change the structure of systems, looking for a way to provide a more stable knowledge transferring and a preservation of the organizational knowledge. Parameters of the SD model will be adjusted in order to run a sensitivity analysis and to test the impact of policies to change the system into a more appropriate behavior of knowledge transferring. The aim of this procedure is to ensure that the FT organization maintains its core competence without affecting its ultimate goals of providing FT campaigns according to the customer's needs.

REFERENCES

Albers JA (2009) A practical approach to implementing knowledge management. Journal of Knowledge Management Practice 10(1):1-14.

Amaral JAA (2012) Desvendando sistemas; [accessed 2015 Mar 6]. http://issuu.com/profjoaoarantes/docs/desvendandosistemasebook

Bendoly E (2014) System dynamics understanding in projects: information sharing, psychological safety, and performance effects. Prod Oper Manag 23(8):1352-1369. doi: 10.1111/poms.12024

Bock F (1999) The intelligent approach to knowledge management: viewing KM in terms of content, culture, process, and infrastructure. Knowl Manag Rev 7:22-25.

Brasil, Ministério da Defesa (2011) Comando da Aeronáutica. RICA 21-99: regimento interno do Instituto de Pesquisas e Ensaios em Voo. São José dos Campos: Ministério da Defesa.

Brasil, Ministério da Defesa (2012) Comando da Aeronáutica. DCA 1-1: Doutrina Básica da Força Aérea Brasileira. Brasília: Ministério da Defesa.

Chen L, Fong PSW (2015) Evaluation of knowledge management performance: an organic approach. Inform Manag 52(4):431-453. doi: 10.1016/j.im.2015.01.005

Coyle RG (1996) System dynamics modelling: a practical approach. Vol. 1. Boca Raton: CRC Press.

Coyle RG (2000) Qualitative and quantitative modelling in system dynamics: some research questions. Syst Dynam Rev 16(3):225-244. doi: 10.1002/1099-1727(200023)16:3<225::AID-SDR195>3.0.CO;2-D

Dangelico RM, Garavelli AC, Petruzzelli AM (2010) A system dynamics model to analyze technology districts' evolution in a knowledge-based perspective. Technovation 30(2):142-153. doi: 10.1016/j.technovation.2009.09.006

Davenport TH, Prusak L (1998) Working knowledge: how organizations manage what they know. New York: Harvard Business School Press.

Fallah-Fini S, Rahmandad H, Huang TTK, Bures RM, Glass TA (2014) Modeling US adult obesity trends: a system dynamics model for estimating energy imbalance gap. Am J Publ Health 104(7):1230-1239. doi: 10.2105/AJPH.2014.301882

Figueiredo JCB (2009) Estudo da difusão da tecnologia móvel celular no Brasil: uma abordagem com o uso de Dinâmica de Sistemas. Prod 19(1):230-245. doi: 10.1590/S0103-65132009000100015

Follador RC, Souza CE, Marto AG, Silva RGA, Góes LCS (2009) Comparison of in-flight measured and computed aeroelastic damping: modal identification procedures and modeling approaches. Paper presented at: IFASD 2009. Proccedings of the International Forum on Aeroelasticity and Structural Dynamics; Seattle, USA.

Follador RC, Trabasso LG (2015) Knowledge Management maturity level in a Brazilian Air Force flight test environment. Proceedings of the 2015 Portland International Conference on Management of Engineering and Technology (PICMET); Portland, USA.

Fonseca GA (2011) Métricas em gestão do conhecimento: modelo para avaliação do impacto de comunidades de prática em uma empresa desenvolvedora e fabricante de produtos complexos (Master's thesis). São José dos Campos: Instituto Tecnológico de Aeronáutica. In portuguese.

Ford AT (1999) Modeling the environment: an introduction to system dynamics — models of environmental systems. Washington: Island Press.

Forrester JW (1971) Counterintuitive behavior of social systems. Theor Decis 2(2):109-140. doi: 10.1007/BF00148991

Forrester JW (1994) System dynamics, systems thinking, and soft OR. Syst Dynam Rev 10(2-3):245-256. doi: 10.1002/sdr.4260100211

Forrester JW (1997) Industrial dynamics. J Oper Res Soc 48(10):1037-1041.

Gonçalves P (2009) Behavior modes, pathways and overall trajectories: eigenvector and eigenvalue analysis of dynamic systems. Syst Dynam Rev 25(1):35-62. doi: 10.2139/ssrn.1131392

Gonçalves P (2011) Balancing provision of relief and recovery with capacity building in humanitarian operations. Operations Management Research 4(1-2):39-50. doi: 10.1007/s12063-011-0045-7

Gratton GB (2010) Flight test reports. In: Blockley R, Shyy W, editors. Encyclopedia of aerospace engineering. Chichester: Wiley. p. 1-10.

Gray III WR (2005) Boundary-avoidance tracking: a new pilot tracking model. Proceedings of the 2005 AIAA Atmospheric Flight Mechanics Conference and Exhibit; San Francisco, USA.

Grundstein M (2008) Assessing the enterprise's knowledge management maturity level. Int J Knowl Learn 4(5):415-426. doi: 10.1504/IJKL.2008.02206

Kimberlin RD (2003) Flight testing of fixed-wing aircraft. Reston: American Institute of Aeronautics and Astronautics.

Liao YW, Wang YM, Wang YS, Tu YM (2015) Understanding the dynamics between organizational IT investment strategy and market performance: a system dynamics approach. Comput Ind 71:46-57. doi: 10.1016/j.compind.2015.02.006

Lloria MB (2008) A review of the main approaches to knowledge management. Knowl Manag Res Pract 6(1):77-89. doi: 10.1057/palgrave.kmrp.8500164

McCarthy JT, Hocum CL, Albright RC, Rogers J, Gallaher EJ, Steensma DP, Dingli D (2014) Biomedical system dynamics to improve anemia control with Darbepoetin Alfa in long-term hemodialysis patients. Mayo Clin Proc 89(1):87-94. doi: 10.1016/j.mayocp.2013.10.022

McNabb DE (2007) Knowledge management in the public sector: a blueprint for innovation in government. Armonk: M.E. Sharpe.

Meadows DH, Wright D (2008) Thinking in systems: a primer. White River Junction: Chelsea Green Publishing.

Morecroft JDW (2007) Strategic modelling and business dynamics: a feedback systems approach. Chichester: John Wiley & Sons.

Nonaka I, Konno N, Toyama R (2001) Emergence of "ba". In: Nonaka I, Nishiguchi T. Knowledge emergence: social, technical, and evolutionary dimensions of knowledge creation. Oxford, New York: Oxford University Press. p. 13-29.

Nonaka I, Takeuchi H (2004) Hitotsubashi on knowledge management. Singapore: John Wiley & Sons.

Prahalad CK, Hamel G (1990) The core competence of the corporation. Harvard Business Review; [accessed 2016 Jul 1]. http://www.expert2business.com/itson/Articles/CoreCompetencies.pdf

Repenning NP, Sterman JD (2002) Capability traps and self-confirming attribution errors in the dynamics of process improvement. Admin Sci Q 47(2):265-295. doi: 10.2307/3094806

Špicar R (2014) System dynamics archetypes in capacity planning. Procedia Engineering 69:1350-1355. doi: 10.1016/j.proeng.2014.03.128

Sterman JD (2002) Systems dynamics modeling: tools for learning in a complex world. IEEE Eng Manag Rev 30(1):42-42. doi: 10.1109/EMR.2002.1022404

Swinerd C, McNaught KR (2014) Simulating the diffusion of technological innovation with an integrated hybrid agent-based system dynamics model. J Simulat 8(3):231-240. doi: 10.1057/jos.2014.2

Teng Y, Kong N, Tu W (2015) Optimizing strategies for population-based chlamydia infection screening among young women: an age-structured system dynamics approach. BMC Publ Health 15(1):639. doi: 10.1186/s12889-015-1975-z

United States, Federal Aviation Administration (2012) Advisory Circular AC25-7C: Flight Test Guide for Certification of Transport Category Airplanes. Washington: Federal Aviation Administration.

Villela PR (2005) Introdução à dinâmica de sistemas. Juiz de Fora: Universidade Federal de Juiz de Fora.

Wolstenholme EF (1999) Qualitative vs quantitative modelling: the evolving balance. J Oper Res Soc 50(4):422-428. doi: 10.2307/3010462

Zaim S, Bayyurt N, Tarim M, Zaim H, Guc Y (2013) System dynamics modeling of a knowledge management process: a case study in Turkish Airlines. Procedia - Social and Behavioral Sciences 99:545-552. doi: 10.1016/j.sbspro.2013.10.524

2

Experimental Investigation on Limit Cycle Wing Rock Effect on Wing Body Configuration Induced by Forebody Vortices

Zhen Rong[1,2], Xueying Deng[2], Baofeng Ma[2], Bing Wang[2]

ABSTRACT: The purpose of this paper is to present the aerodynamic and flow characteristics of a slender body with a 30° swept wing configuration undergoing a limit cycle oscillation using a synchronous measurement and control technique of wing rock/particle image velocimetry/dynamic pressure associated with the time history of the wing rock motion. The experimental investigation was concentrated on 3 main areas: motion characteristics, static and dynamic surface pressures and static and dynamic particle image velocimetry. The tests' results revealed that the lag in asymmetric twin vortices over the forebody switching from the left vortice pattern to the right one exhibits a hysteresis evolvement during the wing rock motion; the asymmetric triple vortices over the forebody interacted with the flowfield over wings appeared to induce the instability and damping moments. The main flow phenomena responsible for wing rock of wing body configuration were completely determined by the forebody vortices. These exhibit apparent dynamic hysteresis in vertical position, which further influences the wing flows, and the dynamic hysteresis of flows yields the damping moments sustaining the oscillations.

KEYWORDS: Wing rock, Asymmetric forebody vortices, Wing body, High angle of attack, Wind tunnel test.

INTRODUCTION

The demand for improved airplane performance occasionally results in maneuvers falling outside the designated flight envelope. One of the limitations to combat effectiveness for all fighter aircraft is the phenomenon of wing rock — a primary motion in roll with, in some cases, a coupled oscillation in yaw. Generally, the onset of wing rock is attributed to a loss of stability in the lateral/directional mode and can be caused by a number of different aerodynamic phenomena.

At high angles of attack, the vortices emanating from the forebody of an aircraft can be very strong. Studies (Katz 1999; Nelson and Pelletier 2003; Ericsson 1989) have shown that interactions between asymmetric forebody vortices and the other surfaces on the aircraft are primarily responsible for wing rock in this flight regime. In a more recent study, Ericsson and Beyers (2003) believed that forebody vortices induce wing rock on a variety of wings with aspect ratios substantially larger than unity. In order to understand the static and dynamic stabilities of an aircraft at high angle of attack, Brandon and Nguyen (1988) found that a forebody cross-sectional shape can strongly effect wing rock behavior at high angle of attack. However, most of the studies focus on the kinematic characteristics of wing rock, and there is little effort to pay close attention to the flow mechanism. Although the mechanism of the wing rock has not been understood so far, some test results indicate that this flow phenomenon does not show repeatability. The HARV F-18 was tested by Quast (1991) in a wind tunnel test using a 2.5%-scale model, and the wing-rock was caused by the interaction of the wing with the forebody vortices. Nevertheless, the limit cycle

1.Zhejiang University – School of Aeronautics and Astronautics – Hangzhou/Zhejiang – China. **2.**Ministry of Education – Beihang University – Key Laboratory of Fluid Mechanics – Beijing/China.

Author for correspondence: Zhen Rong | Zhejiang University – School of Aeronautics and Astronautics | 38 Zheda Rd – Xihu Dist | Hangzhou/Zhejiang – China – 310027 | Email: rongzhen@zju.edu.cn

oscillation on F-18 model was not existent in the test made by Ross and Nguyen (1988). So, it was maybe a determinacy problem for the roll oscillation similar to the asymmetric vortices over a slender body at high angle of attack.

Recently, Deng *et al.* (2008) and Ma *et al.* (2015) have studied the effects of artificial perturbation on the tip of the model on behaviors of the wing rock induced by the forebody vortices. Their results showed determinacy problems for the roll oscillation induced by forebody vortices and concluded that, with natural tip perturbations, the roll motion of the model was not deterministic due to the randomness of these perturbations, but the deterministic roll motion could be predetermined by setting an artificial perturbation on the tip of the model. In addition, the model exhibited a limit cycle oscillation motion as the artificial tip perturbations were set at the azimuth angle of 0° or 180° on the tip of the model.

The experiments and analysis reported in the cited references provide some insight into the kinematic and aerodynamic characteristics of the wing rock phenomenon. However, these experiments produced very little data on the unsteady flowfield associated with the rolling motion. To truly understand the relationship between the vertical flowfield and the unsteady aerodynamic roll moment driving the wing in the limit cycle motion induced by the forebody vortices when the artificial tip perturbation is set at the azimuth angle of 0° or 180°, one needs to have information about the surface pressure distributions and the vortices flow structures.

The purpose of this paper is to present the forebody vortices and flowfield over the wings behavior on a wing body configuration undergoing a limit cycle motion using the synchronous measurement and control technique of wing rock/particle image velocimetry (PIV)/dynamic pressure (Rong *et al.* 2010). The investigations consist of motion history, surface pressure and digital particle image velocimetry (DPIV) experiments with the goal of identifying the primary fluid mechanisms causing the limit cycle oscillation behavior.

The tests' results revealed that the lag in asymmetric twin vortices over the forebody switching from the left vortice pattern (LVP) to the right vortice pattern (RVP) would exhibit a hysteresis evolvement during the wing rock motion, and the asymmetric triple vortices over the forebody which interacted with the flowfield over wings appeared to induce the instability and damping moments. The main flow phenomena responsible for wing rock of wing body configuration were completely determined by the forebody vortices. The forebody vortices exhibit apparent dynamic hysteresis in vertical position, which further influences the wing flows, and the dynamic hysteresis of flows yields the damping moments sustaining the oscillations.

EXPERIMENTAL SETUP

All experiments were conducted in the D4 wind tunnel of Fluid Mechanics Institute of Beihang University. This low-speed low-noise tunnel had a 1.5 × 1.5 × 2.5 m opened test section, with a turbulence level of less than 0.1%; speeds up to 65 $m \cdot s^{-1}$ could be achieved through a closed-circuit continuous flow system.

Two basic models were used in the test. The free-to-roll test model (Fig. 1a) was used to measure the oscillatory motion history, and the pressure test model (Fig. 1b) was used for surface pressure distributions as well as vortex flowfield combined tests. The model had a length-to-base-diameter ratio (L_b/D) equal to 8, and the fore-body of the model was pointed tangent ogive with fineness ratio, L_f/D = 3.0. The wings were sharp edge delta wings with a 30° sweep and had a 45° beveled edge on either sides of leading edges. The model was made of aluminum, with moment of inertia I = 0.007 $kg \cdot m^{-2}$ around the body roll axis and corresponding non-dimensional moment of inertia $\bar{I}$ = 0.63 — $\bar{I} = I/\rho L$ (Brandon and Nguyen 1988), with airflow density ρ = 1.225 $kg \cdot m^{-3}$ and length of wing span L = 390 mm.

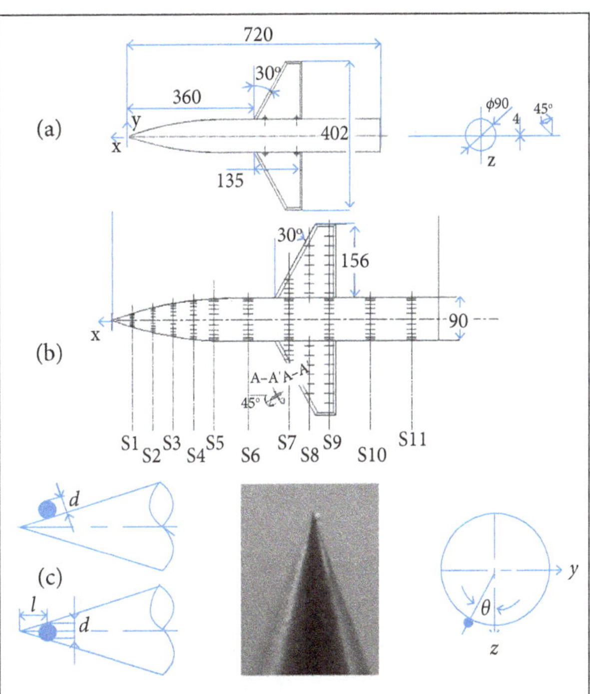

Figure 1. Test models. (a) Free-to-roll model; (b) Pressure test model; (c) Model tip with an artificial perturbation.

The pressure test model had also 11 tapping sections, and pressure taps stations were shown in Table 1. A piece of artificial perturbation (Fig. 1c) made by chinaware was set at the tip of the model at azimuth angle of $\gamma = 0°$, being the windward symmetric plane.

Table 1. Pressure tap stations.

Station	Dimension (mm)	Non-dimensional
S1	45.0	0.5
S2	90.0	1.0
S3	135.0	1.5
S4	180.0	2.0
S5	225.0	2.5
S6	301.5	3.35
S7(W1)	391.5	4.35
S8 (W2)	436.5	4.85
S9 (W3)	481.5	5.35
S10	571.5	6.35
S11	661.5	7.35

FREE-TO-ROLL SYSTEM

The model was mounted on a dynamic roll rig installed in the D4 wind tunnel (Fig. 2a). The model sting was fitted through a low-friction bearing that could allow the model to rotate "freely" around the longitudinal axis. Motion history was measured using a 12-bit high-precision optical encoder, yielding a resolution of ±0.088°.

In order to obtain the results of the surface flow and the spatial flow structure simultaneously, a synchronous measurement and control technique of wing rock/PIV/dynamic pressure was exploited in D4 wing tunnel (Rong *et al.* 2010). The surface pressure was measured using pressure scanners based on pressure tubing. The vortex wakes were determined using a PIV system. For the pressure measurements, the pressure tubes were arranged inside the model. These tubes will interfere with the free oscillations of the model, so a forced-to-roll rig driven by a motor was used to drive another model with pressure tubes, thereby reproducing the time histories obtained previously and measuring the surface pressure and vortex wakes during the prescribed motion to avoid the interference. It provided a useful technique for the research of dynamic flow characteristics as the PIV phase-locking accurately and PIV working with dynamic pressure acquisition in-phase were carried out. Figure 2b was the sketch of this combined measurement system.

Digital Particle Image Velocimetry System (DPIV) measurements were taken using a Dantec PIV system (Dantec Company 2000), incorporating a pair of pulsed Nd:YAG lasers with a maximum energy input of 350 mJ per pulse, in order to create a more detailed velocity picture of the flowfield over the model. To illuminate the desired planes, the PIV system was placed besides the test section of the wind tunnel (Fig. 2c). Images were captured using a high-digital camera with a resolution of 2,048 × 2,048 pixels. The commercial software package Flow-Manager and an auto-correlation algorithm were used to analyze the images, with an interrogation window size of 64 by 64 pixels and to produce velocity vectors for further processing. Sequences of 8 instantaneous frames were taken for each case and the time-averaged (or phase-averaged for dynamic roll motion)

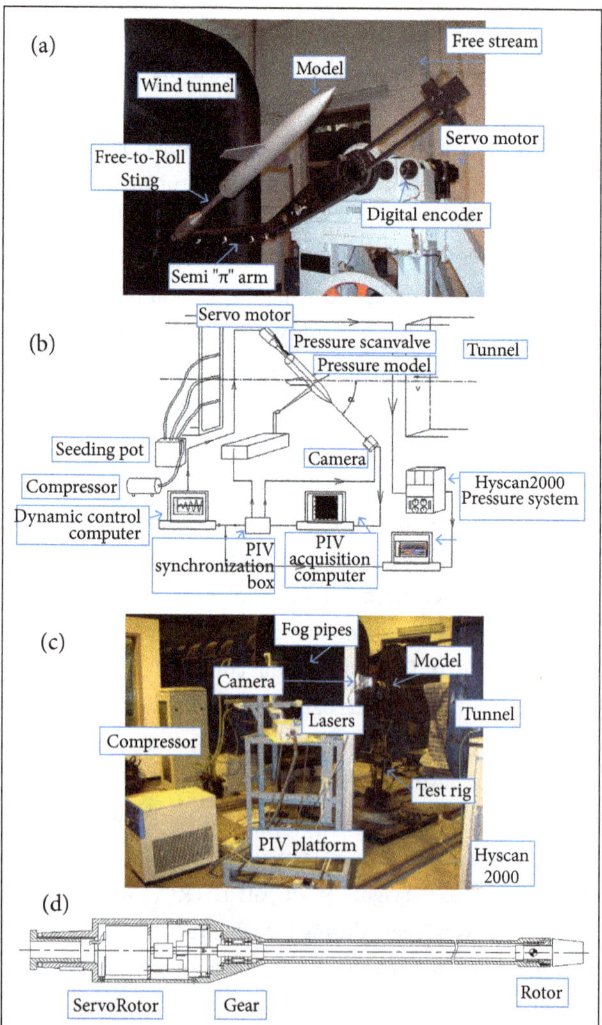

Figure 2. Test facilities. (a) Free-to-roll rig; (b) PIV/dynamic pressure combined test system; (c) Scene of the PIV system;(d) Forced-to-roll rig.

velocity fields were calculated. The flow was seeded with particles of mean diameter of 1 μm compressed by the edible bean oil. For the dynamic case, the PIV system was extra-triggered in order to capture the flow field at a specific roll angle and then perform phase averaging (Fig. 3). An error of ±0.02° in the extra-triggered angle existed in these measurements.

The Hyscan2000 was a data acquisition system designed for high-speed data collection and made up of 5 parts, such as DAQ data acquisition, IFM2000 ZOC Module, ZOC Pressure Scanning Modules, CSM2000 Cable-Service Unit and SPC3000 Servo Pressure Calibration Module (Scanivalve Corporation 1992). This data acquisition system could sample and store pressure at rates up to 100,000 channels per second, and the accuracy of pressure measurement was near ±0.06% F.S. The ZOC contains up to 64 pressure sensors in pressure ranged from 0 to 1 psid, and the frequency of the pressure collection was set at 128 Hz. It was manifested that to meet the frequency response requirements, the pressure pipes connected ZOC with model were shorten than 1 m (Cao 2008). In the tests, 0.5 m long pressure pipes were chosen. The working mode of the Hyscan2000 was extra-trigger and host modes.

The forced-to-roll system was accomplished with a DC servo motor and motion control computer board. Encoder signals were used for feedback (Wang *et al.* 2009), as shown in Fig. 2d. Digital proportional-integral-derivative (PID) with velocity and acceleration feed-forward control was implemented for precision tracking of the time history. Time histories taken with the free-to-roll apparatus were used to provide the input signal to the rock trajectory. The dynamic pressure data and DPIV system was synchronized with the motion control system so that the roll angle and time were known for each sample. The simulation of the time history had a high precision, and the error was limited to 5%, as shown in Fig. 4. The combined measurement data were taken at a free stream velocity $V = 15\ \text{m·s}^{-1}$, corresponding to a Reynolds number, based on base diameter, of $\text{Re} = \rho VD/\mu = 9.0 \times 10^4$ and in sub-critical Reynolds number flow (Lamont 1982).

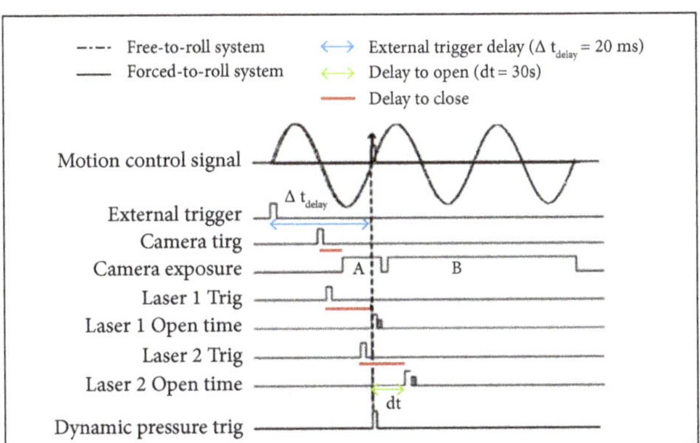

Figure 3. The working principle of synchronous test.

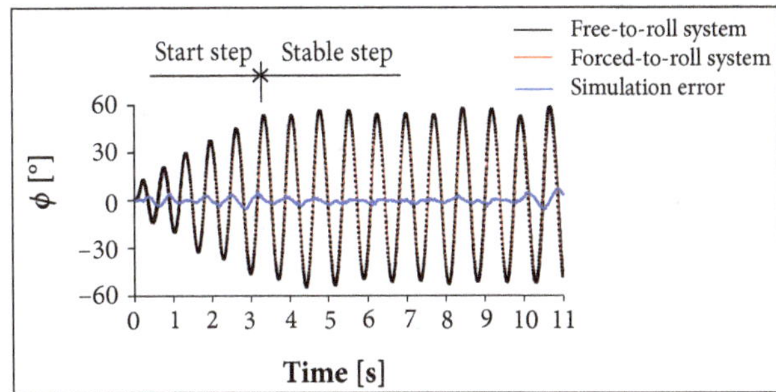

Figure 4. Curve of forced-roll simulation (Wang *et al.* 2009).

RESULTS

Figure 5 is the typical plot of the limit cycle motion time history that is induced by the asymmetric fore-body vortices when the artificial tip perturbations are set at the azimuth angle of 0°. The motion was regardless of the initial conditions. From the plot of time history, the amplitude of the limit cycle motion was about 55° and the frequency of oscillation was 1.35 Hz. Figure 6a shows the variation of limit cycle amplitude with increasing Reynolds number for the model configuration. The trend of the amplitude remained relatively in the same level range of 40° – 60°. The dependence of limit cycle frequency on Reynolds number was also studied, as seen in Fig. 6b. It was revealed that the dominant frequency increased in what appeared to be a linear trend, but the reduced frequency maintained a relatively constant value over the Reynolds number range in sub-critical Reynolds number flow.

In order to investigate the mechanism responsible for the roll moment behavior, experiments were conducted to investigate the behavior of the flowfields over the fore-body and wings of the model using the synchronous measurement and control technique of wing rock/PIV/dynamic pressure, and couple of these information with the behavior of the model motion.

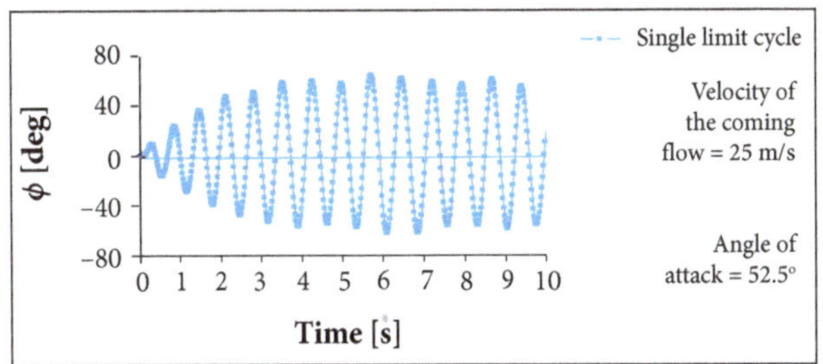

Figure 5. The chart of limit cycle oscillation.

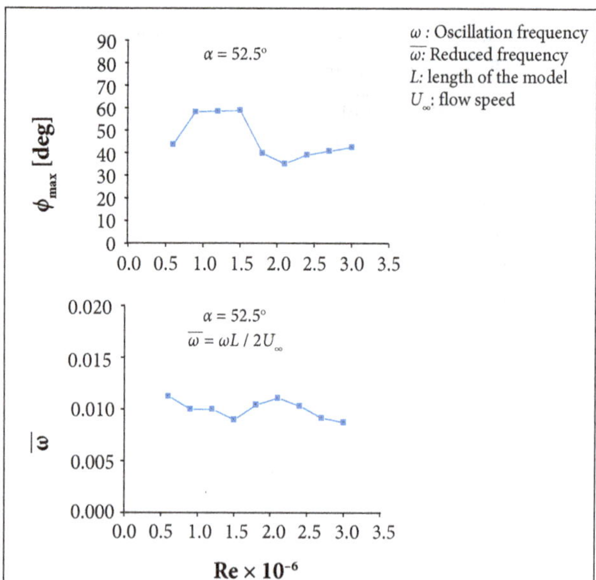

Figure 6. Variations of wing rock with Reynolds number. (a) Limit cycle amplitude; (b) Limit cycle reduced frequency.

EFFECT OF THE TIP PERTURBATION AND FLOW STRUCTURES OVER THE WING BODY

Recently, Deng *et al.* (2002) and Chen (2003) have completed a series of investigations about the tip-perturbation effect and characteristics of asymmetric vortices flow over slender body. Their experiments results showed that the behaviors and structure of asymmetric vortices over a slender body were mainly controlled by artificial perturbation on the tip of slender body compared with geometrical minute irregularities on the model from the machining tolerances. There were 4 sensitive circumferential locations at which bi-stable vortices over slender body were switched by the artificial perturbation (Fig. 7a).

The behaviors of multi-vortices flow structure were investigated by Deng *et al.* (2003) and Wang (2003) in detail. From their discussions, it was clearly shown that the asymmetric vortices flow patterns at regular state were developed and evolved along the slender body axis, which closely correlates with behaviors of sectional side force distribution. The asymmetric vortices flowfield could be zoned into different regions with the specific features, which included asymmetric twin vortices inception region and fully developed region, asymmetric triple vortices region, 4 vortices region, 5 vortices region and Karman vortices street like region shown in Fig. 7b. Consequently, a physical model of asymmetric vortices structure at regular state can be concluded with detailed zonal analysis shown in Fig. 7c.

The Tip Perturbation Effect

Figure 8a presents the longitudinal sectional-side-force distribution of the wing-body model for all the roll angles with an interval of 15 degrees. In the figure, $C_y \sim x/D$ curves — where C_y is the side force of the model, x is the axial distance from the tip of the model, and D means diameter of the afterbody — were collapsed into 2 families which were clearly shown in the 2 mirrored bi-stable states similar to the asymmetric fore-body vortices over the slender body. Left vortices on fore-body (LVF) were referred to that the left one of asymmetric vortices was located in the lower position while the right one was in the higher position when the flow was observed from the x direction, and so was right vortices on fore-body (RVF). The fore-body vortices exhibited twin symmetric vortices at $\phi = 0°$ due to the response between the artificial perturbation and asymmetric vortices. The pressure distributions and PIV pictures shown in Figs. 8b and 8c and also demonstrated the character of the bi-stable states.

Following the sectional-side-force distribution in Fig. 8a, the flow regions over the wing-body could be divided into different regions: asymmetric twin vortices (A-B), asymmetric triple vortices region (B-C), wings region (C-D) and after-body region (D-E). The present paper discusses the flow pattern development for RVF regular state at $\phi = 30°$.

The $C_y \sim x/D$ curves in region A-B presented the asymmetric twin vortices were gradually developing from the A station and fully developed near the B station. The pressure distributions

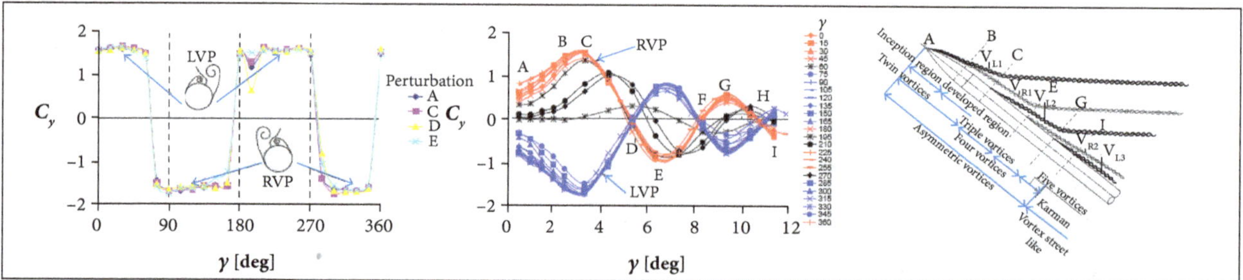

Figure 7. (a) Asymmetric vortices pattern variation with tip perturbation circumferential angle (Deng *et al.* 2002); (b) $C_y \sim x/D$ with tip perturbation of slender body (Deng *et al.* 2003); (c) Physical model of multi-vortices structure (Wang 2003).

and PIV results showed that in this region the flow pattern over the model was clearly twin asymmetric vortices (RVF), as seen in Fig. 9a, and the strength of the asymmetric vortices appeared to increase from $x/D = 0.5$ to $x/D = 2.5$ (Fig. 9b).

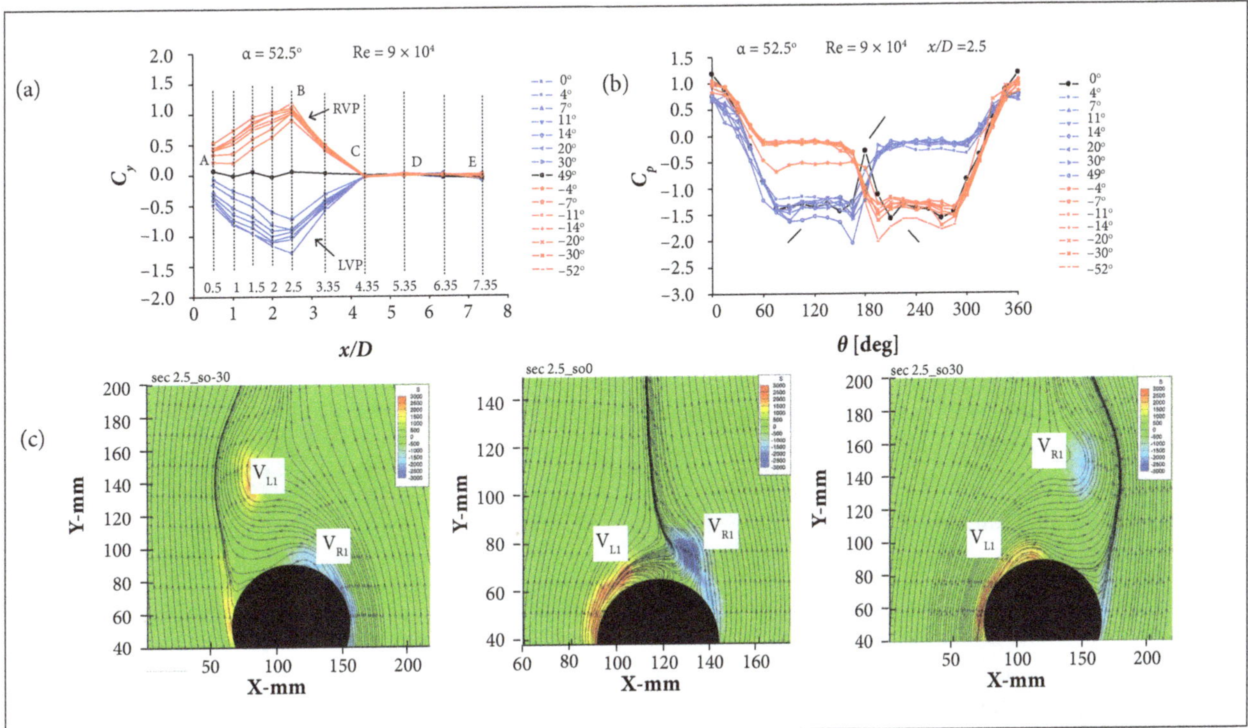

Figure 8. (a) $C_y \sim x/D$ with roll angle of wing body; (b) Pressure distributions with roll angle at $x/D = 2.5$ (C_p is the pressure coefficient); (c) PIV results with roll angle at $x/D = 2.5$.

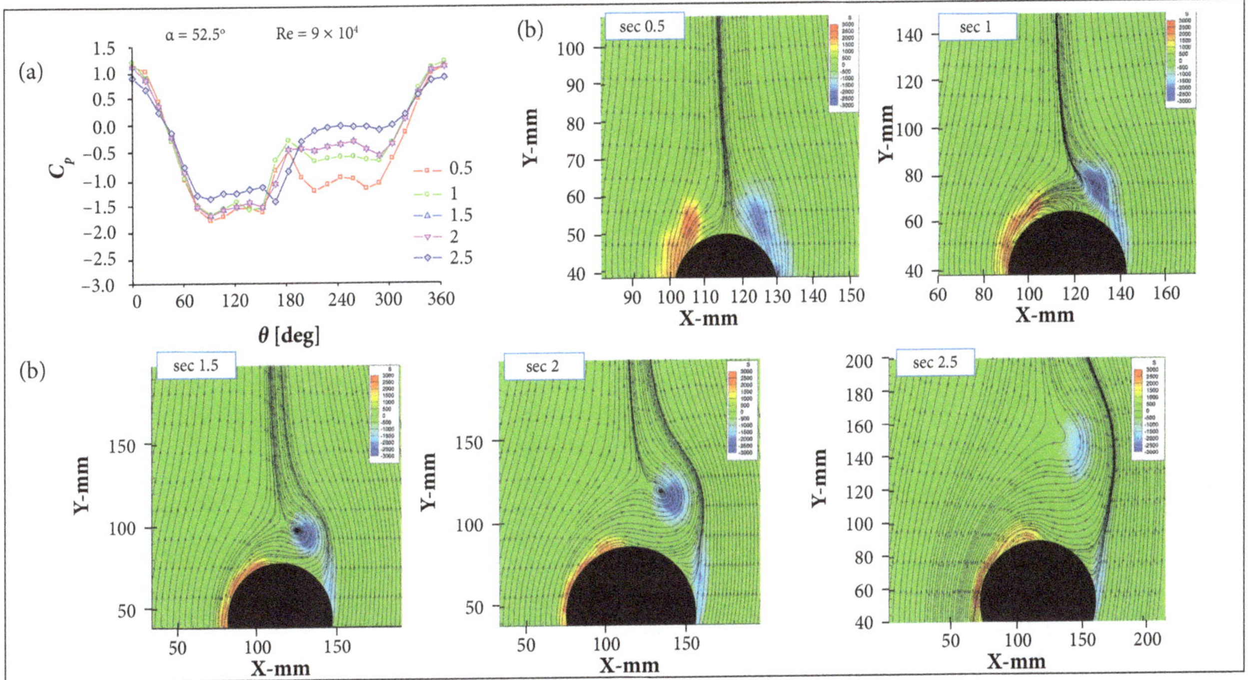

Figure 9. (a) Pressure distributions with different sections at $\phi = 30°$; (b) PIV results with different sections at $\phi = 30°$.

In the region B-C, the sectional side force decreased with x/D from the B station (C_y maximum) to the C station (near zero) shown in Fig. 8a. As the fore-body vortices developed in the downstream direction, a triple vortices system exhibited at x/D = 3.35.The highly located vortices LVF1 was breakaway from the surface and a new vortices LVF2 (the new left vortices on the forebody) created beneath LVF1 indicated by pressure distribution and PIV picture shown in Figs. 10a and 10b.

With the growth of the triple vortices over the wings, the separated shear layer feeding vorticity was blocked by the wing and the left vorticity remained in the flowfield. As was known to all, the leading edge vortices on the lowly-swept wing were already breakdown at this high angle of attack. The flowfield over the S8 section measured by PIV was shown in Figs. 10c and 10d (region C-D). Two strong vortices sheets were induced by the fore-body vortices near the surface of model (LVF1 and RVF2) and an anti-clockwise rolling moment which could drive the model oscillating in negative direction when the model was free to roll was generated.

In afterbody region (D-E), the sectional side force decayed to or near zero shown in Fig. 8a. The pressure distribution and PIV picture appeared similar to the Karman vortices street like the region (Wang 2003) in Figs. 10e and 10f.

From above detailed discussions, the spatial vortices flow over evolved along the wing body was clearly indicated in Fig. 11a, which included asymmetric twin vortices region, asymmetric triple vortices region, unsteady flow region over the wings and unsteady flow region over the afterbody. Consequently, the spatial flow structure rebuilt by 2D-DPIV results has completed, and quasi-quantificational description about the evolution of spatial vortices flow could be obtained in Fig. 11b.

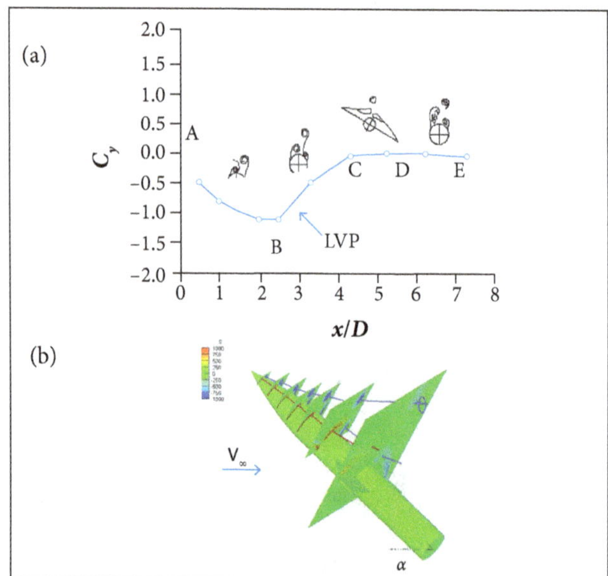

Figure 11. (a) Spatial flow structure shown by Cy distribution; (b) Spatial flow structure shown by PIV.

FLOW CHARACTERISTICS AND MECHANISM

Cross-section Flow and its Evolution

Figure 12 shows the C_y distribution at x/D = 2.5 during the limit cycle motion simulation; the results differed greatly from the static case where the $C_y \sim \phi$ curves were undergoing hysteresis loops near $\phi = 0°$. When the model rolls past $\phi = 0°$ in positive direction, C_y remains positive; as the model continues to roll,

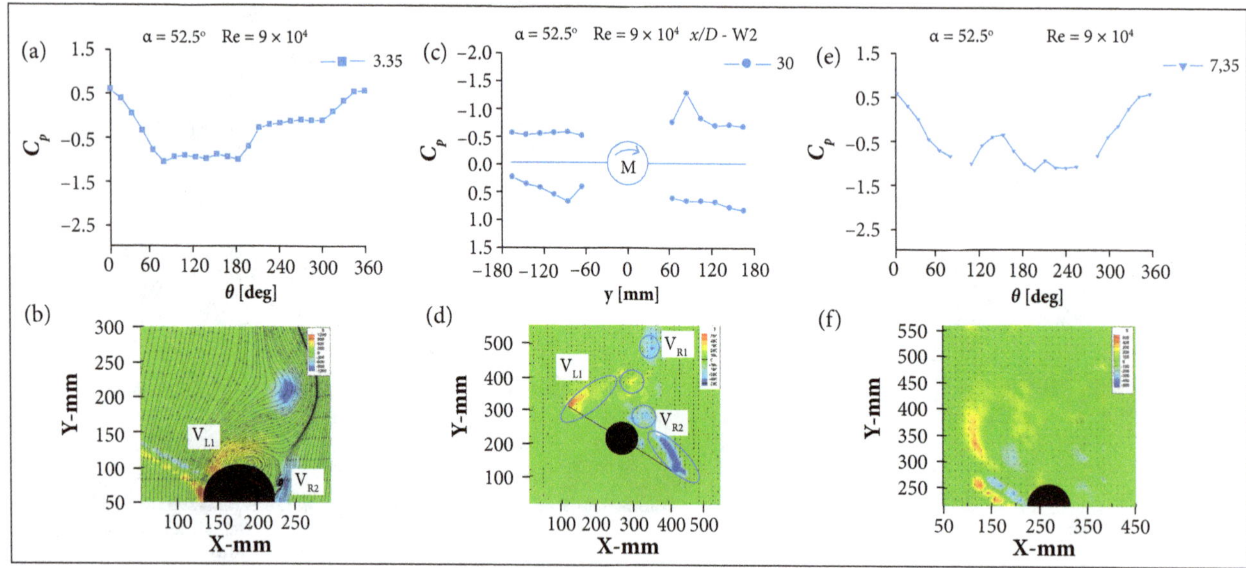

Figure 10. Pressure distribution and PIV results.

C_y changes from a great positive value to 0, and then continues to negative values. As the role of the inertia, C_y returned the positive maximum platform after an overshooting. The model followed to roll from positive maximum platform to $\phi = 0°$, and C_y retained the same value. When the model started to roll in the range of negative angles, the forebody vortices experienced a similar evolution.

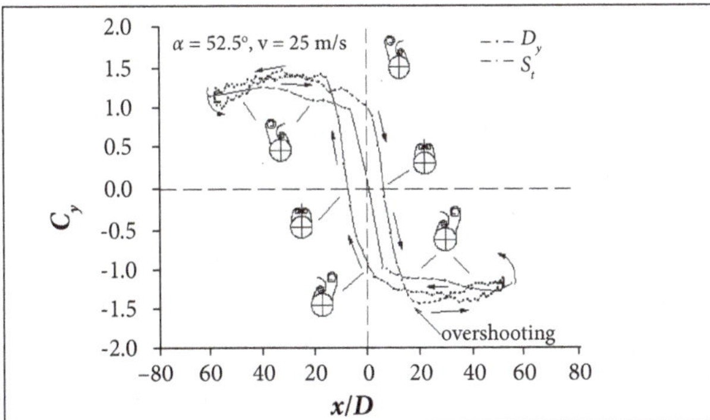

Figure 12. Chart of side force and vortices forms at $x/D = 2.5$ during oscillation.

Further analysis of the forebody vortices characteristics and evolution was to proceed by the combined PIV and dynamic pressure measurements results in Fig. 13. Compared to Fig. 8, as a result of forebody vortices evolvement lagged the rolling motion and the response of perturbation did not immediately waken, the pattern of forebody vortices remains the same as RVF when the model rolls past $\phi = 0°$ and a positive C_y was induced, as shown in Fig. 13a; to continue rolling, the asymmetric degree of vortices came to be weakened, the RVF was lower, and the LVF was upper, until a symmetrical pattern was exhibited. Now C_y equaled to zero. As the model continued to roll, the forebody vortices pattern became LVF-dominated by the tip perturbation until $\phi = 20°$. Rolling to the maximum displacement during the wing rock motion, the forebody vortices were kept to be the LVF. This is the forebody vortices hysteresis process moving from $\phi = 0°$ to the maximum positive displacement at positive direction. There would be undergoing a similar evolution moving from $\phi = 0°$ to the maximum negative displacement in negative direction. However, the forebody vortices would keep the LVF at positive angle phase and RVF at negative angle phase, which is consistent with the response of perturbation. It was concluded that the forebody vortices had a close relationship with the tip perturbation during the wing rock motion. Figure 13b displayed the surface pressure distributions at $x/D = 2.5$ measured in the combined tests synchronized with PIV. The results from the surface pressure can be seen when the rolling angle ϕ is passing through 0° from negative to positive, the pressure distribution was asymmetric, left vortices are higher and right vortices are lower. To continue rolling, pressure distribution tended to be symmetric until completely symmetrical distribution was exhibited and a peak pressure appeared at the leeward; then the pressure distributions

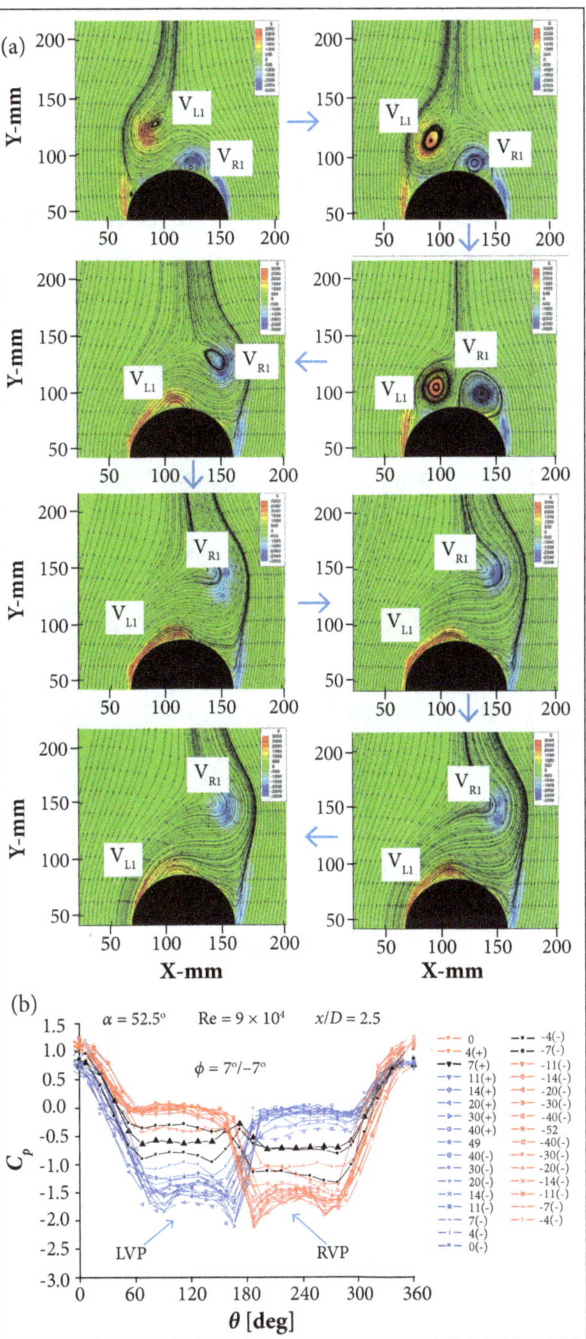

Figure 13. (a) PIV results and dynamic pressure measurement combined technique ($x/D = 2.5$); (b) Dynamic pressure measured by PIV and dynamic pressure measurement combined technique ($x/D = 2.5$).

presented LVF when increasing the roll angle. In conclusion, the forebody vortices characteristics and evolutions shown by pressure distributions were in agreement with the PIV results.

As a result of the pressure taps be on the forebody in the normal direction, roll moment could not be integral by the pressure results. Otherwise, it was found that the forebody vortices could not cause the fuselage rock only in the free-to-roll tests. So, the flowfield over the wings interacted with the forebody vortices was thought to be a driving mechanism of the wing rock motion.

Figure 14 shows that the evolution of the flowfield over wings from $\phi = 0°$ to the maximum positive displacement in positive direction. When the model rolls past $\phi = 0°$, the vortices at forebody remain as the RVF, so in the flowfield over the wing VL1 and VL2 are located at the left side, and VR1 is located at the right side. It was clearly that the vorticity sheet was induced by the VL2 and VR1 with the leading edge of the wing. Due to the VL2 was closer to the wing surface, a larger lift would generate on the left side and

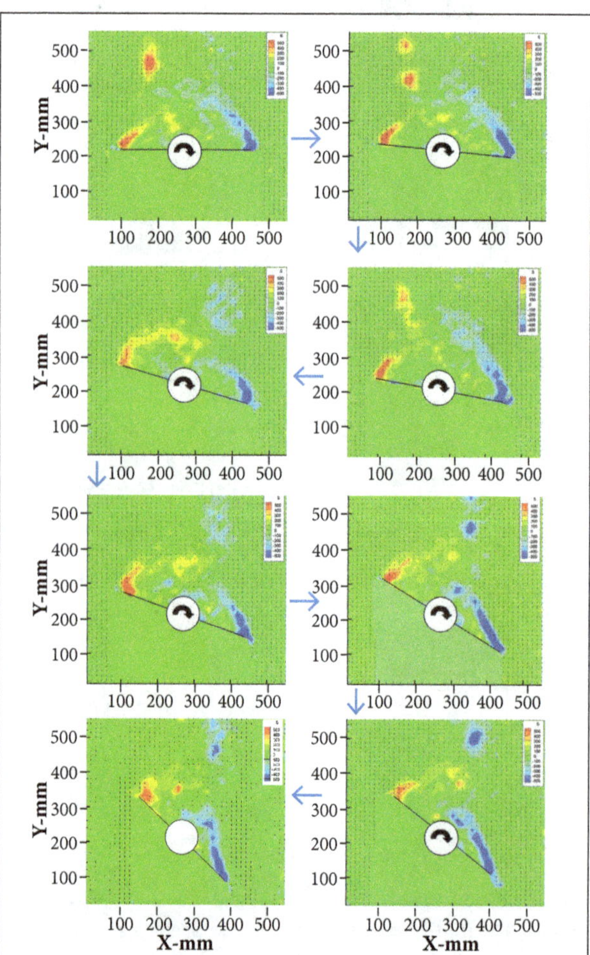

Figure 14. PIV results and dynamic pressure measurement combined technique (S8).

a positive rolling moment would drive the model post rotate at clockwise direction, as the shown pressure distribution. To continue rolling, the new forebody vortice VL2 gradually disappeared and the VL1 moved downward to the surface. However, the VR1 raised upward until the twin forebody vortices became symmetric and the restoring moment was close to 0. Then the forebody vortices pattern switched and in the flowfied over the wing VR2 appeared, so the VR2 was closer to the wing surface and a negative rolling moment would drive the model post rotate at counterclockwise direction; the model was driven to decelerate to roll until the maximum negative displacement. PIV results also revealed a great deal about the behavior of the flowfield on the oscillating wing. However, PIV alone cannot quantify how the unsteady flowfield generates the wing rock motion. For these reasons, unsteady surface pressures were measured on S8 station of wing undergoing wing rock. The pressure results and section rolling moment integrated by the pressure distribution on the S8 station, as shown in Fig. 14. Although the moment coefficient was obtained at only one chord-wise station, all of the non-linearity seen in the actual roll moment curve was captured. The data show the region of instability, the damping lobes and slope depression at the larger roll angles. The pressure distribution of the top surface generated all of the instability in roll moment and very little damping. The bottom surface provided the classical roll damping hysteresis which reduced the instability caused by the flowfied over the top surface.

Spatial Vortices Flow and its Evolution

In Fig. 15, the $C_y \sim x/D$ curves during the limit cycle oscillation show that the main contribution of C_y remained to obtain from forebody sections. The magnitude of C_y increased gradually from the tip to section $x/D = 2.5$, and the maximum of C_y appeared at $x/D = 2.5$ and then C_y gradually reduced to zero. The hysteresis of $C_y \sim x/D$ could be clearly recognized as the result of forebody switching lag. The hysteresis of the spatial flow structure during the whole oscillation measured by PIV was seen in Fig. 16. When the model was rolling past $\phi = 7°$, the forebody vortices and flow over wings were almost symmetrical simultaneously.

Analysis of the Flow Mechanism

The rolling moment $C_L = I \cdot \ddot{\phi}$, where $\ddot{\phi}$ represents angular acceleration, yielded the typical wing rock hysteresis loops when plotted as a function of roll angle (Fig. 17a). Another achievement of rolling moment can be obtained by the integral of ΔC_p at the regions of wings (S7/S8/S9), where $\Delta C_p = C_{pdown} - C_{pup}$ (Fig. 17b). Note the clockwise loop in the plots which indicated a dynamic instability and that the restoring moment was linear with roll angle. Energy was being fed to the system, therefore, the roll angle amplitude was increasing. Two stable damping lobes had formed for larger roll angles. The area of these lobes equaled the area of the unstable portion of the plot such that the net energy exchange was zero. This condition was necessary for the limit cycle oscillation to be sustained (Nguyen *et al.* 1981).

Figure 18a presents the normal position variation of 2 vortices with roll angle, which were near the surface mainly affected by the aerodynamics in triple vortices system at x/D = 3.35.

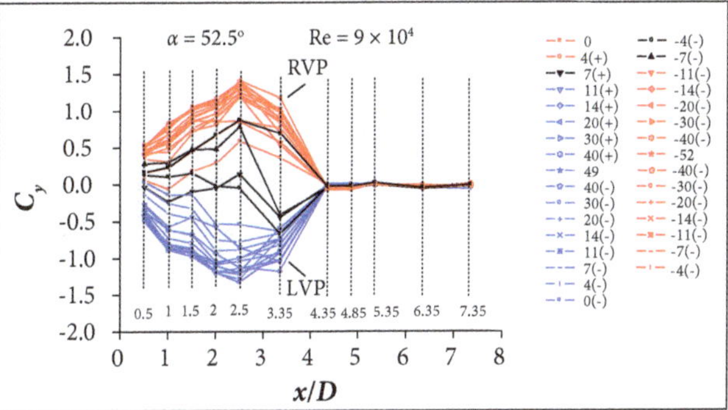

Figure 15. $C_y \sim x/D$ with roll angle during wing rock.

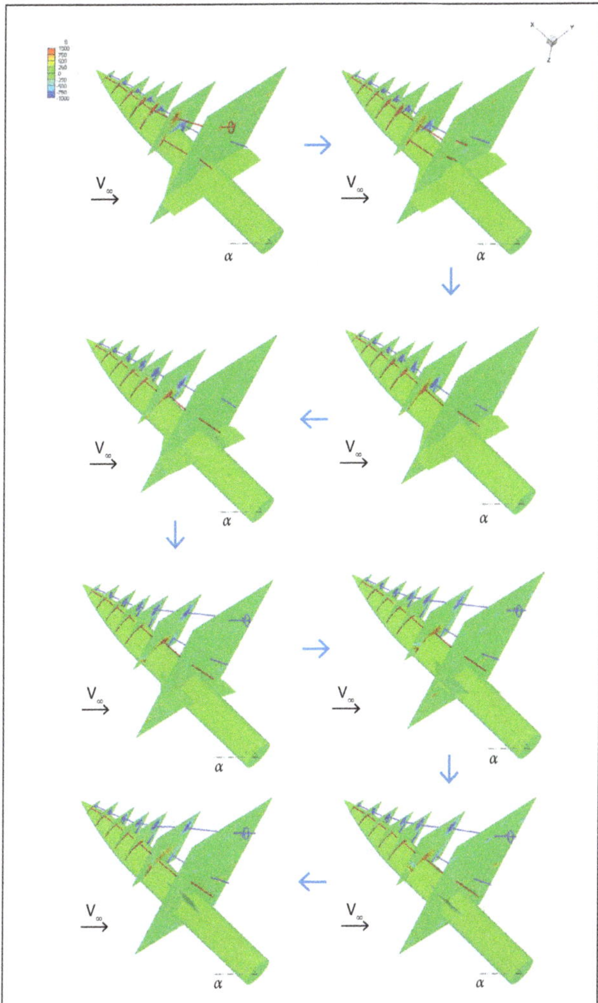

Figure 16. Spatial flow structure shown by PIV during wing rock.

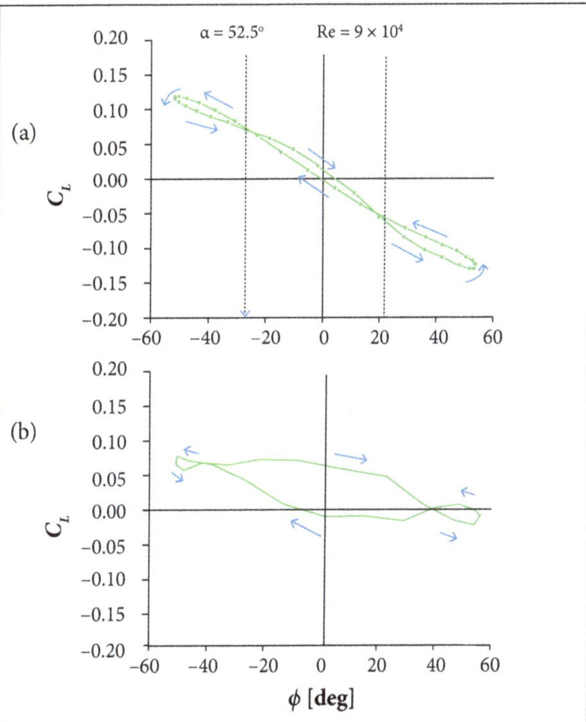

Figure 17. (a) $C_L \sim \phi$ distribution calculated by roll history; (b) $C_L \sim \phi$ distribution obtained by C_p integral at wings.

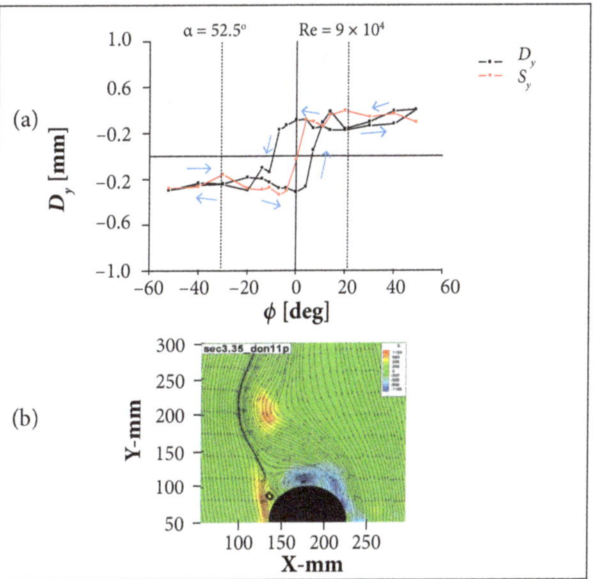

Figure 18. (a) Dynamic vortices asymmetry at x/D = 3.35; (b) Sketch of asymmetric vortices position.

The presence of the hysteresis lobe was more prominent in the normal direction in this plot; d_y was stand by the normal asymmetric in rolling processing between the 2 vortices near the model and defined as $d_y = y_L - y_R$, where L means left and R, right; s_y was stand by the normal asymmetric in static state between the 2 vortices near the model and defined as $s_y = y_L - y_R$. Figure 18b is a sketch of asymmetric vortices position used to define asymmetric parameters. It can be clearly seen that the hysteresis loop lay roughly within the −20 and +30 roll angle position. A similar look at the roll moment plot (Fig. 19a) shows the same beginning and ending points for the roll moment hysteresis loop and the normal position data show the same agreement.

Furthermore, the flow mechanism was studied by the roll motion and aerodynamics coupling. Figure 19 is integrated with the angular displacement, angle velocity and rolling moment in one graph in order to understand the dynamic flow characteristics and the physical evolution more intuitively. Figure 19a shows that, when the model was rolling past $\phi = 0°$, the kinematics and dynamics characteristics of the model ($\phi = 0°$, > 0, $C_L > 0$) were seen in the projection of curves ~ ϕ, and $C_L \sim \phi$ model is an accelerated rolling in positive direction, where C_L is the rolling moment coefficient. Rolling on, the angular velocity turned to maximum and the parameter CL was closed to zero. Then model rolled past $\phi = 7°$; the kinematics and dynamics characteristics of the model ($\phi = 7°$, > 0, and $C_L < 0$) can be found in Fig. 19c. To continue rolling slower by the damping moment, the model was rolling until the largest positive displacement $\phi = 49°$; the kinematics and dynamics characteristics of the model are $\phi = 49°$, $\dot{\phi} = 0$, $C_L < 0$. Then the model would accelerate to roll in negative direction. The above-mentioned oscillation was the evolution of a quarter, like "spring effect", but this was linearization course $\ddot{x} = -kx$ and the roll moment of wing rock possess a hysteresis characteristics.

Above to all, the hysteresis of the forebody vortices and the lag displacement revealed the mechanism of wing rock that was caused by the forebody vortices interacting with the wing flowfield which was controlled by the forebody vortices.

CONCLUSIONS

One of the limitations to the combat effectiveness for most of the fighter aircraft is the phenomenon of wing rock. The purpose of this paper was to present the aerodynamic and flow characteristics of a slender body with a 30° swept wing configuration undergoing a limit cycle oscillation induced by forebody vortices using a synchronous measurement and control technique of wing rock/PIV/dynamic pressure. The flow mechanics of the rolling motion induced by forebody vortices was analyzed associated with the time history of the wing rock motion. The main conclusions of this investigation are summarized as follows:

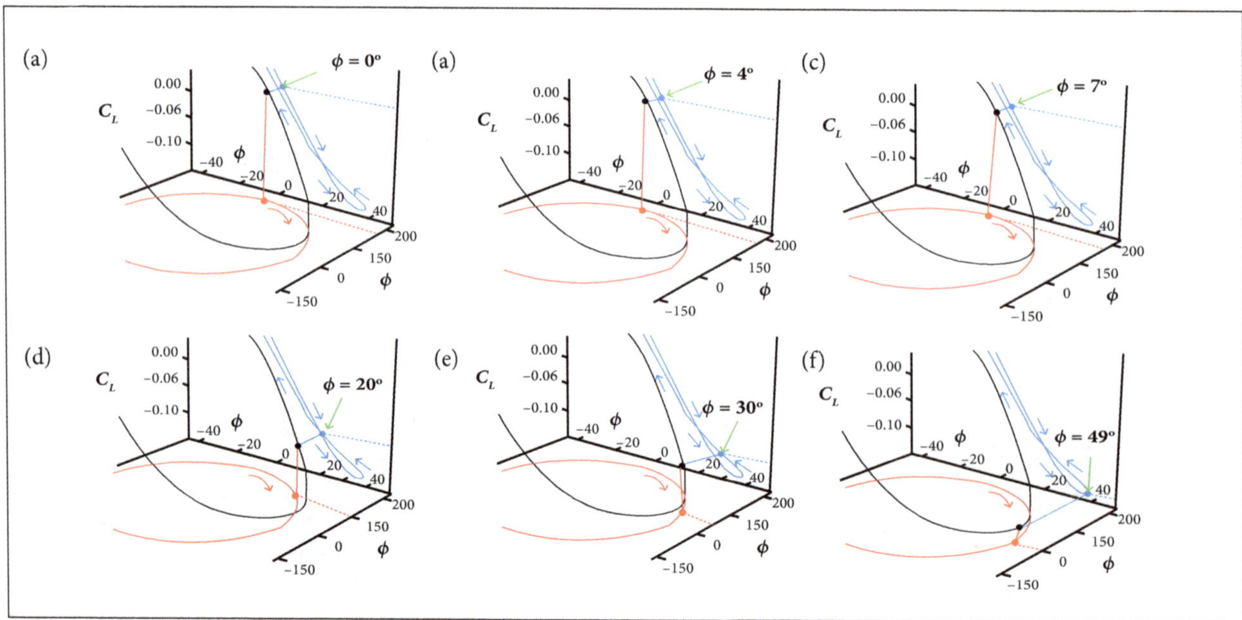

Figure 19. Analysis of motion and flow coupling about wing rock. (a) $\phi = 0°$ (+); (b) $\phi = 4°$ (+); (c) $\phi = 7°$ (+); (d) $\phi = 20°$ (+); (e) $\phi = 30°$ (+); (f) $\phi = 49°$ (+).

- In sub-critical Reynolds number flow at angle of attack of 52.5°, the model exhibits a limited cycle oscillation motion when the artificial tip perturbations locate near the azimuthal angles of 0° and 180° on the tip.
- Asymmetric vortices over forebody of wing body appeared to be 2 mirrored bi-stable states similar to the asymmetric forebody vortices over the slender body, and the flow regions over the wing body along the *x* direction can be divided into different regions: asymmetric twin vortices, asymmetric triple vortices region, wings region and afterbody region.
- Limit cycle oscillation of the wing body configuration is caused by the forebody vortices interacting with the wing flowfield, which was controlled by the forebody vortices. The surface pressure distributions and the DPIV results during the rolling motion revealed that the asymmetric twin vortices over the forebody switching from LVP to RVP would exhibit a hysteresis evolvement during the wing rock motion. At the wing region, the asymmetric triple vortices over the forebody interacted with the flowfield over the wing appeared to induce the instability and damping moments; the flowfield at the bottom of the wing was found to provide a majority of damping moment. In general, the wind tunnel tests confirmed that the main flow phenomena responsible for wing rock of the wing body configuration were determined by the forebody vortices and interacted with these and flowfield on the wing.

ACKNOWLEDGEMENTS

The project is supported by the National Natural Science Foundation of China (NSFC), under Grant No. 10702004, and partly supported by NSFC under Grant No. 11502234, as well as the Zhejiang Provincial Natural Science Foundation under grant No. LY14A020002.

REFERENCES

Brandon JM, Nguyen LT (1988) Experimental study of effects of forebody geometry on high angle-of-attack stability. J Aircraft 25(7):591-597. doi: 10.2514/3.45628

Cao BC (2008) Dynamic pressure measurement technology in the wind tunnel and its application in the investigation of wing-rock phenomenon (Master's thesis). Beijing: Beihang University. In Chinese.

Chen X (2003) Perturbation effect and active control on behaviors of asymmetric vortices over slender body (PhD thesis). Beijing: Beihang University. In Chinese.

Dantec Company (2000) Flow manager software and introduction to PIV instrumentation. Skovlunde: Dantec Company.

Deng X, Chen X, Wang Y, Liu P (2002) Influence of tip perturbations on behaviors of asymmetric vortices over slender body. AIAA 2002-4710.

Deng X, Wang G, Chen X, Wang Y, Liu P, Xi Z (2003) A physical model of asymmetric vortices flow structure in regular state over slender body at high angle of attack. Sci China E 46(6):561-573. doi: 10.1360/02ye0164

Deng XY, Tian W, Ma BF, Wang YK (2008) Recent progress on the study of asymmetric vortex flow over slender bodies. Acta Mech Sin 24(5):475-487. doi: 10.1007/s10409-008-0197-3

Ericsson LE (1989) Wing rock generated by forebody vortices. J Aircraft 26(2):110-116. doi: 10.2514/3.45731

Ericsson L, Beyers M (2003) The challenge of determining combat aircraft wing rock through subscale testing. Proceedings of the 41st Aerospace Sciences Meeting and Exhibit, Aerospace Sciences Meetings; Reno, USA.

Katz J (1999) Wing/vortex interactions and wing rock. Progr Aero Sci 35(7):727-750. doi: 10.1016/S0376-0421(99)00004-4

Lamont PJ (1982) Pressures around an inclined ogive cylinder with laminar, transitional, or turbulent separation. AIAA J 20(11):1492-1499. doi: 10.2514/3.51212

Ma BF, Deng XY, Rong Z, Wang B (2015) The self-excited rolling oscillations induced by fore-body vortices. Aero Sci Tech 47:299-313. doi: 10.1016/j.ast.2015.10.003

Nelson RC, Pelletier A (2003) The unsteady aerodynamics of slender wings and aircraft undergoing large amplitude maneuvers. Progress in Aerospace Sciences 39(2-3):185-248. doi: 10.1016/S0376-0421(02)00088-X

Nguyen LT, Yip LP, Chambers JR (1981) Self-induced wing rock of slender delta wings. Proceedings of the Atmospheric Flight Mechanics Conference; Albuquerque, USA.

Quast T (1991) A study of high alpha dynamics and flow visualization for a 2.5% model of the F-18 HARV undergoing wing rock (Master's thesis). South Bend: University of Notre Dame.

Rong Z, Deng XY, Wang B, Ma BF (2010) The development of a synchronous measurement and control technique of wing rock/PIV/dynamic pressure and application to the study of wing rock. Acta Aerodynamica Sinica 28(2):180-187.

Ross AJ, Nguyen LT (1988) Some observations regarding wing rock oscillations at high angles of attack. AIAA Paper 88-4371-CP.

Scanivalve Corporation (1992) Hyscan 2000 pressure system. Liberty Lake: Scanivalve Corporation.

Wang B, Huang CD, Ma BF, Deng XY (2009) The control method of precise reproduction of the wing rock motion. Journal of Experiments in Fluid Mechanics 23(1):79-104.

Wang G (2003) Experimental study of flowfield structure around an ogive-cylinder (PhD thesis). Beijing: Beihang University. In Chinese

3

Ferrite Quantification Methodologies for Duplex Stainless Steel

Arnaldo Forgas Júnior[1], Jorge Otubo[1], Rodrigo Magnabosco[2]

ABSTRACT: In order to quantify ferrite content, three techniques, XRD, ferritoscope and optical metallography, were applied to a duplex stainless steel UNS S31803 solution-treated for 30 min at 1,000, 1,100 and 1,200 °C, and then compared to equilibrium of phases predicted by ThermoCalc® simulation. As expected, the microstructure is composed only by austenite and ferrite phases, and ferrite content increases as the solution treatment temperature increases. The microstructure presents preferred grains orientation along the rolling directions even for a sample solution treated for 30 min at 1,200 °C. For all solution treatment temperatures, the ferrite volume fractions obtained by XRD measurements were higher than those achieved by the other two techniques and ThermoCalc® simulation, probably due to texturing effect of previous rolling process. Values obtained by quantitative metallography look more assertive as it is a direct measurement method but the ferritoscope technique should be considered mainly for *in loco* measurement.

KEYWORDS: Duplex stainless steels, Ferrite quantification, Ferritoscope, Quantitative metallography, X-ray diffraction, ThermoCalc®.

INTRODUCTION

Duplex stainless steels (DSS) are characterized ideally by equal amounts of ferrite and austenite, which provides increased mechanical resistance (680 to 880 MPa ultimate tensile strength) due to the fine grain size, typical of these steels (Sedriks 1996; Souza *et al.* 2005). They present higher corrosion resistance when compared to ferritic stainless steels, promoted by high content of chromium, nitrogen and molybdenum, and, due to the presence of austenite, the DSS present good ductility and toughness (250 J impact toughness) (Young *et al.* 2007).

This behavior (corrosion resistance, strength) is promoted by the presence of approximately 50% volume fraction of ferrite (Kashiwar *et al.* 2012), and therefore the control of the ferrite content in DSS is necessary to ensure the desired properties.

DSS form a specific group of materials that are used in a variety of industrial applications. The main factor that makes these materials useful for the production of components for the chemical, food and aerospace industries, which requires long-life components, are precisely their high corrosion resistance and mechanical strength (Oliveira *et al.* 2014).

Nowadays, the use of DSS has grown considerably in several applications, such as, for pressure vessels, heat exchangers, pipelines, evaporators, storage towers and pipelines in oil and gas industries for transportation of dry and/or wet carbon dioxide (Kashiwar *et al.* 2012) and rod technologies developed in the aerospace industry (Badoo 2008). DSS are also used on a large scale in the construction of off-shore platforms, in seawater injection systems and systems applied for removal of CO_2 and H_2S (Souza *et al.* 2005).

1.Departamento de Ciência e Tecnologia Aeroespacial – Instituto Tecnológico de Aeronáutica – Divisão de Engenharia Aeronáutica e Mecânica – São José dos Campos/SP – Brazil. **2.**Fundação Educacional Inaciana – São Bernardo do Campo/SP – Brazil.

Author for correspondence: Arnaldo Forgas Júnior | Departamento de Ciência e Tecnologia Aeroespacial – Instituto Tecnológico de Aeronáutica – Divisão de Engenharia Aeronáutica e Mecânica | Praça Marechal Eduardo Gomes, 50 – Vila das Acácias | CEP: 12.228-900 – São José dos Campos/SP – Brazil | Email: forgas.jr@gmail.com

The increase of the demand for DSS in industrial applications made mandatory the control of the manufacturing processes and the necessity to know accurately the quantities of the phases before the final use. Unfortunately, the phase quantification can be influenced by the existing measurement techniques. Previous studies (Magnabosco and Spomberg 2011; Tavares *et al.* 2012) showed quantification of the ferrite volume fraction of DSS using two different techniques: quantitative metallography after Beraha etching and magnetic measurements with a ferritoscope. The results showed considerable discrepancies between the values obtained by these techniques (Magnabosco and Spomberg 2011).

Another technique to evaluate the volume fraction of the phases is by analysis of the peak intensity of X-ray diffraction (XRD) patterns. For a randomly oriented DSS sample, quantitative measurements of the relative ferrite and austenite content can be obtained from XRD patterns taking into account the total integrated intensity of all diffraction peaks for each phase, which is proportional to the volume fraction of that phase. If the crystalline phase or grains of each phase are randomly oriented, the integrated intensity from any single diffraction peak related to a *(hkl)* crystalline plane is also proportional to the volume fraction of that phase (ASTM E 975-13; Cullity and Stock 2001).

The main objective of this research is to compare different methodologies of ferrite phase quantification: quantitative optical metallography, magnetic measurement by ferritoscope and analysis of the peak intensities of the phases by X-ray diffraction, comparing them to the equilibrium of phases predicted by ThermoCalc® simulation.

EXPERIMENTAL PROCEDURES

The studied material is a hot rolled DSS (UNS S31803) plate 300 mm long × 200 mm wide × 3 mm thick with the chemical composition shown in Table 1.

Samples from the original plate were solution-treated under nitrogen atmosphere at three different temperatures, 1,000, 1,100 and 1,200 °C, for 30 min, and then cooled in water. For each temperature, the solution-treated strips samples were cut into specimens of 10 × 10 mm and subsequently embedded in thermosetting resin. The samples were analyzed considering the plan surface of the plate viewing from the rolling direction. Further, the specimens were ground to 500-mesh emery paper and then polished down to 1 μm size diamond paste lubricated with ethanol in a semi-automatic polishing equipment.

After polishing, the microstructures of the specimens were revealed with modified Beraha etchant (20 mL of hydrochloric acid + 80 mL of distilled and deionized water), and to this stock solution 1 g of potassium metabisulphite + 2 g of ammonium bifluoride were added just before etching. The etching time was approximately 30 s, and the etching was interrupted by immersion in water. The surfaces after etching were dried by ethanol evaporation, aided by a cold air jet, enabling the micrograph recording of the samples.

The volume fraction of ferrite was quantified by three different techniques:

- Magnetic measurement by ferritoscope: after polishing of the specimens, ten measurements of the volume fraction of ferrite of each sample were taken with a FISCHER MP30 ferritoscope; the equipment was calibrated with appropriate standards, with detection limit of 0.1% ferrite. The ferritoscope is a method that measures the fraction of the ferromagnetic phase. The measurement method is based on magnetic induction in which a magnetic field is generated by a coil that interacts with the magnetic phase of the sample. The changes in the magnetic field induce a voltage proportional to the ferromagnetic phase content in a second coil. This voltage is then evaluated.
- Quantitative optical metallography: after polishing and etching, measurements of phase contents were made by the point counting technique prescribed in ASTM Standard E562-02, using an image analysis routine with Leica QMetals software, connected to a Leica DMLM microscope. Ten fields per sample were analyzed at 500X magnification.
- X-ray diffraction: to confirm the existence of ferrite and austenite, X-ray diffraction patterns were obtained using a Shimadzu XRD-7000 diffractometer under Cu-Kα radiation; diffraction scans were performed at $30° < 2\theta < 120°$ at a rate of 1°/min and sampling every 0.04°; Cu source was excited at acceleration voltage of 30 kV and current of 30 mA.

Table 1. Chemical composition (mass percentage) of DSS.

Cr	Ni	Mo	Mn	N	C	Si	P	S	Fe
22.07	5.68	3.20	1.38	0.17	0.017	0.34	0.02	0.001	Balance

The method used to determine phase content from XRD tests was described in the study of Moser *et al.* (2014). The quantitative estimation is based on the use of internal ratios. Assuming that the grains are randomly oriented, the integrated intensity of a given phase *i* is proportional to the volume fraction of that phase, *Vi*, as shown in Eq. 1:

$$V_i = \frac{\frac{1}{n}\sum_1^n \frac{I_i^j}{R_i^j}}{\frac{1}{n}\sum_1^n \frac{I_\gamma^j}{R_\gamma^j} + \frac{1}{n}\sum_1^n \frac{I_\alpha^j}{R_\alpha^j}} \quad (1)$$

where: n is the number of peaks examined for each phase (being i the phases γ or α); $j = 1, 2, 3, \ldots, n$; R is the material scattering factor and is described in Eq. 2:

$$R_{hkl} = \left(\frac{1}{V^2}\right)\left[|F|p\left(\frac{1+\cos^2 2\theta}{\sin^2\theta\cos\theta}\right)\right]e^{-2M} \quad (2)$$

where: V is the volume fraction of the unit cell; F is the structure factor; *hkl* are the Miller indexes of the reflection plane; p is the multiplying factor associated to the specific *hkl* plane; θ is the diffraction angle; and e^{-2M} is Debye-Waller Factor (DWF).

Table 2 shows the values of those variables for specific planes of typical austenite and ferrite phases. The volume fraction of the unit cell V is obtained through the lattice parameter a of ferrite and austenite, determined from XRD patterns after the application of the Nelson-Riley extrapolation method for accurate unit-cell dimensions of crystals (Nelson and Riley 1945).

Equilibrium in the solution-treatment temperatures was predicted by ThermoCalc® software using TCFe7 database, inputting in the software:

- Chemical composition according to Table 1.
- Temperature range (1,000 to 1,200 °C).

Table 2. Values for the determination of the material scattering factor for specific planes of ferrite and austenite in stainless steels (ASTM E 975-13; Cullity and Stock 2001).

Austenite (FCC)				Ferrite (BCC)			
hkl	F	p	DWF	*hkl*	F	p	DWF
(111)	17.454	8	0.960	(110)	17.285	12	0.958
(200)	16.460	6	0.947	(200)	14.695	6	0.918
(220)	13.703	12	0.897	(211)	12.994	24	0.880
(311)	12.345	24	0.861	(220)	11.752	12	0.843

FCC: Face centred cubic; BCC: Body centred cubic.

So, it is possible to analyze the measured ferrite contents by the three described methods and the expected equilibrium ferrite fraction from ThermoCalc® simulation.

RESULTS AND DISCUTION

Figures 1, 2 and 3 show, respectively, the micrographs of specimens' solution treated at 1,000, 1,100 and 1,200 °C etched with modified Beraha reagent. It is verified the presence of only two phases: ferrite (dark gray) and austenite (light gray), as expected (Nilsson 1992; Nilsson and Chai 2012).

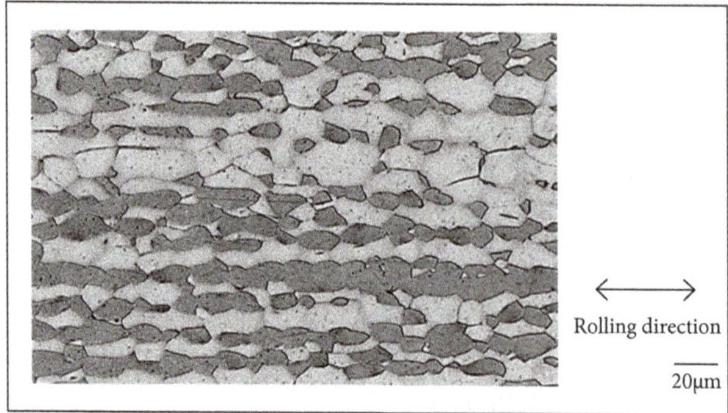

Figure 1. Micrograph of the sample heat-treated at 1,000 °C.

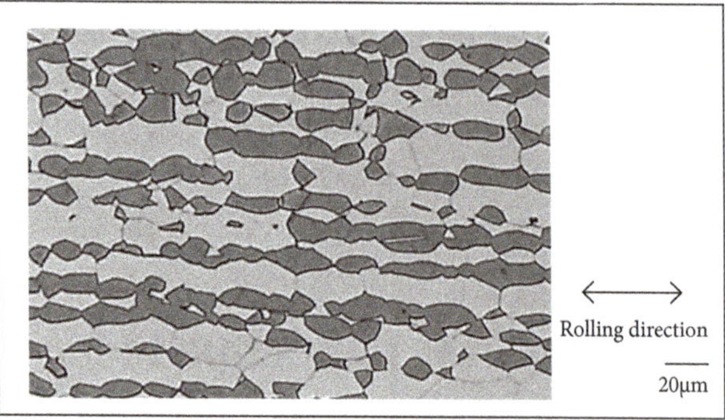

Figure 2. Micrograph of the sample heat-treated at 1,100 °C.

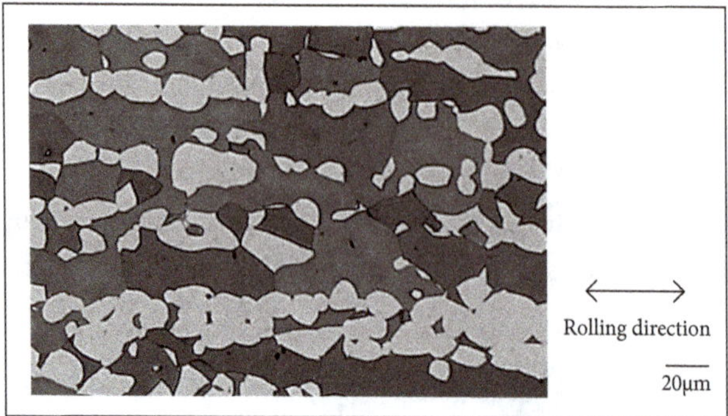

Figure 3. Micrograph of the sample heat-treated at 1,200 °C.

Micrographs showed that ferrite and austenite grains distribution present a preferred orientation in the rolling direction as indicated by double arrows. Those preferred grains orientation decreased but it was not completely eliminated as the solution treatment temperatures increase, and the random structure needed for quantitative optical microscopy and XRD determination of ferrite phase previously discussed is impaired. As expected, the increase in solution treatment temperature leads to larger grain sizes.

Figures 4, 5 and 6 present XRD patterns for samples solution treated at 1,000, 1,100 and 1,200 °C, respectively, showing that only ferrite and austenite phases are identified in the microstructures, confirming the observations of optical micrographs (Figs. 1 to 3). The unit cell parameters (a) for ferrite and austenite (calculated from XRD results) were, respectively, 0.2880 and 0.3601 nm in all solution-treatment temperatures, and the respective unit cell (V) volumes were 0.023879 and 0.046711 nm^3.

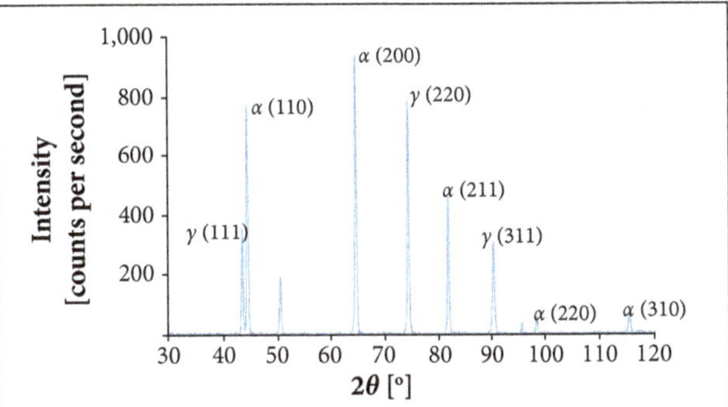

Figure 4. X-Ray diffraction for sample solution treated at 1,000 °C.

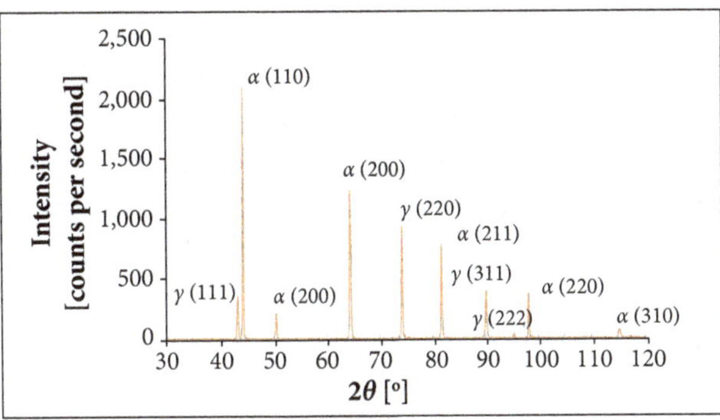

Figure 5. X-ray diffraction for sample solution treated at 1,100 °C.

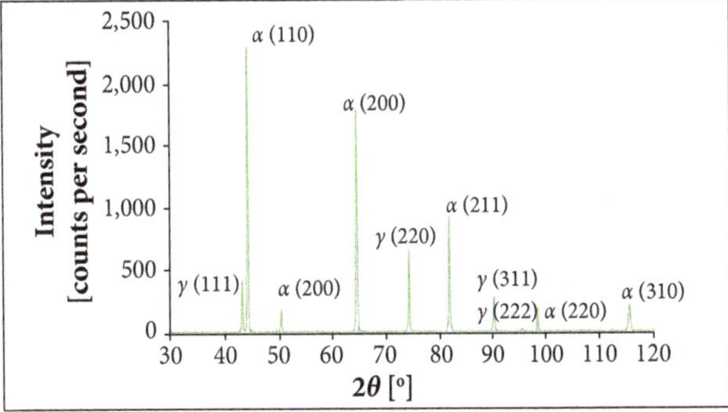

Figure 6. X-ray diffraction for sample solution treated at 1,200 °C.

Duplex stainless steel solidifies from liquid phase through ferritic field, and between 1,200 and 1,000 °C, approximately, ferrite plus austenite area is achieved. In this field, it forms duplex structure. Since the fractions of ferrite and austenite may be calculated, these relative fractions can be controlled by selecting the appropriate heating temperature (Nilsson 1992). Figure 7 illustrates this variation as a function of solution-treated temperature obtained by ThermoCalc® software with TCFE7 database. It is noted that the phases present in this temperature range are only ferrite and austenite, and, as expected, with increasing solution-treated temperature higher is the volume fraction of ferrite.

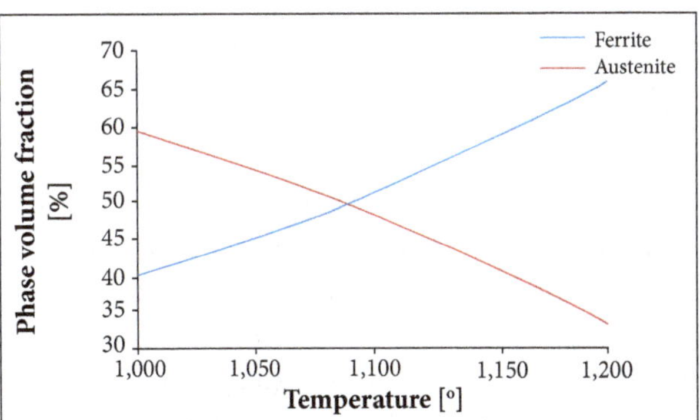

Figure 7. Results from ThermoCalc® software simulation showing only two equilibrium phases, from 1,000 to 1,200 °C temperature range.

Table 3 presents the ferrite volume fraction of the solution-treated samples at 1,000, 1,100 and 1,200 °C for all techniques: XRD, ferritoscope, quantitative optical metallography and also ThermoCalc® prediction. It can be noted that increasing the solution-treatment temperature the ferrite content increases as well, considering all applied techniques.

Figure 8 shows a comparison between the values of ferrite volume fraction obtained by these methodologies, XRD, ferritoscope, and quantitative optical microscopy, all of them compared to ThermoCalc® prediction.

This figure could be separated into two regions: region 1 corresponding to higher ferrite volume fraction varying from

Table 3. Ferrite content by measurement techniques.

Temperature (°C)	Ferrite content (%)			
	XRD	Ferritoscope	Quantitative optical microscopy	ThermoCalc®
1,000	58.4	39.8 ± 0.5	41.5 ± 0.9	40.4
1,100	68.5	46.2 ± 0.5	48.1 ± 0.8	51.3
1,200	87.2	57.0 ± 0.7	58.5 ± 1.5	66.4

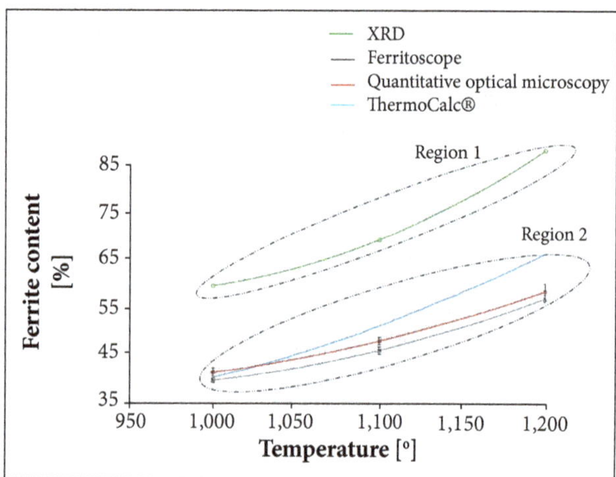

Figure 8. Ferrite content obtained by the four measuring techniques.

56 to 88%, obtained by XRD measurements, and region 2 with lower ferrite content varying from 40 to 66%, obtained by the other methods (quantitative optical metallography, ferritoscope and ThermoCalc® prediction), both taking into account the solution-treated temperatures.

The high values of ferrite content, region 1, presented by XRD measurements, could be attributed to the preferred grains orientations that persist even for the solution-treated sample at 1,200 °C conditioned to the previous hot rolled 3 mm thick plate. Nevertheless, all techniques show the same trend of increasing ferrite volume fraction with increasing solution-treated temperature, as provided in ThermoCalc®.

In region 2, it is verified that solution-treated samples at 1,000 °C produced similar results comparing the two techniques, quantitative microscopy and ferritoscope, to the equilibrium prediction of ThermoCalc® simulation. However, at 1,100 and 1,200 °C, the result obtained by ThermoCalc® simulation was higher than that obtained by quantitative microscopy and ferritoscope, increasing proportionally as the solution treatment temperature increases. This difference could be attributed to the fact that ThermoCalc® simulation does not take into account texture and especially grain size, which can affect the kinetics of phase formation. Quantitative microscopy and ferritoscope showed equivalent results as far as the differences are within the experimental errors. It can be concluded that values obtained by quantitative optical microscopy could be more assertive as far as it is a direct measurement method, but ferritoscope technique could be a fast alternative tool for ferrite quantification *in loco*. All three techniques, although giving different results, showed clearly the tendency of the increase the volume fraction of ferrite as the solution treatment temperature increases, in accordance with the equilibrium prediction.

CONCLUSIONS

Different techniques were used to quantify the volume fraction of ferrite in a duplex stainless steel and were compared with ThermoCalc® simulations. All of them showed clearly the increase in ferrite content as the solution treatment temperature increases.

The plot of ferrite content as a function of solution treatment temperature showed two distinct regions, one related to XRD technique, presenting higher values of volume fraction, and another region with lower values obtained by optical metallography, ferritoscope techniques and ThermoCalc® simulations.

The higher values of ferrite content presented by XRD technique could be attributed possibly to the influence of the texture imposed by hot rolling, as the preferred grains orientation along the rolling direction is not eliminated even in a solution treatment at 1,200 °C.

Quantitative metallography seems to be the most assertive technique to measure volume fraction of the phases in DSS, but ferritoscope technique should be considered for practical applications such as *in loco* measurement.

REFERENCES

Badoo NR (2008) Stainless steel in construction: a review of research, applications, challenges and opportunities. Journal of Constructional Steel Research 64(11):1199-1206. doi: 10.1016/j.jcsr.2008.07.011

Cullity BD, Stock SR (2001) Elements of X-ray diffraction. 3rd ed. New Jersey: Prentice-Hall.

Kashiwar A, Vennela N, Kamath SL, Kathirkar RK (2012) Effect of solution annealing temperature on precipitation on 2205 duplex stainless steel. Mater Char 74:55-63. doi: 10.1016/j.matchar.2012.09.008

Magnabosco R, Spomberg S (2011) Comparative study of ferrite quantification methods applied to duplex stainless steel. Proceedings of the 7th European Stainless Steel Conference — Science and Market; Milan, Italy.

Moser NH, Gross TS, Korkolis YP (2014) Martensite formation in conventional and isothermal tension of 304 austenitic stainless steel measured by X-ray diffraction. Metall Mater Trans 45(11):4891-4896. doi: 10.1007/s11661-014-2422-y

Nelson JB, Riley DP (1945) An experimental investigation of extrapolation methods in the derivation of accurate unit-cell dimensions in crystals. Proc Phys Soc 57(3):160-177. doi: 10.1088/0959-5309/57/3/302

Nilsson JO (1992) Super duplex stainless steels. Materials Science and Technology 8(8):685-700. doi: 10.1179/mst.1992.8.8.685

Nilsson JO, Chai G (2012) The physical metallurgy of duplex stainless steel. Sandvik Materials Technology, R&D Centre, S-81181. Sandviken: Sandvik Materials Technology.

Oliveira CA, Diniz AE, Bertazzoli R (2014) Correlating tool wear, surface roughness and corrosion resistance in the turning process of super duplex stainless steel. J Braz Soc Mech Sci Eng 36(4):775-785. doi: 10.1007/s40430-013-0119-6

Sedriks AJ (1996) Corrosion of stainless steel. 2nd ed. New York: John Wiley.

Souza GC, Pardal JM, Tavares SSM, Fonseca MPCF, Martins JLF, Moura EP, Cardote IF (2005) Evaluation of proportion of phases in joints welded from duplex stainless steel pipes by means of non-destructive testing. Weld Int 29(10):762-770. doi: 10.1080/09507116.2014.932985

Tavares SSM, Pardal JM, Abreu HFG, Nunes CS, Silva MR (2012) Tensile properties of duplex UNS s32205 and lean duplex UNS S32304 steels and the influence of short duration 475 °C aging. Materials Research 15(6):859-864. doi: 10.1590/S1516-14392012005000116

Young MC, Tsay LW, Shin CS, Chan SLI (2007) The effect of short time post-weld heat treatment on the fatigue crack growth of 2205 duplex stainless steel welds. Int J Fatig 29(12):2155-2162. doi: 10.1016/j.ijfatigue.2007.01.004

Immersed Boundary Method Based on the Implementation of Conservation Equations Along the Boundary Using Control-Volume Finite-Element Scheme

Seyedeh Nasrin Hosseini[1], Seyed Mohammad Hossein Karimian[1]

ABSTRACT: In this study conservation equations were implemented along the boundaries via ghost control-volume immersed boundary method. The control-volume finite-element method was applied on a cartesian grid to simulate 2-D incompressible flow. In this approach, mass and momentum equations were conserved in the whole domain including boundary control volumes by introducing ghost-control volume concept. The Taylor problem was selected to validate the present method. Four different case studies of Taylor problem encompassing both inviscid and viscous flow conditions in ordinary and 45° rotated grid were used for more investigation. Comparisons were made between the results of the present method and those obtained from the exact solution. Results of the present method indicated accurate predictions of the velocity and pressure fields in midline, diagonal, and all boundaries. The agreement between the results of the present method and the exact solution was very good throughout the whole temporal domain. Furthermore, comparison of the rate of kinetic energy decay in viscous case showed same level of agreement between the results.

KEYWORDS: Immersed boundary method, Control-volume-based finite element, Sub-control volumes, Conservation of mass and momentum equations, Ghost node, Ghost sub-control volume.

INTRODUCTION

The immersed boundary method (IBM) is known as a powerful approach for simulating flows in moving boundary and complex geometry problems. In this method, discretization of equations is carried out on a Cartesian grid, which is simple to generate. However, the boundary does not conform to the grid lines, and therefore indirect methods are employed to apply the boundary conditions. This creates a range of different methods developed in the context of IBM which are applied to elastic (Peskin 1972, 1982; Beyer Jr 1992; Fauci and McDonald 1995; Zhu and Peskin 2003) and solid (Berger and Aftosmis 1998; Khadra *et al.* 2000; Tseng and Ferziger 2003; Saiki and Biringen 1996) boundaries. The conventional ghost-node method is currently used in problems with solid boundaries, where the value of ghost-node is set as to meet the boundary conditions. In ghost-node methods, finite difference scheme is usually used to simulate the flow field and the value of ghost-node is determined using a kind of interpolation schemes (Mittal and Iaccarino 2005; Majumdar *et al.* 2001; Ghias *et al.* 2004; Mittal *et al.* 2008). While these approaches are considered fairly fast in convergence and simple in application, mass and momentum equations are not conserved in applying boundary conditions. However, the so-called cut-cell method is a complicated approach based on Cartesian grid (Clarke *et al.* 1986; Udaykumar *et al.* 2001, 1999, 1996; Ye *et al.* 1999), which implements conservation laws in boundary cells. In this method the shape of Cartesian cells in the vicinity of the

1.Amirkabir University of Technology – Aerospace Engineering Department – Tehran/Tehran – Iran.

Author for correspondence: Seyed Mohammad Hossein Karimian | Amirkabir University of Technology – Aerospace Engineering Department | 424 Hafez Ave. | 015875-4413 – Tehran/Tehran – Iran | Email: hkarim@aut.ac.ir

boundary is changed to fit the boundary. In cut-cell method, cells are divided by the boundary, and conservation laws are implemented in divided cells conforming to the boundary. Comparing to ghost node methods used in IBM, cut-cell method is extremely complicated. This is because the boundary may cut the Cartesian grids anywhere on the cells and create new arbitrary shape. It would make it more difficult to discretize the equations and calculation of fluxes particularly in 2- and 3-D and moving problems.

In the present study, an immersed boundary method based on CVFE scheme is proposed in the context of ghost node concept in which conservation of conserved quantities is enforced. Importantly, the present method has the capability to conserve mass and momentum equations along the boundary. The present approach is different from the cut-cell method such that boundary cell shapes remain unchanged.

NUMERICAL ALGORITHM

The governing equations in the present method are solved via CVFE scheme, which was presented by Minkowycz *et al.* (1988) to discretize governing equations. Sub-control-volume (SCV) and node types are further explained to implement boundary conditions.

CONTROL-VOLUME FINITE-ELEMENT METHOD

In this scheme, solution domain is always discretized into a number of Cartesian elements. As shown in Fig. 1a, a local coordinate system (s,t) is defined in the middle of each element. This local coordinate system divides each element into 4 SCVs. Each SCV is associated with an element node at its vertex. Therefore, as shown in Fig. 1c, the grey area represents a control volume made from surrounding 4 SCVs neighbour elements. All primitive variables are located at the vertices of the elements, placing them in the middle of each control volumes.

Although governing equations are finally conserved on control volumes such as the one shown in Fig. 1b, their formation are done through the assembly of elemental equations (Minkowycz *et al.* 1988). Elemental equations of each element include conservation of governing equations on the 4 SCVs of that element. Variables and their gradients should be evaluated at the integration points (Fig. 1a) to determine the flux at each sub-control surface. Variables with elliptic nature or of diffusion type such as pressure and diffusion can be calculated using bilinear interpolation. Minkowycz *et al.* (1988) presented a bilinear shape function to determine the value of variables everywhere in the element (Fig. 1a). Accordingly the value of variable φ and its gradients can be determined by:

$$\varphi(s,t) = \sum_{i=1}^{4} N_i(s,t)\Phi_i \tag{1}$$

$$\frac{\partial\varphi(s,t)}{\partial x} = \sum_{i=1}^{4} \frac{\partial N_i(s,t)}{\partial x}\Phi_i = \sum_{i=1}^{4} D_{x,i}\,\Phi_i \tag{2}$$

where ϕ_i is the value of φ at the vertices of each element; N_i is the ith bilinear shape function.

Modelling of other variables without elliptic nature or diffusion type such as velocity components in mass fluxes and convection terms will be discussed in more details later. Details of the CVFE method and the formation of the system of governing equation were presented by Minkowycz *et al.* (1988).

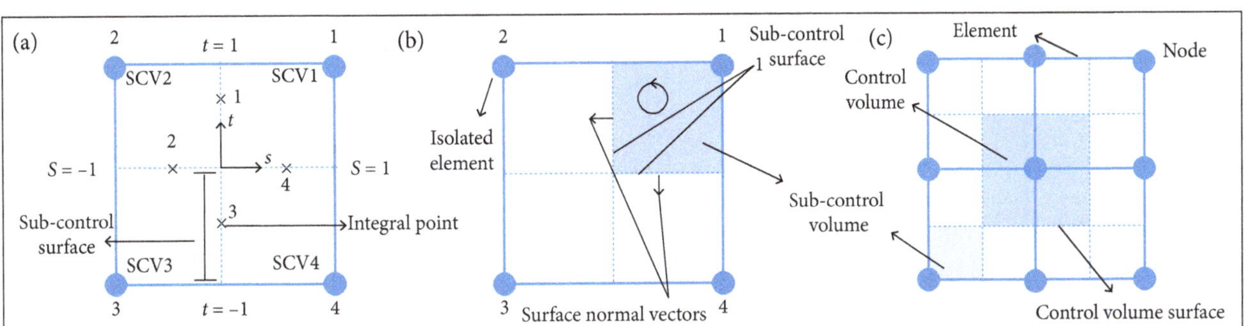

Figure. 1. (a) Definitions of the element. Local coordinate system of (*s,t*) is located in the middle of the element, sub-control surface is indicated, and integral points are shown via cross symbols in the middle of sub-control surfaces; (b) The grey area is the SCV, and surface normal vectors are indicated in its outward direction; (c) The dark grey area in the center of the figure is control volume made up of 4 surrounding SCVs and the light grey area is SCV.

SUB-CONTROL VOLUMES AND NODE TYPES

Discretization of governing equations and calculation of fluxes are done on SCVs; hence, the classification of different sub-control volumes and nodes is described here. There are 4 SCVs in each element as previously explained according to Fig. 1a. Depending on the location of elements in the domain, the SCVs and nodes are classified into 3 types in this paper. The first type of SCV is the "ordinary" or "fluid" one that is in the middle of the solution domain and it has no boundary in its SCV or in its related element (Fig. 2). An ordinary node is assigned to each related ordinary SCV. In the second type the boundary has crossed the SCV. This type of SCV and its pertaining node are called ghost SCV type I and ghost node type I, respectively (Fig. 3). Lastly, as shown in Fig. 4, the third type is defined when the boundary is placed in the SCVs of fluid nodes in the element. These SCVs are called ghost SCV type II and accordingly each related node is called ghost node type II (Fig. 4). To conclude, in this method, whenever the immersed boundary is placed within an element, nodes outside of the flow field are called ghost nodes (nodes 2 and 3 in Figs. 3 and 4) and their corresponding SCVs (SCVs 2 and 3 in Figs. 3 and 4) are called ghost SCVs. In this paper boundary conditions are applied via ghost SCVs (Fig. 5). Note that SCVs of both fluid and ghost nodes are always considered as ordinary SCVs or ghost SCVs, respectively, regardless of the boundary location. As noted earlier in conventional IBMs (sharp interface methods — Seo and Mittal 2011; Ghias *et al.* 2007) boundary conditions are applied via the assignment of appropriate values for the flow variables to the ghost nodes. These values are mostly assigned by a kind of interpolation scheme (Mittal and Iaccarino 2005; Majumdar *et al.* 2001; Ghias *et al.* 2004, 2007). In the present method, however, flow variables on the ghost nodes are determined by implementation of conservation laws and the boundary condition on ghost SCVs. Details of the method will be discussed in following section.

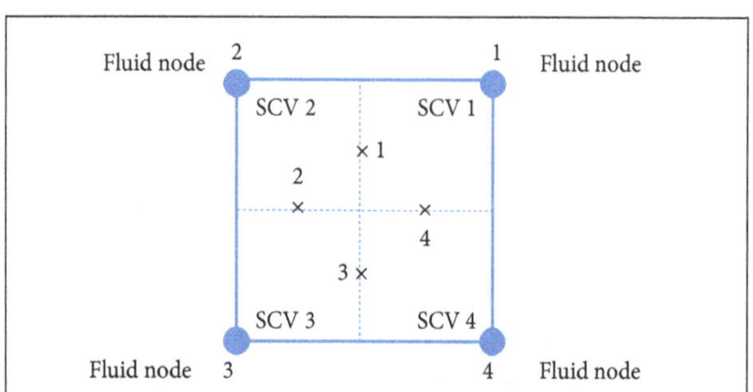

Figure 2. Ordinary SCV and node. All 4 SCVs and nodes are ordinary (fluid).

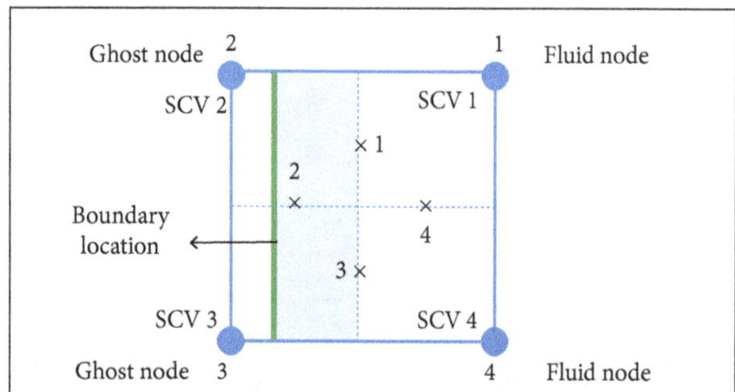

Figure 3. Ghost SCV and node type I; grey area indicated in SCVs 2 and 3 are ghost SCVs type I, nodes 1 and 4 are fluid nodes and nodes 2 and 3 are ghost node type I.

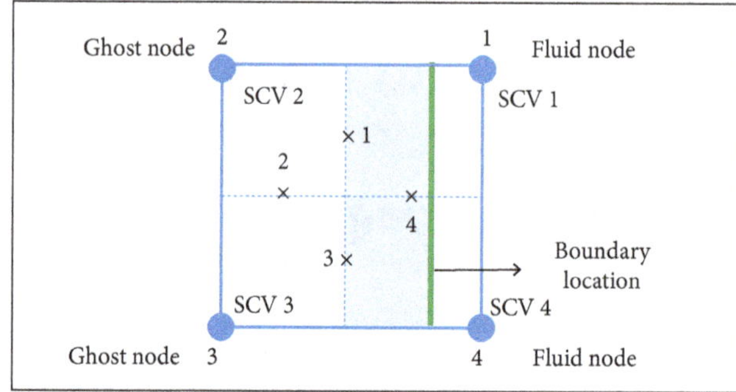

Figure 4. Ghost SCV and node type II. The grey area indicated in SCVs 1 and 4 are ghost SCV type II related to nodes 2 and 3 (ghost nodes type II). Nodes 1 and 4 are fluid ones.

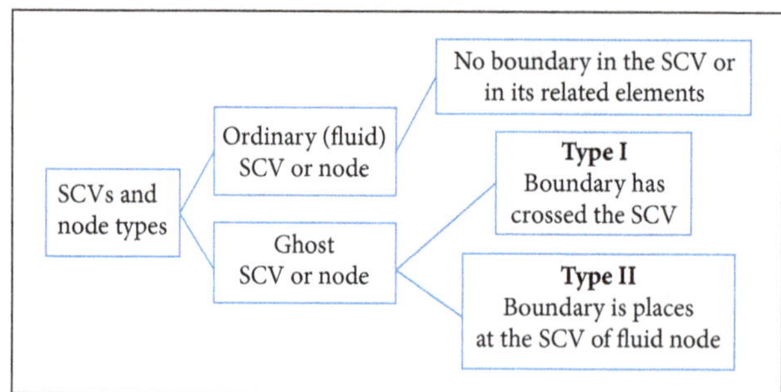

Figure 5. SCV and node types classification.

GOVERNING EQUATIONS AND DISCRETIZATION

In Eq. 3 there is a detail analysis of how Navier-Stokes equations were discretized. The integral form of the incompressible Navier-Stokes equations for 2-D flow is given by

$$\int_V \frac{\partial Q}{\partial t} dV + \int_S (E ds_x + F ds_y) = \int_S (G ds_x + H ds_y) \tag{3}$$

where Q is the vector of conserved quantities; E and F are convection flux vectors; G and H are diffusion flux vectors; V is volume.

The extended form of these vectors is:

$$Q = \begin{pmatrix} o \\ \rho U \\ \rho V \end{pmatrix}; \ E = \begin{pmatrix} (\rho u) \\ (\rho u)u + p \\ (\rho u)v \end{pmatrix};$$

$$F = \begin{pmatrix} (\rho v) \\ (\rho v)u \\ (\rho v)v + p \end{pmatrix}; \ G = \begin{pmatrix} 0 \\ \tau_{xx} \\ \tau_{xy} \end{pmatrix}\tau; \ H = \begin{pmatrix} 0 \\ \tau_{yx} \\ \tau_{yy} \end{pmatrix} \quad (4)$$

$$\tau_{xx} = 2\mu\frac{\partial u}{\partial x}; \tau_{yy} = 2\mu\frac{\partial v}{\partial y};$$

$$\tau_{xy} = \tau_{yx} = \mu(\frac{\partial u}{\partial y} + \frac{\partial v}{\partial x})$$

where ρ represents density; u and v are velocity in x and y directions, respectively; τ is shear stress; μ is viscosity; p means pressure.

Upper-case letters were used to indicate nodal values and the lower-case ones, to show the values of variables on integral points (*ip*). After substituting stress tensor within G and H, the simplified form would be as shown in Eq. (5).

$$G = \begin{pmatrix} 0 \\ \mu\frac{\partial u}{\partial x} \\ \mu\frac{\partial v}{\partial x} \end{pmatrix}; \ H = \begin{pmatrix} 0 \\ \mu\frac{\partial u}{\partial y} \\ \mu\frac{\partial v}{\partial y} \end{pmatrix} \quad (5)$$

Firstly the ordinary SCV is explained. Navier-Stokes equations should be discretized in all of the four SCVs of each element in order to form element-level equations. In a case of ordinary SCV, the process of discretizing is straightforward as described in Karimian and Schneider (1994a). This process is explained in more details as follows. SCV 1 in Fig. 1a is considered here where Eq. 3 is written for this SCV as follows:

$$\int_{SCV1}\frac{\partial Q}{\partial t}dV + \int_{SS1\ \&\ SS4}(Eds_x + Fds_y) = = \int_{SS1\ \&\ SS4}(Gds_x + Hds_y) \quad (6)$$

where SS stands for the inner sub-control surface shown in Fig. 1a; ds_x and ds_y are the components of normal surface vector in the outward direction.

The volume integral of the transient term is estimated using a lumped approach. Surface-integrals of E, F, G, and H are calculated by their average values over SS at the midpoint location, which is the *ip*. For the diffusion flux vectors G and H, bi-linear interpolation (Eq. 2) is used to directly evaluate the components of stress tensor (Karimian and Schneider 1994b). In the convection flux vectors E and F, pressure is evaluated using bilinear interpolation (Eq. 1), and the momentum fluxes are linearized with respect to mass fluxes and. Velocity components u and v in mass fluxes are called integral-point convecting velocities and have been previously denoted by (ρu) and (ρv) (Karimian and Schneider 1994a). Other values of u and v in the momentum fluxes, which are convected by the mass fluxes through the control-volume surfaces, are called convected velocities. Convecting and convected velocities are cell-face, which are modelled in terms of nodal values of velocity and pressure.

Karimian and Schneider (1994a) reported the implementation of the corresponding governing equations of flow to derive cell-face velocities (convected and convecting velocities) (Karimian and Schneider 1994a). In this method convected velocity is obtained from the following equation:

$$\rho\frac{\partial u}{\partial t} + \rho q\frac{\partial u}{\partial s} + \frac{\partial p}{\partial x} - \mu\nabla^2 u = 0 \quad (7)$$

In Eq. 7 the convection term is represented in stream wise direction and $q = (u^2 + v^2)^{1/2}$. Expression for convected velocity u is obtained on integration points which encompass all relevant variables related to flow condition. The convecting velocity $\hat{u}$ on *ip* is obtained from Eq. 8 as follows:

$$\left(\rho\frac{\partial u}{\partial t} + \rho q\frac{\partial u}{\partial s} + \frac{\partial p}{\partial x} - \mu\nabla^2 u\right) - - u\left(\rho\frac{\partial u}{\partial x} + \rho\frac{\partial v}{\partial y}\right) = 0 \quad (8)$$

For details about the modeling of cell-face velocities and their role in resolving pressure velocity decoupling in incompressible flow, see (Karimian and Schneider 1994a).

In the current research after completing the discretization of Navier-Stokes equations, a fully coupled algorithm is used to solve the resulted system of equations to obtain the flow variables (pressure and velocity components: p, u, and v). This system of equations is solved simultaneously using a band solver.

BOUNDARY CONDITIONS AND GHOST SUB-CONTROL VOLUMES

In IBM, flow variables are assigned so that their value guarantees satisfaction of boundary condition on the immersed boundary. As mentioned before, in the present method flow, variables on the ghost nodes are determined by implementation of conservation laws and the boundary condition on ghost SCVs. Therefore, the key-point in the present method is to clearly implement conservation laws on ghost SCVs along the boundaries. This process is explained here for the ghost SCVs types one and two.

Ghost Sub-Control Volume Type I

In Fig. 6 an element with ordinary SCVs 1 and 4, and ghost SCVs 2 and 3 is presented. Implementation of Eq. 3 on ordinary SCVs 1 and 4 is done as described in previous section. Thus, mass and momentum conservation equations on ordinary SCV 1 would be:

$$\rho\hat{u}_1 ds_{x_1^1} + \rho\hat{v}_4 ds_{y_4^1} = 0 \tag{9}$$

$$\frac{\rho V_1(U_1 - U_1^o)}{\Delta t} + (\rho\hat{u}_1 ds_{x_1^1})u_1 + (\rho\hat{v}_4 ds_{y_4^1})u_4 + + p_1 ds_{x_1^1} - \mu\frac{\partial u}{\partial x}ds_{x_1^1} - \mu\frac{\partial u}{\partial y}ds_{y_4^1} = 0 \tag{10}$$

$$\frac{\rho V_1(V_1 - V_1^o)}{\Delta t} + (\rho\hat{u}_1 ds_{x_1^1})v_1 + (\rho\hat{v}_4 ds_{y_4^1})v_4 + + p_4 ds_{y_4^1} - \mu\frac{\partial v}{\partial x}ds_{x_1^1} - \mu\frac{\partial v}{\partial y}ds_{y_4^1} = 0 \tag{11}$$

where: V_1 is the volume of ordinary SCV 1; the superscript "o" denotes value from the previous time step; Δt is the time step. Lower and upper numeric indices in the normal surface vector components, for instance $ds_{x_1^2}$, denote that ds_x is calculated on sub-surface 1 for the SCV 2. Similar equations can be obtained for other ordinary SCVs in the domain, *e.g.*, SCV 4 in this element.

Next the implementation of Eq. 3 on ghost SCVs is explained. Ghost SCVs 2 and 3 are type I. The grey area in Fig. 6 represents the ghost SCV 2 in the flow field. This is an "effective" volume of ghost SCV 2 denoted by V_2 this part. Substituting these parameters in Eq. 3 for SCV 2 it results in:

$$\int_{SCV2}\frac{\partial Q}{\partial t}dV + \int_{SS1\,\&\,SS2_r\&\,SSb}(Eds_x + Fds_y) - - \int_{SS1\,\&\,SS2_r\&\,SSb}(Gds_x + Hds_y) = 0 \tag{12}$$

On SSb, flux vectors E, F, G, and H are evaluated on ip_b. These flux vectors are evaluated for SS2 on ip_2. The discrete form of Eq. 12 is given by

$$\rho\hat{u}_1 ds_{x_1^2} + \rho\hat{v}_2 ds_{y_{2r}^2} + (\rho\overrightarrow{\hat{q}_b}.\overrightarrow{ds_b}) = 0 \tag{13}$$

$$\frac{\rho V_2(U_2 - U_2^o)}{\Delta t} + (\rho\hat{u}_1 ds_{x_1^2})u_1 + (\rho\hat{v}_2 ds_{y_{2r}^2})u_2 + + (\rho\overrightarrow{\hat{q}_b}.\overrightarrow{ds_b})u_b + p_1 ds_{x_1^2} + p_b ds_{x_b} - -\mu\left.\frac{\partial u}{\partial x}\right|_1 ds_{x_1^2} - \mu\left.\frac{\partial u}{\partial x}\right|_b ds_{x_b} - \mu\left.\frac{\partial u}{\partial y}\right|_b ds_{y_b} - - \mu\left.\frac{\partial u}{\partial y}\right|_2 ds_{y_{2r}^2} = 0 \tag{14}$$

$$\frac{\rho V_2(V_2 - V_2^o)}{\Delta t} + (\rho\hat{u}_1 ds_{x_1^2})v_1 + (\rho\hat{v}_2 ds_{y_{2r}^2})v_2 + + (\rho\overrightarrow{\hat{q}_b}.\overrightarrow{ds_b})v_b + p_2 ds_{y_2^2} + p_b ds_{y_b} - -\mu\left.\frac{\partial v}{\partial x}\right|_1 ds_{x_1^2} - \mu\left.\frac{\partial v}{\partial x}\right|_b ds_{x_b} - - \mu\left.\frac{\partial v}{\partial y}\right|_b ds_{y_b} - \mu\left.\frac{\partial v}{\partial y}\right|_2 ds_{y_{2r}^2} = 0 \tag{15}$$

where $\hat{q}_b = (\hat{u}_b^2 + \hat{v}_b^2)^{1/2}$ is the convecting velocity vector; $\overrightarrow{ds_b}$ is the normal surface vector in the outward direction and

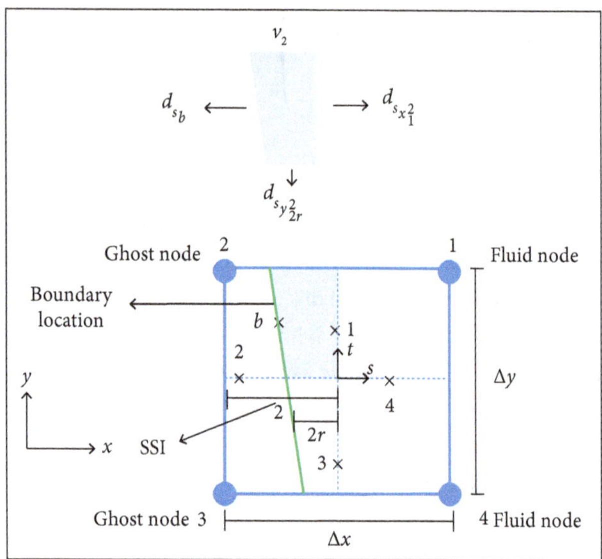

Figure 6. Ghost SCV type I (grey area in SCV 2 is considered); SSI is the left part of sub-surface 2; SS2r is the right part of sub-surface 2 along the grey area; SSb is the boundary portion in SCV 2; V_2 is the volume of the grey area; ds_b is normal surface vector of boundary in SCV 2 in direction to outward of the grey area; $ds_{x_1^2}$ is the normal surface vector of sub-surface1; $ds_{y_{2r}^2}$ is normal surface vector related to right part of sub-surface 2; Δx and Δy are grid dimensions; points 1, 2, 3, and 4 indicated with cross symbols are *ip*.

ds_{xb} and ds_{yb} are the components of $\overrightarrow{ds_b}$ in x and y directions, respectively.

Depending on the boundary condition, appropriate constraints can be forced in Eqs. 13 to 15. For instance, if the boundary b is solid, then $(\overrightarrow{p\hat{q}_b} \cdot \overrightarrow{ds_b}) = 0$, $u_b = 0$ and $v_b = 0$; p_b is described based on the nodal pressures of element using bi-linear interpolation. Moreover, velocity gradients of $\left.\frac{\partial u}{\partial x}\right|_b$, $\left.\frac{\partial u}{\partial y}\right|_b$, $\left.\frac{\partial v}{\partial x}\right|_b$ and $\left.\frac{\partial v}{\partial y}\right|_b$ are evaluated using bilinear interpolaion defined in Eq. 2.

Ghost Sub-Control Volume Type II

In Fig. 7 an element with 2 fluid nodes and two ghost nodes is shown. As mentioned in the section "Sub-Control Volumes and Node Types", SCVs 1 and 4 are considered ordinary SCVs, and SCVs 2 and 3 are ghost SCVs type II.

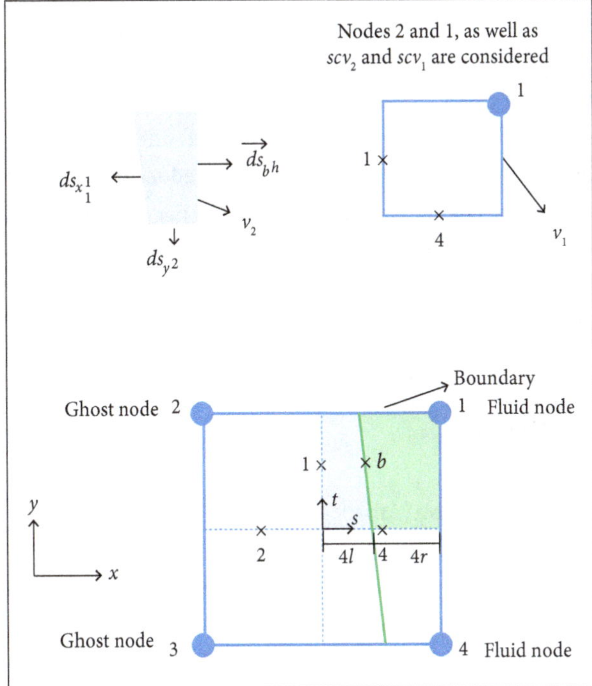

Figure 7. Ghost SCV type II (SCV 1 and SCV 2 are considered); the grey area is ghost-SCV type II assigned to node 2; the dotted area is the difference area between complete SCV 1 area and the grey area which contains fluid; 4r is the right part of sub-surface 4; 4l is the left part of sub-surface 4 along the grey area; SSb is the boundary portion in SCV 1; v_2 is volume of the grey area; v_1 is complete volume of SCV 1; ds_{bh} is normal surface vector of boundary in SCV 1 in the direction outward from the grey area; ds_{bd} is the normal surface vector on SSb in the direction outward from the dotted area; $ds_{x_1^1}$ is normal surface vector of sub-surface 1; $ds_{y_{4l}^1}$ is normal surface vector related to the left part of sub-surface 4; 1,2,3, and 4 points indicated with cross symbol are *ip*.

Since it is important to remain in the IBM general framework, any point within the flow field (*i.e.* inside the boundary) and its SCV are considered ordinary. Here Eq. 3 is applied to the whole area of SCV 1, *i.e.* the area between SS1, SS4, and node 1. Conservation laws for an ordinary SCV were introduced by Eqs. 9-11 in section ghost SCV type I. The actual area within the flow field is the dotted area between *SSb*, *SS4*, and node 1 which is shown by grey area in Fig. 6. This area is assigned to ghost node 2 and is called ghost SCV 2. Conservation laws (Eq. 3) are written for this ghost SCV, and boundary condition is applied in these equations. In the present study, boundary condition is applied via ghost SCVs, and not necessarily via the SCVs containing the boundary. Combination of conservation laws for ordinary SCV 1 and ghost SCV 2 will result in the conservation of conserved quantities for the dotted area in SCV 1, which is actually within the flow filed. Implementation of Eq. 3 on ghost SCVs is explained next. Mass conservation equation for the grey area is written as:

$$\rho\hat{u}_1 ds_{x_1^1} + \rho\hat{v}_4 ds_{y_{4l}^1} + \left(\overrightarrow{\rho\hat{q}_b}.\overrightarrow{ds}_{bh}\right) = 0 \tag{16}$$

where: $\overrightarrow{ds}_{bh}$ is the normal surface vector on *SSb*. As shown in Fig. 6 surface 4 is divided into 2 parts where the left side is denoted by *4l* and the right side denoted by *4r*. Mass conservation equations of SCV 1 and SCV 2 are written in the system of equations, and solved simultaneously. In order to obtain the final solution of this method, the 2 following equations are combined:

$$\begin{aligned} \rho\hat{v}_4 ds_{y_{4r}^1} - (\overrightarrow{\rho\hat{q}_b}.\overrightarrow{ds}_{bh}) &= 0 \\ \rho\hat{v}_4 ds_{y_{4r}^1} + (\overrightarrow{\rho\hat{q}_b}.\overrightarrow{ds}_{bd}) &= 0 \end{aligned} \tag{17}$$

where $\overrightarrow{ds}_{bd}$ is the normal surface vector on *SSb* and is equal to $-\overrightarrow{ds}_{bh}$.

Equation 17 is in fact the mass conservation equation for the dotted area with the actual SCV related to node 1. Depending on the boundary condition of the problem, appropriate constraints can be forced in Eq. 16. For instance, if boundary b is solid, then $(\overrightarrow{p\hat{q}_b} \cdot \overrightarrow{ds}_{bh}) = 0$.

Similar procedure is applied for momentum conservation equations. The x-momentum conservation equation for the grey area in Fig. 6 is written as follows:

$$\frac{\rho \mathcal{V}_2(U_2 - U_2^o)}{\Delta t} + (\rho \hat{u}_1 ds_{x_1^1})u_1 + \left(\rho \hat{v}_4 ds_{y_{4l}^1}\right)u_4 + \\ + (\rho \overrightarrow{\hat{q}_b} . \overrightarrow{ds}_{b^h})u_b + p_1 ds_{x_1^1} + p_b ds_{x_b^h} - \\ -\mu \left.\frac{\partial u}{\partial x}\right|_1 ds_{x_1^1} - \mu \left.\frac{\partial u}{\partial x}\right|_{b^h} ds_{x_b^h} - \\ -\mu \left.\frac{\partial u}{\partial y}\right|_{b^h} ds_{y_b^h} - \mu \left.\frac{\partial u}{\partial y}\right|_2 ds_{y_{4l}^1} = 0 \tag{18}$$

Here $\mathcal{V}_2$ is the volume of grey area in Fig. 6. As mentioned for the mass conservation equation, Eqs. 10 and 18 are written in the system of equations and solved simultaneously. In order to obtain the final solution of this method, the 2 following equations are combined:

$$\frac{\rho \mathcal{V}_1(U_1 - U_1^o)}{\Delta t} - \frac{\rho \mathcal{V}_2(U_2 - U_2^o)}{\Delta t} + (\rho \hat{v}_4 ds_{y_{4r}^1})u_4 - \\ - (\rho \overrightarrow{\hat{q}_b} . \overrightarrow{ds}_{b^h})u_b - p_b ds_{x_b^h} - \mu \left.\frac{\partial u}{\partial x}\right|_{b^h} ds_{x_b^h} - \\ -\mu \left.\frac{\partial u}{\partial y}\right|_{b^h} ds_{y_b^h} - \mu \left.\frac{\partial u}{\partial y}\right|_2 ds_{y_{4r}^1} = 0 \tag{19}$$

After substitution, Eq. 19 will become:

$$\left\{\frac{\rho \mathcal{V}_1(U_1 - U_1^o)}{\Delta t} - \frac{\rho \mathcal{V}_2(U_2 - U_2^o)}{\Delta t}\right\} + (\rho \hat{v}_4 ds_{y_{4r}^1})u_4 + \\ + (\rho \overrightarrow{\hat{q}_b} . \overrightarrow{ds}_{b^d})u_b + p_b ds_{x_b^d} + \mu \left.\frac{\partial u}{\partial x}\right|_{b^h} ds_{x_b^h} + \\ +\mu \left.\frac{\partial u}{\partial y}\right|_{b^h} ds_{y_b^h} - \mu \left.\frac{\partial u}{\partial y}\right|_2 ds_{y_{4r}^1} = 0 \tag{20}$$

This is the equation of x-momentum conservation for the dotted area which is the actual SCV of node 1, if $\boldsymbol{U}_1$ and $\boldsymbol{U}_2$ are assumed to be the same in transient term. Here $(\mathcal{V}_1 - \mathcal{V}_2)$ is equal to the volume of the dotted area. Conservation equation of y-momentum for the grey area in Fig. 6 can be obtained from the following equation:

$$\frac{\rho \mathcal{V}_2(V_2 - V_2^o)}{\Delta t} + (\rho \hat{u}_1 ds_{x_1^1})v_1 + \left(\rho \hat{v}_4 ds_{y_{4l}^1}\right)v_4 + \\ + (\rho \overrightarrow{\hat{q}_b} . \overrightarrow{ds}_{b^h})v_b + p_4 ds_{y_4^1} + p_b ds_{y_b^h} - \\ -\mu \left.\frac{\partial v}{\partial x}\right|_{b^h} ds_{x_b^h} - \mu \left.\frac{\partial v}{\partial y}\right|_{b^h} ds_{y_b^h} - \mu \left.\frac{\partial v}{\partial y}\right|_2 ds_{y_{4l}^1} = 0 \tag{21}$$

Since the boundary b is solid, then momentum flux $(\overrightarrow{p\hat{q}}_b \cdot \overrightarrow{ds}_{b^h})u_b$ would be 0; p_b is evaluated using bilinear interpolation, and velocity gradient terms of $\left.\frac{\partial u}{\partial x}\right|_{b^h}$ and $\left.\frac{\partial u}{\partial y}\right|_{b^h}$ are evaluated using bilinear interpolation. Identical procedure can be applied to obtain a similar equation to Eq. 20 for y-momentum conservation equation.

The key-point of the present method is to remain in the IBM context while implementing conservation laws all over the domain including boundary control volumes. Therefore control volumes of nodes within the flow field are always considered ordinary and complete. In this method, boundary conditions are implemented via the ghost control volumes where conservation laws are applied to determine variables values on ghost nodes. This is in contrast to other IBM methods in which interpolation functions (Mittal and Iaccarino 2005; Majumdar *et al.* 2001; Ghias *et al.* 2004), or cut cell methods (Clarke *et al.* 1986; Udaykumar *et al.* 2001, 1999, 1996; Ye *et al.* 1999) are used to determine variables values of ghost nodes to satisfy boundary conditions. The application of this method can be extended to moving boundary problems in IBM context to reduce the spurious pressure oscillations. This is due to local mass conservation errors observed in simulations of moving boundary problems with typical immersed boundary methods (Seo and Mittal 2011).

RESULTS

Taylor problem (Alisadeghi 2012; Alisadeghi and Karimian 2011; Mahesh *et al.* 2004; Darbandi and Vakilipour 2008) is selected to evaluate the present method. This problem corresponds to periodic and counter-rotating vortices whose strength decays in time at a rate determined by the viscosity. The Exact solution for Taylor problem (Alisadeghi 2012; Alisadeghi and Karimian 2011; Mahesh *et al.* 2004; Darbandi and Vakilipour 2008) for velocity components, pressure, and kinetic energy at square domain of unit size is given as:

$$u(x,y) = \sin(\pi x)\cos(\pi y)\, e^{\left(-2\pi^2 t/_{Re}\right)} \tag{22}$$

$$v(x,y) = -\cos(\pi x)\sin(\pi y)\, e^{\left(-2\pi^2 t/_{Re}\right)} \tag{23}$$

$$P(x,y) = 0.25[\cos(2\pi x) + \\ + \cos(2\pi y)]e^{\left(-4\pi^2 t/_{Re}\right)} \tag{24}$$

$$KE(x,y) = 0.5[sin^2(\pi x)cos^2(\pi y) + + sin^2(\pi y)cos^2(\pi x)]\, e^{\left(-4\pi^2 t/_{Re}\right)} \tag{25}$$

In inviscid case where Reynolds number is infinite, the exponential terms in Eqs. 22-25 can be ignored. Pressure contours and streamlines of the exact solution for inviscid Taylor problem are plottet in Fig. 8. The solution of the flowfield will be constant with time if numerical solution is used in inviscid case. For the first test problem, the present method is applied to investigate this fact. Figure 9 shows grid structure and domain boundary used for the present method. Grid spacing in both directions are uniform and equal to 0.05.

Domain boundary, denoted by blackline, is immerssed within the elements close to boundary. Thus, domain size will be 0.93 by 0.93, which is less than unity. In this case, the outer grid nodes, depicted by the black squares, are ghost nodes. SCVs of these ghost nodes are type II (as Fig. 4). Boundary conditions are implemented in conservation equations of mass and momentums for ghost control volumes. In the mass conservation equation (Eq. 16), boundary mass flux $(\overrightarrow{p\hat{q}}_d \cdot \overrightarrow{ds}_{bh})$ is known. In x-momemtum conservation equation (Eq. 18), $(\overrightarrow{p\hat{q}}_b \cdot \overrightarrow{ds}_{b\,h})u_b$ is also calculated from known values of the exact solution; p_b is evaluated from bilinear interpolation from the nodal values of pressure. The same method is used for y-momemtum conservation equation. In the inviscid Taylor problem, solution is initially proceeded for the fist hundered time using $\Delta t = 1e{-6}$ s. Then velocity components calculated by the present method are compared with those of the exact solution along the midlines and diagonal of the domain shown in Fig. 10. As can be observed the present results have remained constant throughout the time and exactly the same as the results of the exact solution.

To obtain the difference between the results of the exact solution and the present method we introduce the following equations for P and KE:

$$E_P = \sum_{1}^{N} \left(\frac{\sqrt[2]{(P_n - P_e)^2}}{P_e} / N \times 100\right) \tag{26}$$

$$E_{KE} = \sum_{1}^{N} \left(\frac{\sqrt[2]{(KE_n - KE_e)^2}}{KE_e} / N \times 100\right) \tag{27}$$

where N is the number of fluid nodes in the domain; subscripts n and e indicate numerical and the exact solutions, respectively.

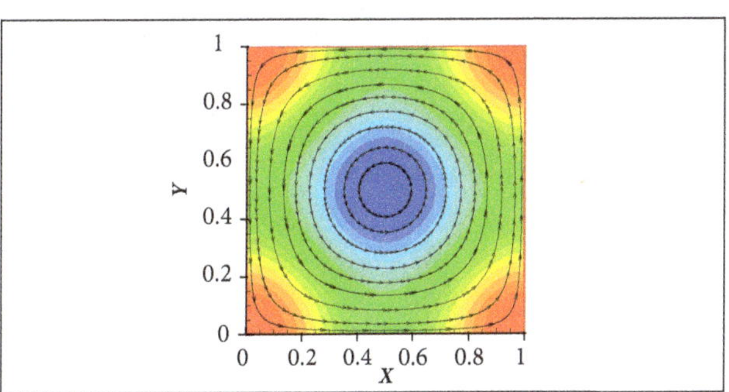

Figure 8. Pressure contours and streamlines of the exact solution for inviscid Taylor problem.

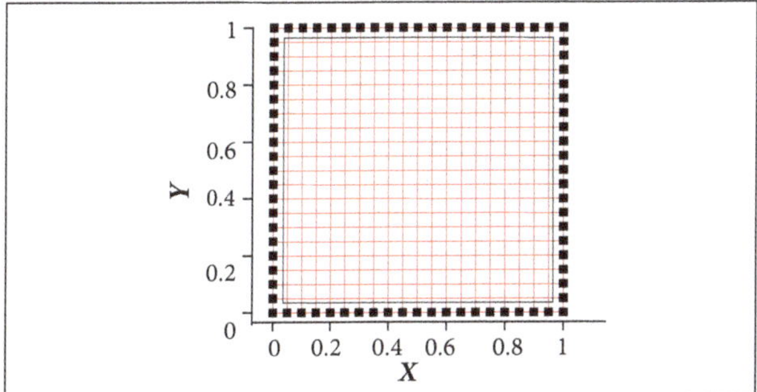

Figure 9. Grid structure, domain boundary (blackline), and ghost nodes (black nodes) for the solution of Taylor problem.

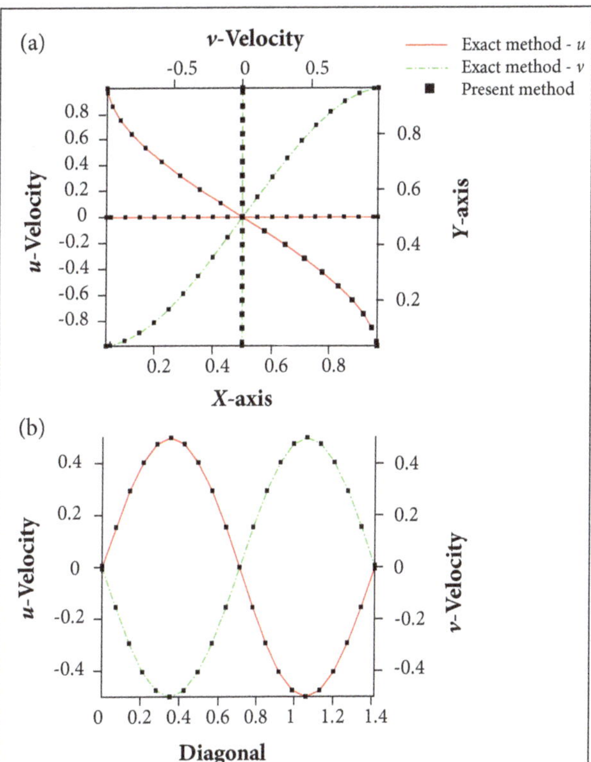

Figure 10. Velocity diagrams of Taylor problem. Comparison with the exact solution: (a) u and v along the horizontal and the vertical midlines; (b) u and v along the diagonal from south west to northeast.

P and *KE* errors reached 0.004 and 9.7e−7%, respectively, and their values did not change during the first 100 time iterations. Error between the results of numerical and the exact solutions shows the high strenght of the present method.

To present the capability of the present method in handling immersed boundaries, Taylor problem is solved on a grid rotated 45° with respect to the solution domain as depicted in Fig. 11. Uniform grid spacing in both directions equal to 0.0707 m. Domain boundary, denoted by the black line in Fig. 11, is immerssed within the elements. Therefore, domain size will be 0.85 by 0.85, which is less than 1 unity. In this case, the outer grid nodes, denoted by the black color in Fig. 12, are ghost nodes. SCVs of these ghost nodes (Fig. 13) are type I and II, as previously defined.

Boundary conditions are implemented using conservation equations of mass and momentums of ghost control volumes. Implementation of conservation equations for ghost SCV type II (Fig. 13b) is similar to the one in the previous test case. For the ghost subcontrol volume type I (Fig. 13a) a similar procedure is followed. This means that boundary mass flux ($\overrightarrow{p\hat{q}}_b \cdot \overrightarrow{ds}_{b\,h}$) in Eq. 13 is calculated by known values of the exact solution. In *x*-momemtum conservation equation introduced by Eq. 14, ($\overrightarrow{p\hat{q}}_b \cdot \overrightarrow{ds}_b$)$u_b$ is calculated using known values of the exact solution; p_b is calculated by bilinear enterpolation using the element nodal values; 1e−6 s is used as time step for the first 100 time steps.

Streamlines obtained from the present method are plotted in Fig. 14. Note that 4 triangles around the central square domain in Fig. 14 are outside the flowfield. In general, streamlines in Fig. 14 are similar to those of Fig. 8. Errors for pressure and kinetic energy, E_p and E_{KE}, were 2.24 and 4.5e−6%, respectively. Similar to the previous test case, KE keeps constant in time as the problem is non-viscous.

Velocity profiles derived from the present method are compared with the results of the exact solution in the middle of the flow field and along the diagonal from south west to north east in Fig. 15, and along left, right, up, and down boundaries in Fig. 16. The results show very good agreement with the results of the exact solution. Therefore it is seen that, in cases

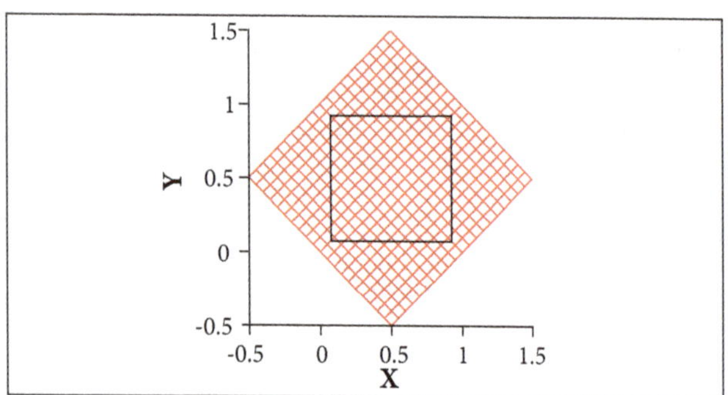

Figure 11. Schematic of Taylor problem (blackline) on a grid rotated 45°.

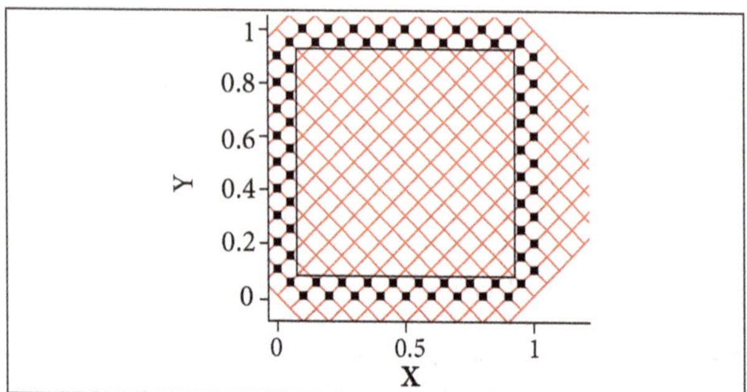

Figure 12. Positions of ghost nodes (blacknodes) with respect to solution domain (blackline).

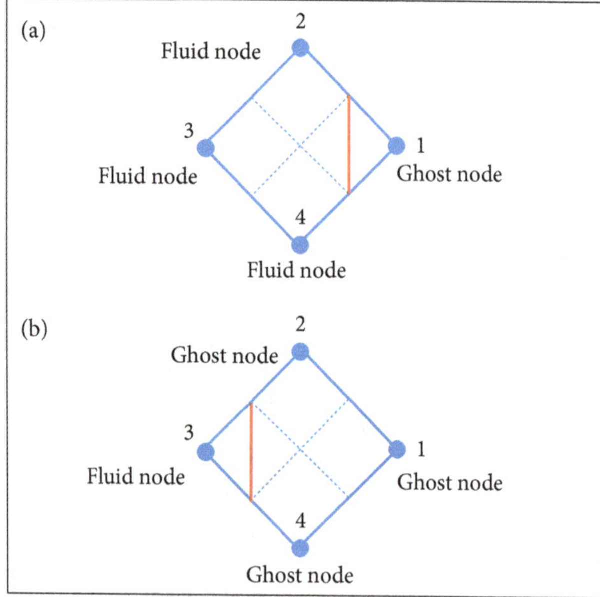

Figure 13. SCVs of ghost nodes of Fig. 12. (a) Type I; (b) Type II.

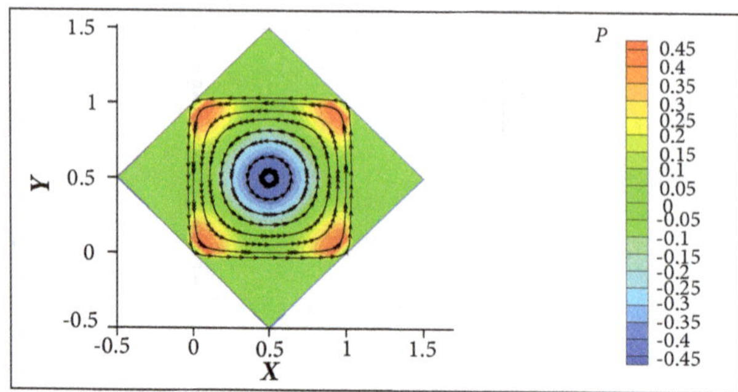

Figure 14. Pressure contours and streamlines for the proposed IBM (Taylor problem with 45° grid rotation).

where grid structure is not aligned with domain boundary, the present method is still able to solve flowfield accurately. The differences between profiles in first and second test cases in Figs. 10 and 14 are due to the difference between sizes of solution domains in the 2 cases.

As our third test case, Taylor problem with Reynolds number of 1,000 is solved using the present method on regular grid domain. This case is chosen to investigate the capability of the present method to solve viscous transient problems. Schematic of the solution domain is similar to Fig. 9, but the location of the boundary is different as shown in Fig. 17. In the viscous case, pressure and velocity fields decay in time at a rate determined by the viscosity. As verified in Eq. 25, *KE* is a function of *t*/*Re*. So the higher the Reynolds number, the lower the rate of *KE* decay.

This test is solved using the present method. Ghost control volumes in Fig. 17 is type I and is considered as described before. In this case, viscosity terms should be calculated at the boundary. Therefore, $\mu \frac{\partial u}{\partial x}\Big|_{b^h} ds_{x_b^h}$ and $\mu \frac{\partial u}{\partial y}\Big|_{b^h} ds_{y_b^h}$ can be found using known values of the exact solution. All the other terms in mass and momentum conservetion equations are calculated in a similar way as explained in the previous test case, although grid configuration is different from the

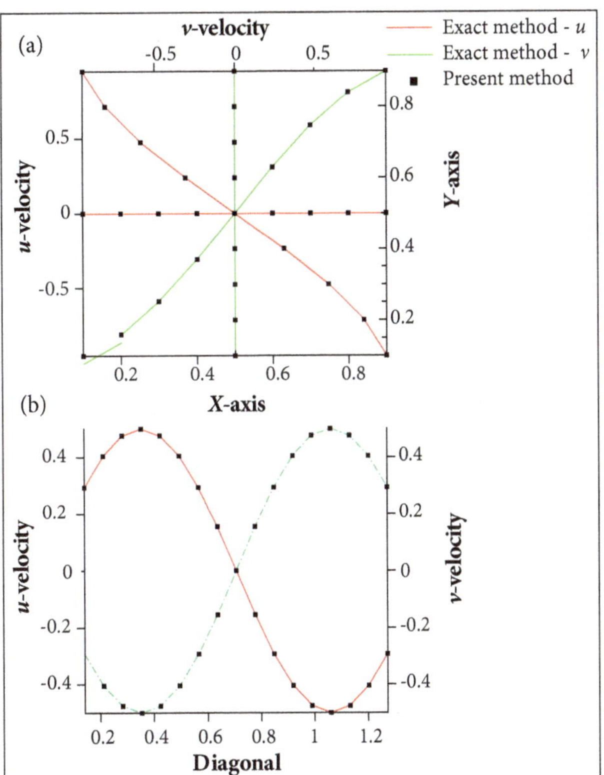

Figure 15. Velocity diagrams of Taylor problem with 45° grid rotation. Comparison with the exact solution: (a) *u* and *v* along the horizontal and the vertical midlines; (b) Velocity components along the diagonal from southwest to northeast.

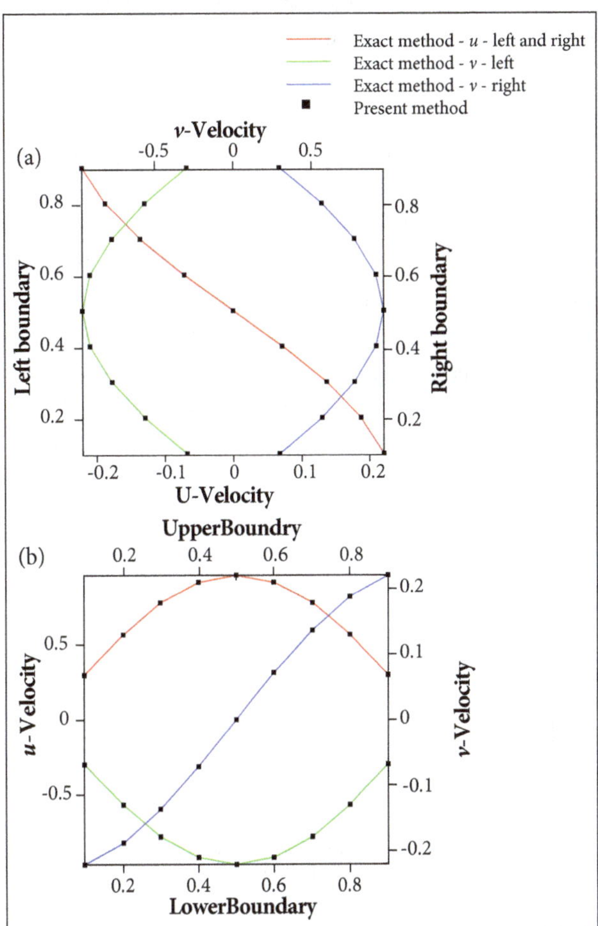

Figure 16. Velocity diagrams of Taylor problem with 45° grid rotation. Comparison with the exact solution: (a) *u* and *v* along the left and right boundaries; (b) *u* and *v* along the up and down boundaries.

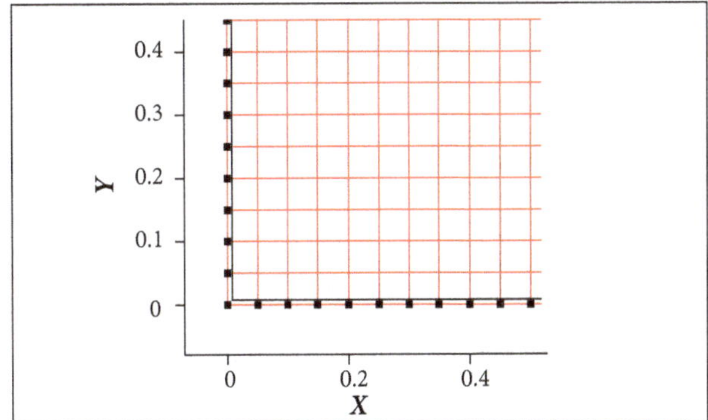

Figure 17. Grid structure for the solution of viscous Taylor problem on regular grid, domain boundary by black line, and ghost nodes by black nodes.

previous test case. Solution proceeds until the pressure and velocity fields are completely decayed by viscosity. In this case, $\Delta t = 0.005$ s and the solution is itereted in every time step until the solution converges. Decay of the temporal kinetic energy is compared with the exact solution in Fig. 18. The result of the present method has excellent agreement with that of the exact solution. Grid independency study was carried out in present research. Viscous Taylor problem on regular grid was solved on 21 × 21, 31 × 31, and 41 × 41 grids. As a sample chosen from the results, pressure variations along the diagonal of the solution domain on 3 grids are compared with each other in Fig. 19. As can be seen, after a small change between results of grid 21 × 21 and grid 31 × 31, no significant changes can be notified between the results of grids 31 × 31 and 41 × 41. The same trend was observed on other results of these 3 grids. So all results are obtained on grid 31 × 31.

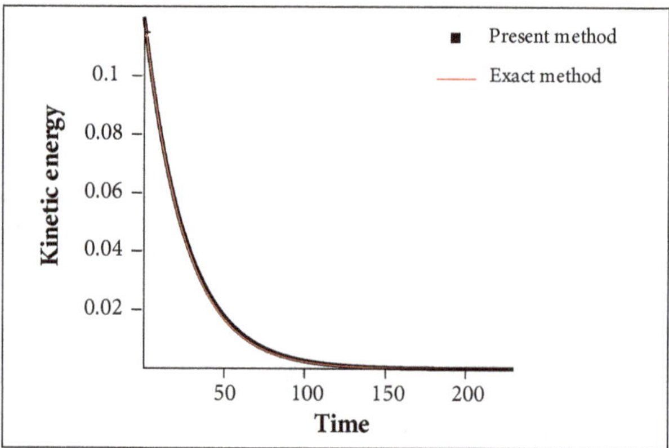

Figure 18. Decay of temporal *KE*: comparison between results of the present method and the exact solution for Reynolds number of 1,000 on a regular grid.

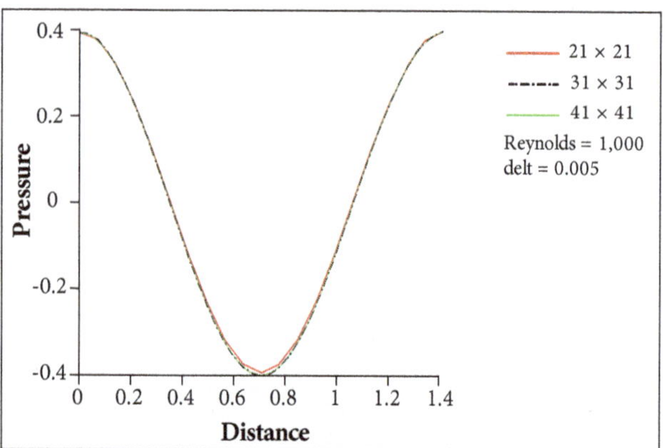

Figure 19. Pressure variation along the diameter for the purpose of grid convergence study. Grids with 21 × 21, 31 × 31 and 41 × 41 nodes.

KE decays faster as Reynolds number decreases. Decay of temporal *KE* are compared with the exact solution for 2 Reynolds numbers of 100 and 1,000 in Fig. 20. The results show very good agreement with the exact solution. As observed, fast decay of *KE* clearly occurs for Reynolds number of 100.

Velocity profiles resulted from the present method are compared with the results of the exact solution in the middle of the flow field, and along the diagonal from southwest to northeast in Fig. 21. Results are presented at different times of of 0.0, 25, and 50 s. Again, as observed, all of the results match excellently with the exact solution obtained from Eqs. 22 and 23.

In addition to the above comparisons, rates of pressure at the center of the solution domain and average of pressure over the solution domain with time are very similar to results of the exact solution as shown in Fig. 22.

Here again, viscous Taylor problem is solved on a 45° rotated grid to present capability of present immersed boundary method on grids skewed with respect to solution boundary. Solution domain and grid structure of this fourth test case are shown in Fig. 23. The only difference between Figs. 12 and 23 is the location of the boundary. In Fig. 23 the domain size is exactly 1 × 1 square, which is smaller in Fig. 12. In both cases the type of the ghost control volmes are the same as shown in Fig. 13, and a 31 × 31 grid is used in both cases. The outer grid nodes, denoted by black color in Fig. 23, are ghost nodes.

Boundary conditions are implemented through conservation equations of mass and momentums of ghost control volumes. Implementation of conservation equations for ghost subcontrol volumes in this case is exactly the same as the one applied in the second case with its grid in Fig. 13. The only

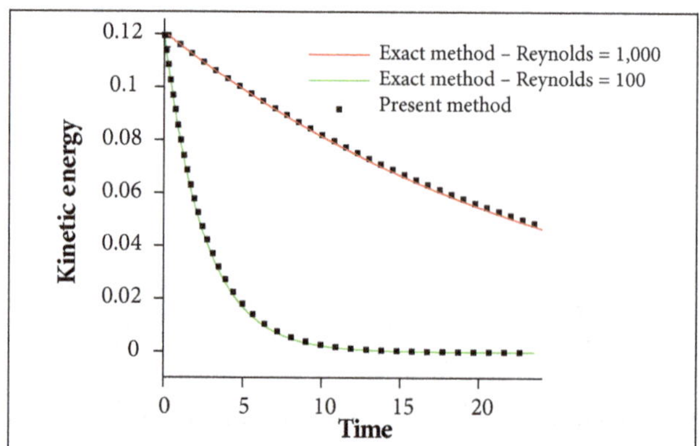

Figure 20. Decay of temporal *KE* for 2 Reynolds numbers of 100 and 1,000. Comparison with the exact solution.

difference in this case is the fact that viscosity terms are calculated at the boundary. Therefore, $\mu \frac{\partial u}{\partial x}\Big|_{b^h} ds_{x_b^h}$ and $\mu \frac{\partial u}{\partial y}\Big|_{b^h} ds_{y_b^h}$ are calculated from known values of the exact solution. Solution is proceeded until velocity and pressure fields completely decay with time and is proceeded in time using time steps of 1e−3 s.

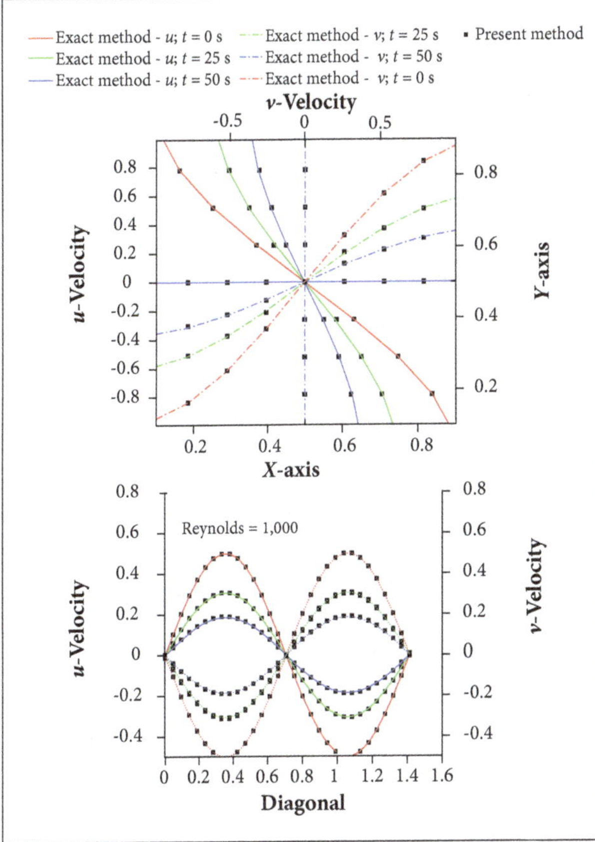

Figure 21. Velocity profiles of viscous Taylor problem in 3 instances of 0.0, 25, and 50.0 s. Comparison with the exact solution: (a) u and v along the horizontal and vertical midlines; (b) u and v along the diagonal from southwest to northeast.

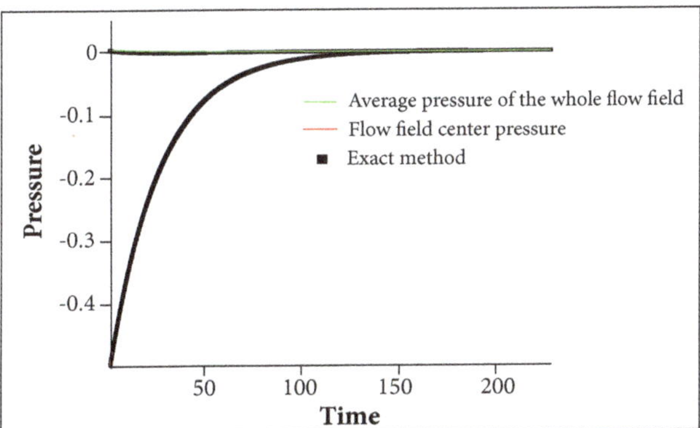

Figure 22. Time variation of pressure at the center of the solution domain and average of pressure over the solution domain: comparison of the present method with the exact solution.

Decay of temporal *KE* and its comparison with the exact solution are illustrated in Fig. 24. Result of the present method follows the exact solution of *KE* decay. Velocity profiles derived from the present method are compared with the results of the exact solution in the middle of the flow field and along the diagonal from southwest to northeast in Fig. 25. Results are presented at various times of 0.0, 25, and 50 s. In all profiles excellent agreement is verified between results of the present method and those of the exact solution. Here again, rates of pressure at the center of the solution domain and the pressure average variations over the solution domain with time also agree very well with the results of the exact solution as shown in Fig. 26.

Now it is possible to claim with confidence that the present method can solve the flow filed in both viscous and non-viscous cases with high accuracy in comparison with the exact solution. The accuracy of the present method is not challenged even on grids not aligned with the boundary domain.

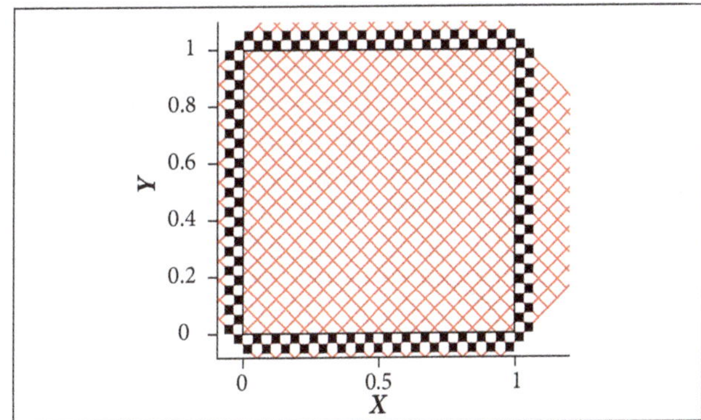

Figure 23. Grid structure for the solution of viscous Taylor problem on 45° rotated grid: domain boundary by black line and ghost nodes by black nodes.

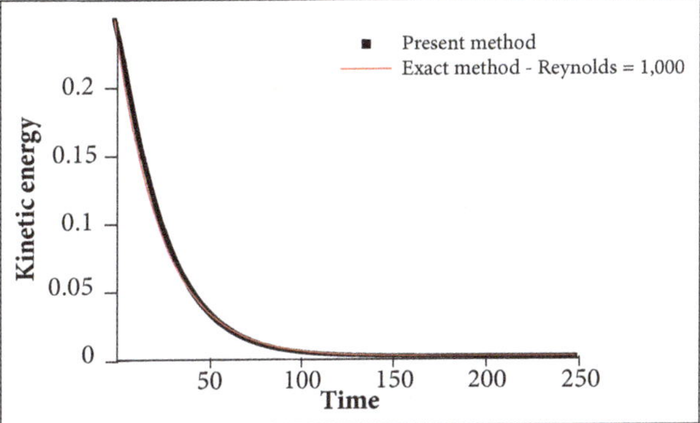

Figure 24. Decay of temporal *KE*: comparison between results of the present method and the exact solution for Reynolds number of 1,000 on a 45° rotated grid.

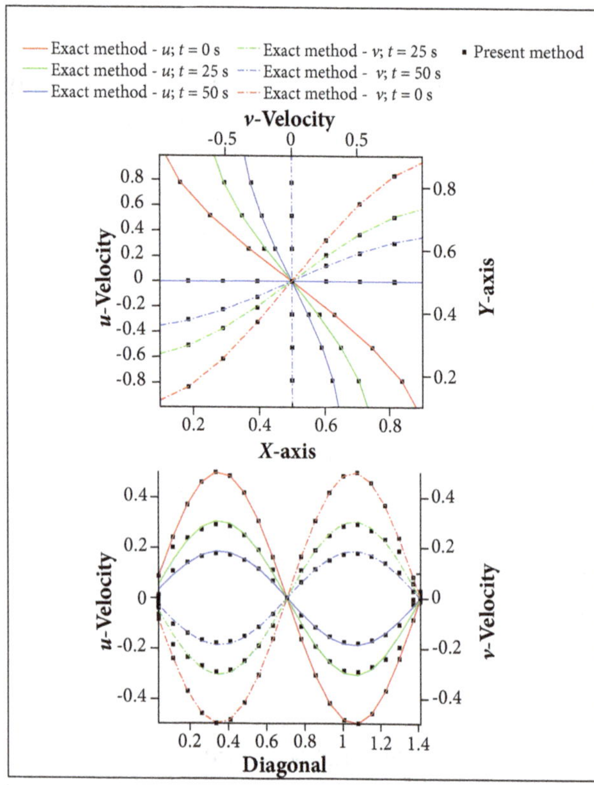

Figure 25. Velocity profiles of viscous Taylor problem in 3 instances of 0.0, 25, and 50 s. Comparison with the exact solution: (a) *u* and *v* along the horizontal and vertical midlines; (b) *u* and *v* along the diagonal from southwest to northeast.

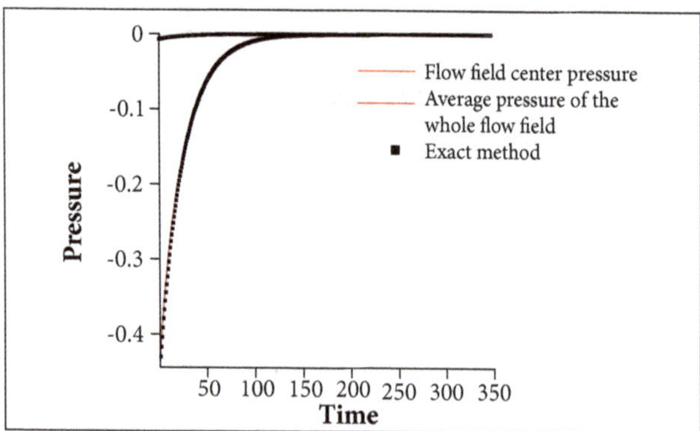

Figure 26. Time variation of pressure at the center of the solution domain and average of pressure over the solution domain: comparison of the present method with the exact solution.

CONCLUSION

In this paper a new immersed boundary method using control volume finite element scheme was introduced for discretization of governing equations. The advantage of this method is that boundary conditions are implemented by conservation of conserved quantities along the boundaries. In typical immersed boundary methods conservation equations are only satisfied within the boundary but not necessarily along the boundary. However, in this study, a new approach for the implementation of boundary conditions was presented in which mass and momentum conservation laws are fully conserved along the boundary as well as inside the domain. In the present method, in addition to the use of ghost node value applied in typical IBM, a new concept of ghost SCV was introduced. This new concept makes the implementation of the conservation laws in the vicinity of the boundary possible. The present method is validated by solving Taylor problem in both non-viscous and viscous cases with Reynolds numbers of 100 and 1,000. Steady and unsteady cases of Taylor problem on regular and 45° rotated grids were also solved for further investigation. Results both in pressure and velocity diagrams show an excellent agreement between the present method and of the exact solution in all cases even at the sharp corners. Based on these results, accurate simulation of the flow fields in physically complex problems can be expected from the present IBM method.

AUTHOR'S CONTRIBUTION

The idea of the present study has been initiated and developed by both authors, as well as the algorithm of the method and its details; Hosseini SN wrote the computer code to implement the algorithm, verified the results, and prepared the manuscript; Karimian SMH supervised the whole research, reviewed and revised the manuscript. Both authors discussed and confirmed the results.

REFERENCES

Alisadeghi H (2012) Modification and improvement of a pressure-velocity coupling algorithm for the numerical solution of two dimensional incompressible flows (PhD thesis). Tehran: Amirkabir University of Technology. In Persian.

Alisadeghi H, Karimian SMH (2011) Different modelings of cell-face velocities and their effects on the pressure-velocity coupling, accuracy and convergence of solution. Int J Numer Meth Fluid 65(8):969-988. doi: 10.1002/fld.2224

Berger M, Aftosmis M (1998) Aspects (and aspect ratios) of Cartesian mesh methods. In: Bruneau CH, editor. Sixteenth International Conference on Numerical Methods in Fluid Dynamics. Vol. 515. Berlin: Springer.

Beyer Jr RP (1992) A computational model of the cochlea using the immersed boundary method. J Comput Phys 98(1):145-162. doi: 10.1016/0021-9991(92)90180-7

Clarke DK, Hassan HA, Salas MD (1986) Euler calculations for multielement airfoils using Cartesian grids. AIAA J 24(3):353-358. doi: 10.2514/3.9273

Darbandi M, Vakilipour S (2008) Developing implicit pressure-weighted upwinding scheme to calculate steady and unsteady flows on unstructured grids. Int J Numer Meth Fluid 56(2):115-141. doi: 10.1002/fld.1451

Fauci LJ, McDonald A (1995) Sperm motility in the presence of boundaries. Bull Math Biol 57(5):679-699. doi: 10.1016/0092-8240(95)00022-I

Ghias R, Mittal R, Dong H (2007) A sharp interface immersed boundary method for compressible viscous flows. J Comput Phys 225(1):528-553. doi: 10.1016/j.jcp.2006.12.007

Ghias R, Mittal R, Lund TS (2004) A non-body conformal grid method for simulation of compressible flows with complex immersed boundaries. Proceedings of the 42nd AIAA Aerospace Sciences Meeting and Exhibit; Reno, USA.

Karimian S, Schneider G (1994a) Numerical solution of two-dimensional incompressible Navier-Stokes Equations: treatment of velocity-pressure coupling. Proceedings of the 25th AIAA Fluid Dynamics Conference; Colorado Springs, USA.

Karimian S, Schneider G (1994b) Pressure-based computational method for compressible and incompressible flows. Journal of Thermodynamics and Heat Transfer 8(2):267-274. doi: 10.2514/3.533

Khadra K, Angot P, Parneix S, Caltagirone JP (2000) Fictitious domain approach for numerical modeling of Navier-Stokes equations. Int J Numer Meth Fluid 34(8):651-684. doi: 10.1002/1097-0363(20001230)34:8<651::AID-FLD61>3.0.CO;2-D

Mahesh K, Constantinescu G, Moin P (2004) A numerical method for large-eddy simulation in complex geometries. J Comput Phys 197(1):215-240. doi: 10.1016/j.jcp.2003.11.031

Majumdar S, Iaccarino G, Durbin PA (2001) RANS solvers with adaptive structured boundary non-conforming grids. Annual Research Briefs. Stanford: Center for Turbulence Research

Minkowycz WJ, Sparrow EM, Murthy JY (1988) Handbook of numerical heat transfer. Vols. 379-421. New York: Wiley.

Mittal R, Dong H, Bozkurttas M, Najjar FM, Vargas A, von Loebbecke A (2008) A versatile sharp interface immersed boundary method for incompressible flows with complex boundaries. J Comput Phys 227(10):4825-4852. doi: 10.1016/j.jcp.2008.01.028

Mittal R, Iaccarino G (2005) Immersed boundary methods. Annu Rev Fluid Mech 37:1-487. doi: 10.1146/annurev.fluid.37.061903.175743

Peskin CS (1972) Flow patterns around heart valves: a numerical method. J Comput Phys 10(2):252-271. doi: 10.1016/0021-9991(72)90065-4

Peskin CS (1982) The fluid dynamics of heart valves: experimental, theoretical, and computational methods. Annu Rev Fluid Mech 14:1-442. doi: 10.1146/annurev.fl.14.010182.001315

Saiki EM, Biringen S (1996) Numerical simulation of a cylinder in uniform flow: application of a virtual boundary method. J Comput Phys 123(2):450-465. doi: 10.1006/jcph.1996.0036

Seo JH, Mittal R (2011) A sharp-interface immersed boundary method with improved mass conservation and reduced spurious pressure oscillations. J Comput Phys 230(19):7347-7363. doi: 10.1016/j.jcp.2011.06.003

Tseng YH, Ferziger JH (2003) A ghost-cell immersed boundary method for flow in complex geometry. J Comput Phys 192(2):593-623. doi: 10.1016/j.jcp.2003.07.024

Udaykumar HS, Mittal R, Rampunggoon P, Khanna A (2001) A sharp interface Cartesian grid method for simulating flows with complex moving boundaries. J Comput Phys 174(1):345-380. doi: 10.1006/jcph.2001.6916

Udaykumar HS, Mittal R, Shyy W (1999) Computation of solid-liquid phase fronts in the sharp interface limit on fixed grids. J Comput Phys 153(2):535-574. doi: 10.1006/jcph.1999.6294

Udaykumar HS, Shyy W, Rao MM (1996) Elafint: a mixed Eulerian-Lagrangian method for fluid flows with complex and moving boundaries. Int J Numer Meth Fluid 22(8):691-712. doi: 10.1002/(SICI)1097-0363(19960430)22:8<691::AID-FLD371>3.0.CO;2-U

Ye T, Mittal R, Udaykumar HS, Shyy W (1999) An accurate Cartesian grid method for viscous incompressible flows with complex immersed boundaries. J Comput Phys 156(2):209-240. doi: 10.1006/jcph.1999.6356

Zhu L, Peskin CS (2003) Interaction of two flapping filaments in a flowing soap film. Phys Fluid 15(7):128-136. doi: 10.1063/1.1582476

5

Finite Element Analysis of Pilot's Helmet Design Using Composite Materials for Military Aircraft

Puran Singh[1], Debashis Pramanik[2], Ran Vijay Singh[2]

ABSTRACT: The objective of this research was to design pilot helmets and to perform analysis of designed ballistic helmet against impact strength of bullet in Solidworks and Laminator software. The material used for construction of the helmet is fiber reinforced polymer matrix composite in which polymer matrix is made of nylon, a thermoset resin, and the fibers are aramid, an aromatic polymide resin developed by E.I. duPont de Nemours and Company and sold under the trademarks "Kevlar®" and "Nomex®". The design of the helmet is done by deciding the stacking sequence of various laminae which are oriented with main material directions at different angles to the global laminate axes in order to produce a structural element in the form of a shell. The simulation of the helmet in Solidworks and Laminator is done with an 8-g AK 47 bullet, hitting it with a velocity of 710 m/s. The model is validated against published data and a good correlation is observed. The result of this project is that a 1.30 kg helmet with shell thickness of 7 mm is obtained, which is economical, light weight and is able to give high-performance protection against ballistic shrapnel and bullets.

KEYWORDS: Finite element analysis, Pilot helmets, Thermoplastic aramid, Composite materials.

INTRODUCTION

The sizing, fit, and comfort have been the most frequent concern of the aviators in each of the different types of aircraft squadrons. Pilot helmet has been used as protective equipment in order to shield human head against serious injuries from shrapnel and bullets. Most modern ballistic helmets are made from a plurality of plies of ballistic material which are laid up in a mold and shaped to the configuration of the helmet. Pilot helmet made of composite materials has become a better equipment compared to traditional steel helmet in terms of the reduction in weight and the improvement in ballistic resistance. Therefore, finite element analysis can be used as a method to characterize the response of composite pilot helmet and to obtain valuable information on parameters affecting impact phenomena (Othman 2009). The first dimension of the helmet is decided according to the average size of human head (Figure 1).

In general, there are two ballistic test standards that are used to determine the quality of protection of the helmet: (1) NIJ-STD-0106.01 Type II and (2) MIL-H-44099A. Nevertheless, different helmet manufacturers may have different ballistic test methods.

Helmet improvement around the head and over the eyes during air combat maneuvers (ACM) or sharp turns in flight involves positive "G" forces in excess of 2 "G". This is attributable to the poor profile of the helmet and its misplaced center of gravity (CG).

The main focus of this research is to study the response of pilot helmet made of composite materials when impacted

1.Amity School of Engineering & Technology – Mechanical & Automation Engineering Department – FEM Laboratory – New Delhi/Delhi – India. **2.**Manav Rachna International University – Faculty of Engineering Technology – Mechanical Engineering Department – Faridabad/Haryana – India.

Author for correspondence: Puran Singh | Amity School of Engineering & Technology – Mechanical & Automation Engineering Department – FEM Laboratory | 580, Najafgarh Kapashera Rd, Dalmia Vihar, Bijwasan | 110061 – New Delhi/Delhi – India | Email: puran.singh910@gmail.com

at high velocity for different sequences of lamina or plies by using finite element analysis. The objectives of this research are:

- To design a pilot helmet that provides high performance protection against ballistic shrapnel and bullets.
- To design a light-weight and economical pilot helmet.
- To analyze deformation as well as stress distribution of the helmet when struck by a bullet at a velocity of 710 m/s.
- To evaluate the failure mechanism occurred on pilot helmet after the impact.

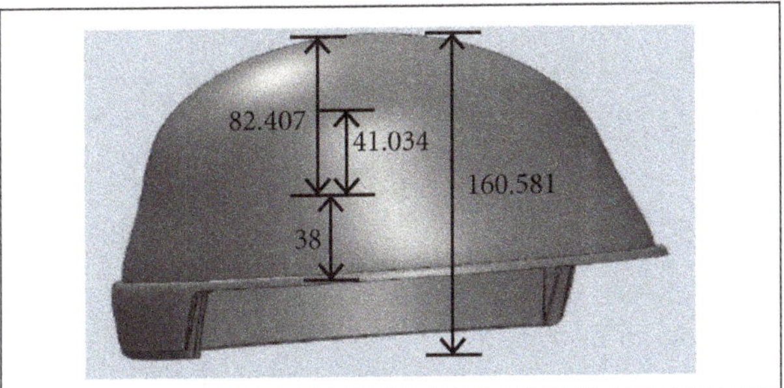

Figure 1. Front view showing dimensions of the helmet.

HELMET DESIGN AND MATERIAL CONSIDERATIONS

Helmet materials and designs have evolved primarily in light of prevailing threats and the invention of new and improved ballistic materials. Figure 2 is a basic summary of U.S. helmet designs and materials since World War I (WWI). For example, the helmet design in WWI was significantly different than in World War II (WWII). WWI was characterized by an unprecedented amount of trench warfare, and the hot and sharp debris falling from relatively high angles were typical. This gave rise to the fairly wide "brim" that gave the WWI helmet its distinctive look.

The combined shells provided higher protection levels over a greater coverage area than the previous M1917 copy of the British Mk I "Brodie" helmet of WWI. The one-size M1 helmet weighed 1.55 kg, had 0.12 m^2 of surface coverage, and protected against the 0.45 caliber round at 244 m/s with a 50% ballistic limit of 396 m/s against the standard North Atlantic Treaty Organization (NATO) 1.1-gram fragment simulator. Thermoplastic aramid matrix systems, one of the most commom materials used, have excellent, mass-efficient ballistic properties. However, the thermoplastic matrix is typically 30 to 60% less rigid than even the toughened thermoset (e.g. phenolic) matrix. This has significant implications for the overall static structural stability and resilience of the thermoplastic aramid shell, as well as the dynamic deflections associated with a ballistic event. To illustrate this phenomenon, consider Fig. 3. A Phantom v.7 high-speed digital camera was used to capture the effects of a simulated ballistic fragment impact on the back side of a flat thermoplastic aramid panel.

This panel had an areal density that was nearly 50% of that recommended for producing a helmet shell. As can be seen in

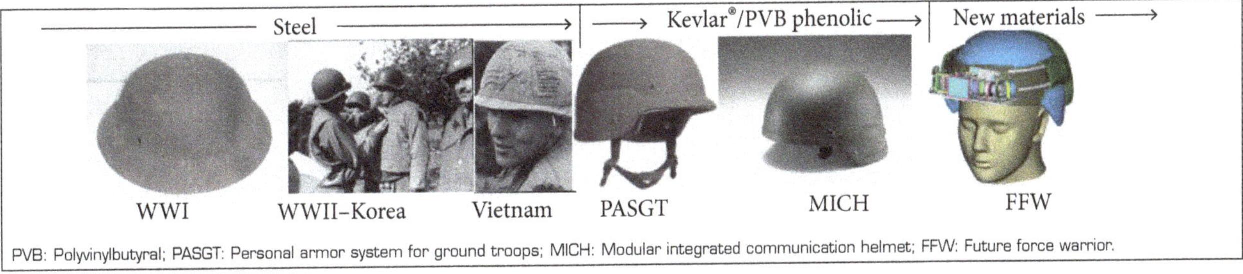

Figure 2. Historical perspective of U.S. Army helmet design and materials (Walsh et al. 2005).

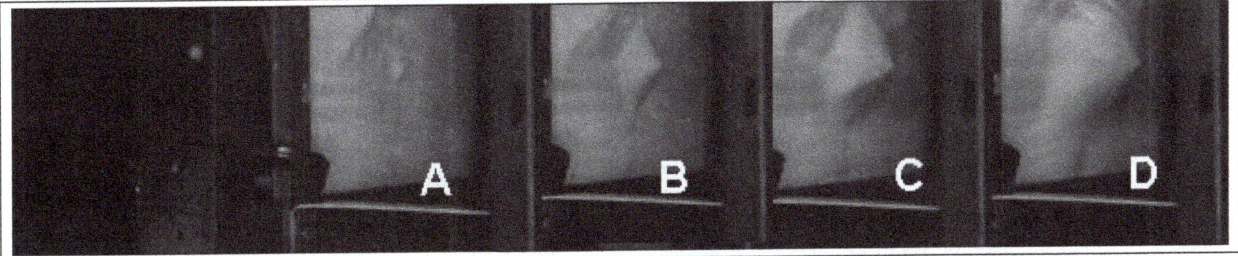

Figure 3. Still photographs from high-speed digital imaging of thermoplastic aramid panel (Walsh et al. 2005).

the sequence of images, the fragment is effectively contained and stopped but not before it induced significant deformation to the overall panel.

A thermoplastic Kevlar® shell at this low areal density may be well-suited for certain applications, but given that the deformation is well over 1 in, it could cause severe skull fracture (and possibly death).

The primary goal of the helmet shell is to protect the pilot from a variety of threats. First, the requirement is to limit the perforation by fragments or bullets through the helmet. Even if the fragment is stopped, the deflection of the shell can engage the skull and cause injury. The current PASGT uses an effective air gap of approximately 13 mm between the inner shell wall and the soldier's head to accommodate any deflection during projectile arrest.

Transient deformation is a direct result of the kinetic energy being dissipated within the ballistic material. Fabrics, although extremely ballistically resilient at real densities around 0.975 g per cm^2, tend to deform significantly. The fragment or bullet could conceivably be arrested by the fabric, but the resulting deformation could still result in a fatal injury by adversely engaging the skull. By contrast, thermoset composites, such as polyvinylbutyral (PVB) phenolic aramid systems, reduce the transient deformation, even though their ballistic performance may be less than that of a pure fabric system. Thermoplastic composite materials offer a compromise of fabric and thermoset composite performance. That is, the thermoplastic tends to deform but not as much as pure fabric, and it tends to have better ballistic resistance than a thermoset-based composite material (Campbell and Cramer 2008).

Practical durability is a necessary trait for any article used in combat. Helmets must also pass static structural tests as well. "Ear-to-ear" loads of 2,000 to 3,500 kPa must be withstood by the helmet for several cycles without any permanent deformation in its structure. Thermoset composites tend to do well, given the higher matrix modulus (as compared to a thermoplastic matrix).

Fully realizing the material and performance benefits of thermoplastic aramids and hybridized solutions will require the rethinking of the manufacturing processes currently in use by most of the U.S. helmet manufacturers. Current processes are configured for mass production of thermoset-based, monolithic Kevlar® helmets. These manufacturing systems typically use expensive, matched steel tools to consolidate the materials. Cold helmet pre-forms are placed in a hot mold and held under pressure until fully cured. Figure 4 is a conceptual schema of such a process.

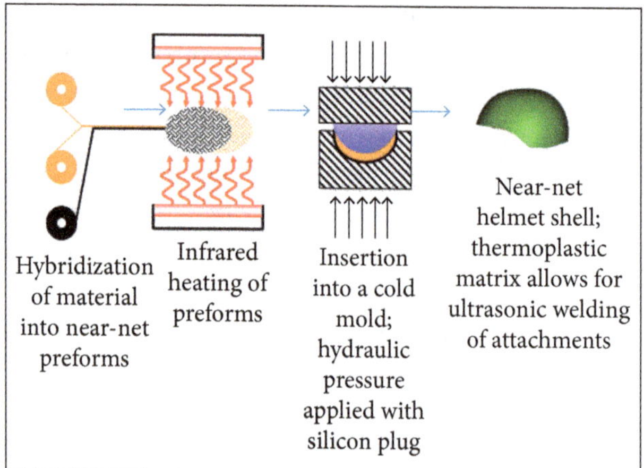

Figure 4. Conceptual schema of the production process of thermoset-based, monolithic Kevlar® helmets (Walsh et al. 2005).

The composite consists of two or more constituents. One is called matrix and the other, reinforcement. The main functions of matrix are:

- It binds the fibers (reinforcement) together and transfers the load to fibers. It provides rigidity and shape to the structure.
- It isolates the fibers so that an individual fiber can act separately. This helps to stop propagation of cracks.
- It provides good surface finish quality and protection to reinforcement fibers against chemical attack and mechanical damage (wear).
- Failure mode of composite is strongly affected by the type of matrix material,and performance characteristics, such as ductility, impact strength etc., are also influenced.

Reinforcement is an important constituent of composite materials. Fiber reinforcement is a thin rod-like structure (Fig. 5). The main functions of fiber reinforcement are:

- It carries the load. In structural composites, 70 to 90% of the load is carried by fiber reinforcements.
- It provides stiffness, strength and thermal stability to the composite.
- It provides electrical conductivity or insulation, depending on the fiber used in the composite.

The general properties of composite materials are light weight, low thermal expansion, high stiffness, high strength and high fatigue resistance.

The matrix in a reinforced plastic may be either thermoset or thermoplastic. In the early days nearly all the thermoset moulding materials were composites in that they contained fillers such as wood flour, mica, cellulose etc. to increase their strength. However, these were not generally regarded as reinforced materials in the sense that they did not contain fibers. Nowadays the major thermoset resins, used in conjunction with glass fiber reinforcement, are unsaturated polyester resins and, to a lesser extent, epoxy resins (Piggott 1980). The most important advantages which these materials can offer are: (i) they do not liberate volatiles during cross-linking and (ii) they can be moulded using low pressures at room temperature.

A wide variety of thermoplastics have been used as the base for reinforced plastics. These include polypropylene, nylon, styrene-based materials, thermoplastic polyesters, acetyl, polycarbonate, polysulfone etc. The choice of a reinforced thermoplastic depends on a wide range of factors which include the nature of the application, the service environment and costs. In many cases conventional thermoplastic processing techniques can be used to produce moulded articles.

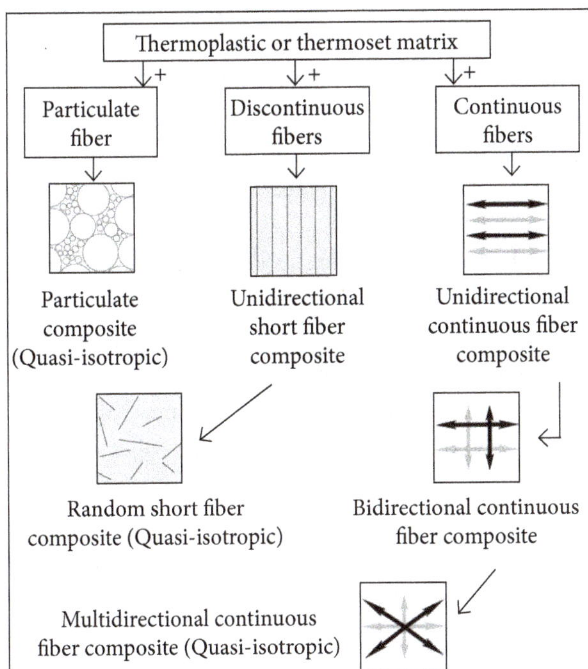

Figure 5. Types of reinforcement.

DESIGN CALCULATIONS AND ANALYSIS

The following assumptions have been taken while designing and analyzing this project:

- It is assumed that the velocity at which the bullet hits the helmet surface is the same as the muzzle velocity.
- It is assumed that the impact time of the bullet on the helmet is 1.5 ms, i.e. the bullet comes to rest 1.5 ms after hitting the surface of the helmet.
- The impact of the bullet on the helmet is assumed to be uniaxial (along *x*-axis).
- For applying the boundary conditions of the helmet during analysis, the bottom part of the helmet is fixed.
- The shape of the bullet is not taken into account while calculating the force of the bullet on the helmet.

The dimensions of the helmet are: width — 180.00 mm, height — 160.58 mm and length — 180.00 mm (Fig. 6).

The composite material chosen is Kevlar® 149/epoxy (fiber/matrix) whose material properties are mentioned in Table 1. Various laminae are prepared out of this composite material,

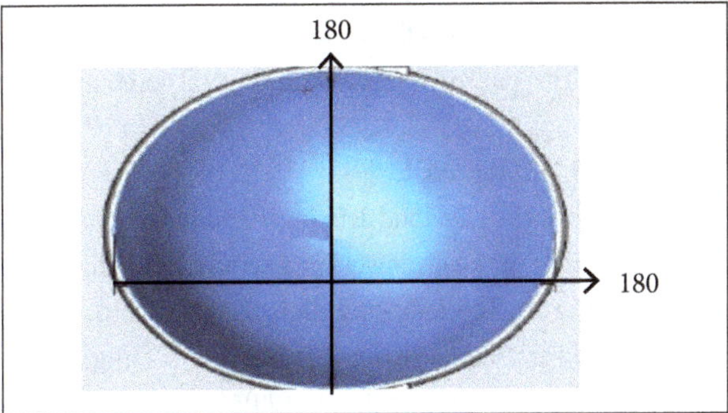

Figure 6. Top view showing the dimensions of the helmet.

Table 1. Properties of Kevlar® 149/epoxy.

Property	Value	Units	Value type
Elastic modulus	1.26e+007	Psi	Constant
Elastic modulus	8e+005	Psi	Constant
Poisson's ratio	0.34	NA	Constant
Shear modulus	3.1e+005	Psi	Constant
Mass density	0.05299	lb/in^3	Constant
Tensile strength	1.85e+005	Psi	Constant
Compressive strength	49,000	Psi	Constant
Yield strength	1.8e+005	Psi	Constant
Thermal expansion coefficient	1.1e–006	F	Constant
Thermal expansion coefficient	3.3e–005	F	Constant
Tensile strength	4,200	Psi	Constant
Compressive strength	22,900	Psi	Constant
Shear strength	7,100	Psi	Constant

and the stacking of various laminae is done, which is oriented with principal material directions at different angles to the global laminate axes producing a structural element in the form of a shell. The strength of this shell is analyzed when a bullet, at a certain velocity, hits it (Silva *et al.* 2005).

The specification of the bullet is as follows: AK 47; caliber: 7.62; mass: 8 g; muzzle velocity: 710 m/s. The force exerted on the helmet is: $m \times a = 8 \times 10^{-3} \times (710/0.0015) = 3{,}786.667$ N. The force with which the bullet hits the helmet is approximately taken as 4,000 N along the x-axis.

Here a software called Laminator is used, which takes the abovementioned material properties, load applied (4,000 N) and different angles of each lamina as the input. It calculates the [*A*], [*B*] and [*D*] matrices and the inverse of these matrices, as well as laminate stiffness properties. Each stacking sequence has 24 laminae. The material of each ply is the same and their thickness is taken as 0.21 mm. This process is repeated for several combinations.

STACKING SEQUENCE 1

After applying the load and defining the boundary conditions to the helmet with the given stacking sequence, a simulation was made to run. The following plots in the form of results were obtained and studied to find the optimum configuration.

Stacking sequence 1: [(45)4, (0)4, (90)4] s with each lamina thickness of 0.21 mm. [*A*], [*B*] and [*D*] matrices are:

$$[A] = \begin{bmatrix} 2.910e{+}007 & 6.300e{+}006 & 4.993e{+}006 \\ 6.300e{+}006 & 2.910e{+}007 & 4.993e{+}006 \\ 4.993e{+}006 & 4.993e{+}006 & 6.481e{+}006 \end{bmatrix}$$

$$[B] = \begin{bmatrix} -4.889e{-}009 & -2.328e{-}010 & -2.328e{-}010 \\ -2.328e{-}010 & -9.546e{-}009 & -4.657e{-}010 \\ -2.328e{-}010 & -4.657e{-}010 & -6.985e{-}010 \end{bmatrix}$$

$$[D] = \begin{bmatrix} 6.412e{+}007 & 2.490e{+}007 & 2.231e{+}007 \\ 2.490e{+}007 & 3.594e{+}007 & 2.231e{+}007 \\ 2.231e{+}007 & 2.231e{+}007 & 2.529e{+}007 \end{bmatrix}$$

The inverse of these matrices is:

$$[A]^{-1} = \begin{bmatrix} 3.998e{-}008 & -3.884e{-}009 & -2.780e{-}008 \\ -3.884e{-}009 & 3.998e{-}008 & -2.780e{-}008 \\ -2.780e{-}008 & -2.780e{-}008 & 1.971e{-}007 \end{bmatrix}$$

$$[B]^{-1} = \begin{bmatrix} 4.872e{-}024 & -3.116e{-}024 & -2.562e{-}024 \\ -3.423e{-}024 & 2.372e{-}023 & -1.668e{-}023 \\ -1.750e{-}024 & -1.825e{-}023 & 2.165e{-}023 \end{bmatrix}$$

$$[D]^{-1} = \begin{bmatrix} 2.339e{-}008 & -7.510e{-}009 & -1.401e{-}008 \\ -7.510e{-}009 & 6.394e{-}008 & -4.979e{-}008 \\ -1.401e{-}008 & -4.979e{-}008 & 9.583e{-}008 \end{bmatrix}$$

The laminate stiffness properties are:

Ex: 4.963e+006

Ey: 4.963e+006

Gxy: 1.006e+006

Vxy: 0.097

We have similar stacking sequence for 2, 3, 4, 5, and 6. After applying the load and defining the boundary conditions to the helmet with the given stacking sequence, a simulation was made to run. The following plots in the form of results were obtained and studied to find the optimum configuration.

Firstly, taking the dimensions for the ballistic helmet, three circular sketches were made. By using those circular sketches as a guiding profile and carefully adjusting various parameters, a shell was formed with the help of surface loft feature (Fig. 7a).

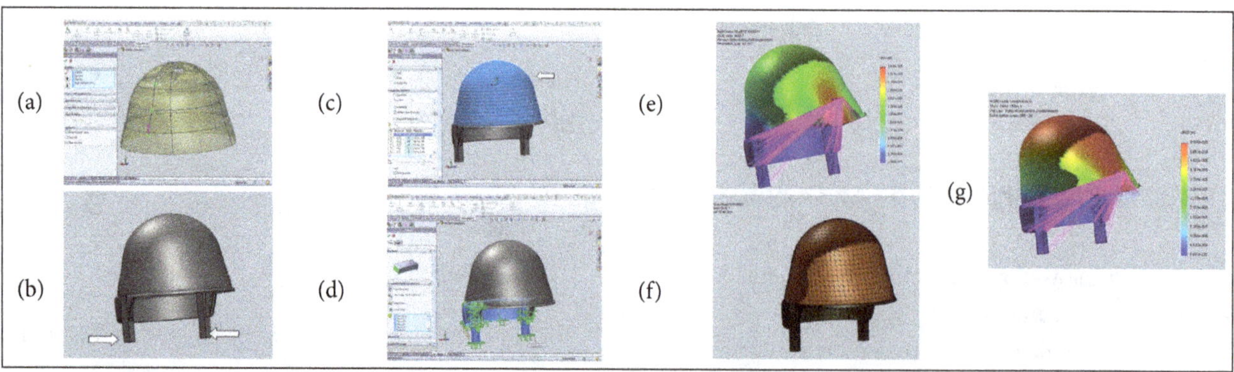

Figure 7. Helmet configuration. (a) Shell forming; (b) Helmet with the support structure; (c) Defining composite layers; (d) Fixed support at the bottom; (e) Application of force; (f) Meshing; (g) Displacement distribution.

For analysis purposes, a support structure was designed to provide the fixture support at the sides of pilot helmet (Fig. 7b). This support works similarly to the straps provided for the support of the helmet in real time.

Linear elastic orthotropic material Kevlar® 149 is defined for all the layers (Fig. 7c). The bottom part of the helmet along with the designed vertical fixtures are fixed and used as boundary conditions for analysis purposes (Fig. 7d). Now, after defining the boundary conditions, the calculated force is applied (Fig. 7e) at the target point taken (where the bullet will hit the helmet). A fine high-quality mesh (with approximate element size equal to 7.9 mm) is generated having parabolic triangular elements with 10,354 nodes and total number of elements equal to 5,058 (Fig. 7f). Results in the form of stress distribution, strain distribution, displacement distribution and factor of safety distribution are obtained and analyzed (Fig. 7g).

ANALYSIS RESULTS

Displacements are measured in meters from the position where the bullet hits the helmet surface before any deformation occurs (Fig. 8a).

Distributions of von Mises stresses are shown in the stress plot obtained after running the simulation (in N/m^2). In this case, a material is said to start yielding when its von Mises stress reaches a critical value known as yield strength. The von Mises stress is used to predict yielding of materials under any loading condition from results of simple uniaxial tensile tests (Fig. 8b).

The plot in Fig. 8c shows the distribution of strain in the top part of pilot helmet when the bullet hits the surface. The plot in Fig. 8d shows the factor of safety at every node. Factor of safety (FoS) is a term describing the structural capacity of a system beyond the applied or actual loads. Tsai-Hill failure criterion is used to evaluate the FoS at the top part of pilot helmet (composite shell). This criterion considers the distortion energy portion of the total strain energy that is stored due to loading. The reaction forces are shown in Table 2. It has proposed that

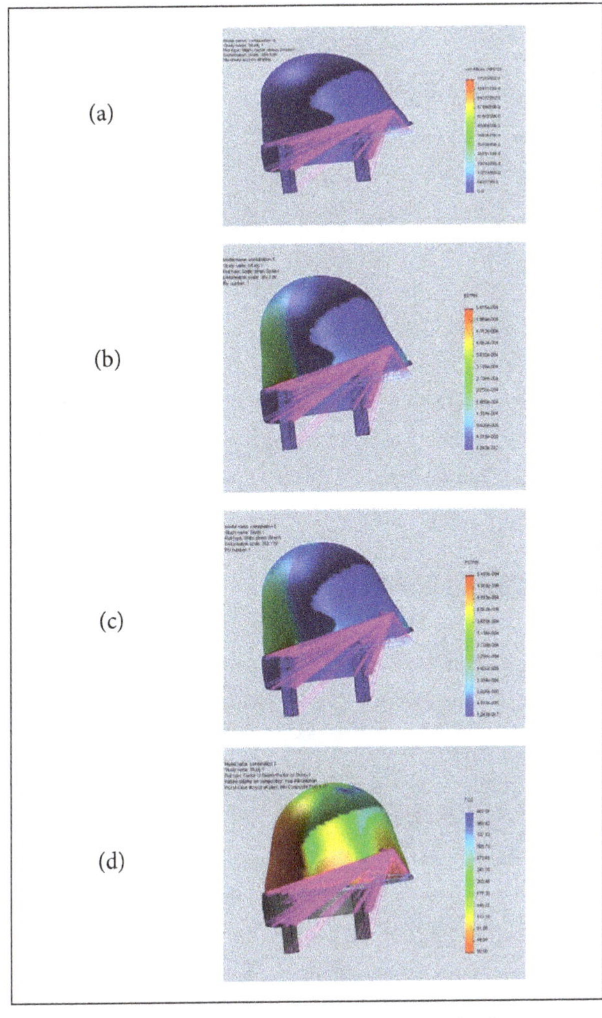

Figure 8. Analysis results. (a) Displacement plot for [(90)4, (0)4, (90)4] s; (b) Stress plot for [(90)4, (0)4, (90)4] s; (c) Strain plot for [(90)4, (0)4, (90)4] s; (d) Factor of safety plot for [(90)4, (0)4, (90)4] s.

Table 2. Reaction forces.

Selection set	Unit	Sum of X	Sum of Y	Sum of Z	Resultant
Entire body	N	−2,403.53	0.100185	−0.00871483	2,403.53
Entire body	N	−2,403.56	−0.0747453	−0.00140771	2,403.56
Entire body	N	−2,403.56	−0.0443136	−0.00578333	2,403.56
Entire body	N	−2,403.49	0.140461	0.0142988	2,403.49
Entire body	N	−2,403.57	−0.00499602	0.000274185	2,403.57
Entire body	N	−2,403.53	0.0195969	−0.00771621	2,403.53

Table 3. Configuration by analysing the results of both Laminator and Solidworks simulations.

Serial number	Sequence	Ex	Ey	Gxy	Vxy	FoS
1	[(45)4, (0)4, (90)4] s	4.963e+006	4.963e+006	1.006e+006	0.097	21
2	[(0)4, (90)2, (0)2, (45)2, (0)2] s	8.857e+006	2.959e+006	7.074e+005	0.152	14
3	[(90)2, (0)2, (45)2, (0)2, (90)2, (0)2] s	6.903e+006	4.934e+006	7.263e+005	0.092	17
4	[(90)2, (0)2, (45)2, (0)2, (90)2, (0)2] s	4.821e+006	1.033e+006	8.338e+005	0.454	6.8
5	[(90)4, (0)4, (90)4] s	4.760e+006	8.715e+006	3.100e+005	0.031	17
6	[(0)2, (45)2, (90)2, (0)2, (45)2, (90)2] s	4.963e+006	4.963e+006	0.097	0.097	20

there should be four helmets to suit flight requirements. After the careful study and comparison of different composite layer combinations in Solidworks and Laminator, the most optimum configuration is shown by stacking sequence 1 — [(45)4, (0)4, (90)4] s, as can be seen in Table 3.

CONCLUSIONS

It has been concluded that the pilot helmet with 24 layers (each layer with 0.21 mm) with total thickness equal to 5.04 mm provides the optimum configuration with the best combination of laminate stiffness properties and FoS (Chawla 1998) under the specified testing conditions (8 g bullet travelling with 710 m/s velocity). The weight of the helmet shell comes out to be 300 g but, in order to give stability and comfort, certain features like foam padding, straps etc. are provided, which increases the weight of the helmet to approximately 1 kg.

The stacking sequence 1 [(45)4, (0)4, (90)4] s, is found as the optimum configuration by analysing the results of both Laminator and Solidworks simulation.

REFERENCES

Othman RB (2009) Finite element analysis of composite ballist helmet subjected to high velocity impact (Master's thesis). Penang: Universiti Sains Malaysia.

Chawla KK (1998) Composite materials: science and engineering. New York: Spinger-Verlag.

Silva MAG, Cismaşiu C, Chiorean CG (2005) Numerical simulation of ballistic impact on composite laminates. Int J Impact Eng 31(3):289-306. doi: 10.1016/j.ijimpeng.2004.01.011

Piggott MR (1980) Load bearing fibre composites. Oxford: Pergamon Press.

Campbell DT, Cramer DR (2008) Hybrid thermoplastic composite ballistic helmet fabrication study. Glenwood Springs: Fiberforge Corporation.

Walsh SM, Scott BR,Spagnoulo DM (2005) The Development of a Hybrid Thermoplastic Ballistic Material With Application to Helmets [accessed 2016 jan 21]. http://www.dtic.mil/dtic/tr/fulltext/u2/a441165.pdf

6

Real-Time Gas Turbine Model for Performance Simulations

Henrique Gazzetta Junior[1], Cleverson Bringhenti[2], João Roberto Barbosa[2], Jesuíno Takachi Tomita[2]

ABSTRACT: Industry and universities around the world invest time and money to develop digital computer programs to predict gas turbine performance. This study aims to demonstrate a brand new digital model developed with the ability to simulate gas turbine real time high fidelity performance. The model herein described run faster than 30ms per point, which is compatible with a high-definition video refresh rate: 30 frames per second. This user-friendly model, built in Visual Basic in modular structure, can be easily configured to simulate almost all the existing gas turbine architectures (single, 2 or 3 shaft engines mixed or unmixed flows). In addition, its real time capability enables simulations with the pilot in the loop at earlier design phases when their feedback may lead to design changes for improvements or corrections. In this paper, besides the model description, it is presented the model run time capability as well as a comparison of the simulated performance with a commercial gas turbine tool for single, 2 and 3 shaft engine architecture.

KEYWORDS: Propulsion, Gas turbines, Aircraft engines, Performance, Computer simulation.

INTRODUCTION

SAE AIR4548 defines a real-time digital engine model as a mathematical performance computer model whose outputs are generated at a rate compatible with the response of the physical system that it represents and with the time requirements of the simulation loop where it is inserted. The early developed models were relatively simple using analog devices and they were firstly used in hardware and software development for aircraft and engine control systems. As the model complexity increased to meet more demanding requirements, analog models became too costly and difficult to use. The early mathematical models, developed to make simulations less expensive, were simply a digital implementation of the analog models and, as digital computers capabilities increased and costs reduced, the engine digital models became very popular. As listed in Bringhenti (1999), some efforts in engine analog, digital or mixed simulation development can be acknowledged through the years notably by Mckinney (1967), Koenig and Fishback (1972), Fishback and Koenig (1972), Szuk (1974), Palmer and Yang (1974), Macmillan (1974), Sellers (1975), Wittenberg (1976), Flack (1990), Stamatis *et al.* (1990), Ismail and Bhinder (1991), Korakianitis and Wilson (1994), Baig and Saravanamuttoo (1997), Bringhenti (1999), Grönstedt (2000), Saravanamuttoo *et al.* (2001), Walsh and Fletcher (2004) and ASME 95-GT-147.

Nowadays, it is wide spread in the aeronautic industry the usage of simulation models for engine or aircraft development and its systems. Most of those models can run steady state simulations only and represents specific engine architecture; others are capable of simulating also the transient states with variable geometry and are flexible to represent almost all types

1.Empresa Brasileira de Aeronáutica – São José dos Campos/SP – Brazil. 2.Departamento de Ciência e Tecnologia Aeroespacial – Instituto Tecnológico de Aeronáutica – Divisão de Engenharia Aeronáutica e Mecânica – São José dos Campos/SP – Brazil.

Author for correspondence: Henrique Gazzetta Junior | Avenida Cassiano Ricardo, 1.411 – Apto. 84B | CEP: 12.240-540 – São José dos Campos/SP – Brazil | Email: henrique.gazzetta@gmail.com

of gas turbines as per Bringhenti (1999, 2003), Grönstedt (2000), and Silva (2011). However, due to the number of iterations and map data readings required in the gas turbine engine simulation process, a long time is required to output the simulation results, what is not compatible with a real-time application.

The real-time engine simulation tool can enhance the simulation activities at early phases of a product design, identifying potential improvements or issues at early phases of an aircraft design when there is room for changes or even step back at virtually no cost.

In addition, a digital real-time engine model could be used for development and testing of control systems, flight simulators, and engine integration with airframe in several aspects.

METHODOLOGY

A brand new engine model was generated to provide high fidelity and real-time gas turbine performance simulation. The model is representative of a three shaft engine, which is the most complex engine architecture. Other existing jet engines architectures (single and 2 shaft engines, mixed and unmixed flows) can be simulated by activating or deactivating components or entire shafts, by defining pressure ratios and efficiencies equal to 1. Additionally, several bleed configurations were modeled in order to give the user the ability to configure the bleed port extraction position, the amount of bleed extraction and the destination of the air being bled from the compressors: outboard bleed (for engine operability, aircraft air conditioning, pressurization, and anti-ice) as well as turbine cooling. In the case of the air being bled for turbine cooling purposes the user can select where exactly the cooling flow will be inserted in the cycle: stators or rotors of the turbine stages. At last, the model can deal with power extraction from all the shafts for aircraft systems. The schematics in Fig. 1 shows the engine model architecture with the airflow paths, power extractions, and power links (components mechanically linked through the shaft) following the proposed nomenclature from SAE AS755. This diagram represents the most complex engine architure to be simulated.The model was built based on blocks with will calculate each engine module separately. The blocks developed for this model are:

- AMB (Standard Atmosphere): this block reads the Altitude, Flight velocity or Mach Number, Ambient temperature or deviation from standard day and air humidity and calculates the engine air inlet properties based on the U.S. Standard Atmosphere 1976, Antoine (1888), and Gordon (1982).
- Air Inlet: this block reads the ambient properties calculated by the Standard Atmosphere block, the input pressure recovery factor and calculates the air intake performance based on the MIL-E-5007D.
- Splitter: the mass flow splitter block is used in several different places in the model, such as bypass and bleed

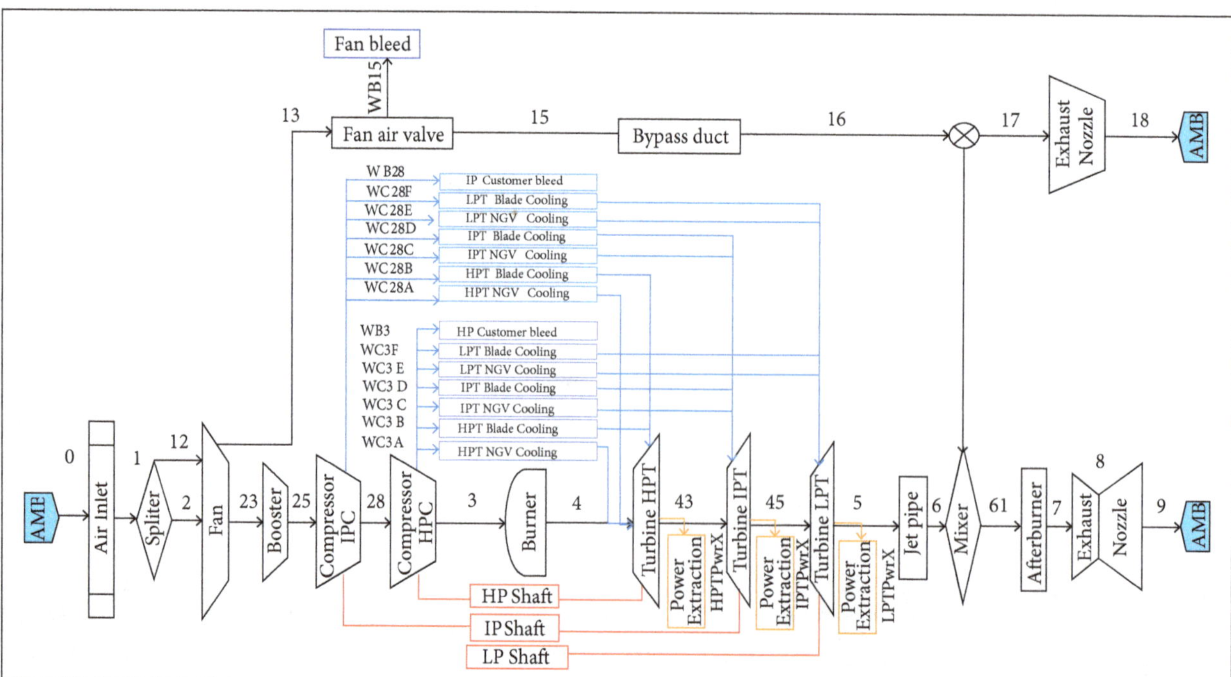

Figure 1. Three shaft direct drive engine model diagram.

extractions, and its basic function is to split the inlet flow in two outlet flows with the same gas properties.

- Compressor: this block reads the compressor characteristics, such as pressure ratio and isentropic efficiency and calculates the gas outlet properties based on the inlet properties and gas compression as described in Saravanamuttoo *et al.* (2001).
- Burner (Combustion Chamber): this block reads the fuel characteristics, such as lower fuel heating value and hydrogen/carbon ratio, and combustion chamber characteristics, such as pressure ratio and exit temperature or fuel flow and calculates the combustion gases properties based on inlet air properties as per Gordon(1982).
- Turbine: the turbine block calculates the gas expansion based on the turbine isentropic efficiency and inlet properties as described in Saravanamuttoo *et al.* (2001).
- Duct losses (Bypass duct and jet pipe): this block calculates the pressure loss through a duct given the pressure recovery factor.
- Mixer: this block calculates the resulting gas properties based on the 2 inlet gas flows. The calculation is based on the chemical composition, pressure, and temperature of each gas flow.
- Exhaust Nozzle: this block calculates the exhaust gas properties and velocity, based on the nozzle inlet gas properties and nozzle coefficients and geometry (convergent or convergent-divergent), as well as gross thrust.

Figure 2 shows the model simulation main process diagram. The flowchart represents all the engine blocks, libraries, input data and iterations necessary to simulate the engine performance. The main steps necessary to perform the simulation are:

- Design Point input read. This block reads all input data necessary to characterize the engine modules and calculate each block at design point.
- Calculate each engine module at component level in the sequence of the gas flow in order to reach the Design Point performance at engine level.
- Read the components maps for off-design performance simulations.
- Scale the components maps based on each module performance previously calculated at Design Point.
- Output the simulation results and the components scaled maps for off-design simulation.
- After finishing the Design Point calculation read the of design inputs, such as operating condition and power setting.
- Set the iterative process starting point. In this model the starting point can be the set equal to the last successfully converged point or a pre-defined starting point calculated based on the flight condition and power setting.
- Calculate the engine components performance and overall performance.
- Check if all the energy balances, mass flow balance and power settings are respected. If so output the calculated engine performance else a new iteration shall be performed with the new operation condition calculated by Broyden or Newton-Raphson method for non-linear system of equation solving.

MODEL DESCRIPTION

The mathematical model described herein are simplified for the sake of the reader clarity. More details can be obtained in the open literature as Mckinney (1967), Koenig and Fishback (1972), Fishback and Koenig (1972), Szuk (1974), Palmer and Yang (1974), Macmillan (1974), Sellers (1975), Wittenberg (1976), Flack (1990), Stamatis *et al.*(1990), Ismail and Bhinder (1991), Korakianitis and Wilson (1994), Baig and

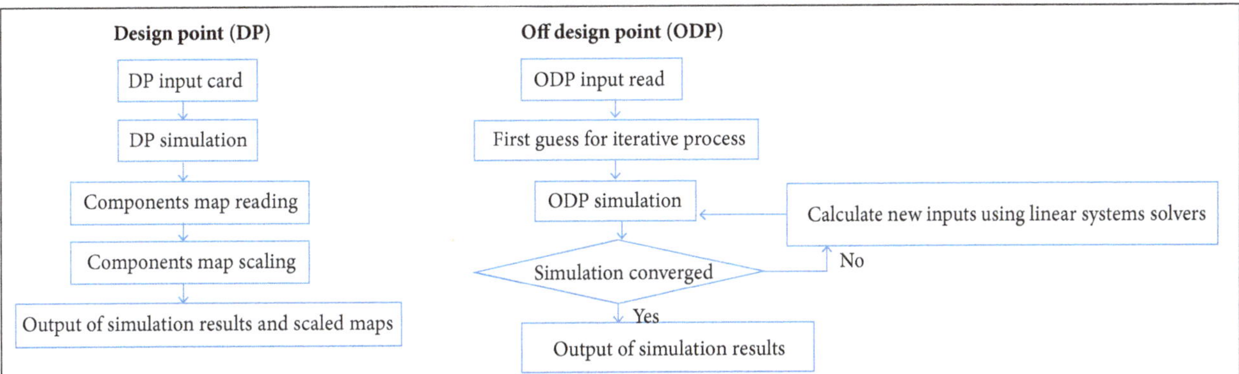

Figure 2. Engine simulation process diagram.

Saravanamuttoo (1997), Bringhenti (1999), Saravanamuttoo *et al.* (2001), Walsh and Fletcher (2004).

STANDARD ATMOSPHERE

The Standard Atmosphere definition implemented in the model described in this paper is based on the U.S. Standard Atmosphere 1976. It splits the atmosphere in 5 different levels up to 85 km grouping in the same level altitudes with similar characteristics of temperature and pressure variation as the altitude increases.

A summary of the atmosphere properties calculation for each atmosphere layer and the parameters to be used in static temperature and pressure calculation are described in Table 1.

$$T = T_b + L_b \times (ALT - H_b) \tag{1}$$

$If \rightarrow L_b \neq 0$

$$P = P_b \times \left[\frac{T_b}{T_b + L_b \times (ALT - H_b)} \right]^{\left[\frac{g_0' \times M_0}{R^* \times L_b} \right]} \tag{2}$$

$If \rightarrow L_b = 0$

$$P = P_b \times e^{\left[\frac{-g_0' \times M_0 \times (ALT - H_b)}{R^* \times T_b} \right]} \tag{3}$$

where: g_0' = 9.80665 m/s^2 is the geopotential gravity; M_0 = 28.96443 kg/mol is the air molecular weight; and R^* = 8,314.62 J/mol·K is the universal gas constant.

HUMIDITY

At all altitudes it is possible to set the humidity contained in the air. For this calculation the Antoine equation (Antoine 1888) determines the saturation vapor pressure for a given temperature for pure components. The Antoine equation and constants for water are:

$$P_{SAT} = 10^{A - \frac{B}{C + T_{water}}} \tag{4}$$

where: P_{SAT} is the saturation pressure in mmHg; T_{water} is the water static temperature in °C; A, B, and C are constants that are specific for each substance. The constants for water are shown in Table 2.

Table 2. Constants for water saturation vapor pressure in Antoine equation.

Temperature	A	B	C
From 1 to 100 °C	8.07131	1,730.63	233.426
From 100 to 374 °C	8.14019	1,810.94	244.485

INTAKE

The engine air inlet simulation was implemented following the MIL-E-5007D, which describes the pressure recovery factors for subsonic, supersonic, and hypersonic flows. Once the engine air inlet does no thermodynamic work and the flow is considered adiabatic, the stagnation temperature through the duct remains constant. Air mass flow and chemical composition also remain the same. The stagnation pressure downstream the air inlet is calculated as follows:

Subsonic flight (Mach < 1)

$$P_{T_{OUT}} = RAMREC \times P_{T_{IN}} \tag{5}$$

Supersonic flight (1 ≤ Mach < 5)

$$P_{T_{OUT}} = RAMREC \times P_{T_{IN}} \times \left[1 - 0.75 \times (MN - 1)^{1.35}\right] \tag{6}$$

Hypersonic flight (Mach ≥ 5)

$$P_{T_{OUT}} = RAMREC \times P_{T_{IN}} \times \left(\frac{800}{MN^4 + 935} \right) \tag{7}$$

Table 1. Standard atmosphere properties calculation summary table (U.S. Standard Atmosphere 1976).

Index (b)	Layer	Geopotential altitude Hb (km)	Thermal gradient Lb (K/km)	Reference temperature Tb (K)	Reference pressure Pb (Pa)
0	Troposphere	0	-6.5	288.15	101,325.0000
1	Tropopause	11	0.0	216.65	22,631.9500
2	Stratosphere	20	+1.0	216.65	5,475.0960
3		32	+2.8	228.65	868.0107
4	Stratopause	47	0.0	270.65	110.9002
5	Mesosphere	51	-2.8	270.65	66.9383
6		71	-2.0	214.65	3.9563

where Pt_{IN} is the inlet stagnation pressure in Pa; Pt_{OUT} is the outlet stagnation pressure in Pa; *MN* is the Mach Number; *RAMREC* the engine air inlet pressure recovery (Pt_{OUT}/Pt_{IN}).

COMPRESSOR

The axial flow compressor was implemented following the classic formulation described by Saravanamuttoo *et al.* (2001), Walsh and Fletcher (2004), and Kurzke (2007). The main equations in the compressor model are described as:

$$CPR = \frac{Pt_{OUT}}{Pt_{IN}} \tag{8}$$

The increase in the stagnation temperature due to work added to the airflow is calculated by:

$$\frac{Tt_{OUT}}{Tt_{IN}} = 1 + \frac{1}{\eta_c} \times \left[\left(\frac{Pt_{OUT}}{Pt_{IN}} \right)^{\frac{(\gamma-1)}{\gamma}} - 1 \right] \tag{9}$$

and the thermodynamic specific work is calculated by:

$$w_{Comp} = h_{OUT} - h_{IN} \tag{10}$$

where: *CPR* is the compressor pressure ratio; Tt_{IN} is the inlet stagnation temperature in K; Tt_{OUT} is the outlet stagnation temperature in K; γ is the specific heat ratio (*Cp/Cv*, being *Cp* and *Cv* the specific heat at constant pressure and volume respectively); η_c is the compressor isentropic efficiency; w_{Comp} is the compressor specific work in W/kg; h_{IN} is the inlet stagnation specific enthalpy in J/kg; h_{OUT} is the outlet stagnation specific enthalpy in J/kg.

COMBUSTION CHAMBER

The combustion chamber model calculates the amount of burnt fuel considering the amount of air and the equivalence ratio. Equivalence ratio is the ratio between the actual fuel air ratio and stoichiometric fuel air ratio, so equivalence ratio equal to 1 means stoichiometric burn, while lower and higher values mean lean and rich burns respectively. The chemical composition of the burnt gases is determined by the following equation, for equivalence ratio (*ER*) ≤ 1, as proposed by Gordon (1982):

$$\begin{aligned} &CH_Y + \alpha(O_2 + 3.727587 \cdot N_2 + 0.0447068 \cdot Ar + \\ &\quad + 0.0015228 \cdot CO_2 + \beta \cdot H_2O) \rightarrow \\ &(1 + 0.0015228 \cdot \alpha) \cdot CO_2 + \left(\frac{Y}{2} + \alpha \cdot \beta \right) \cdot H_2O + \\ &+ 3.727587 \cdot \alpha \cdot N_2 + \left(\alpha - 1 - \frac{Y}{4} \right) \cdot O_2 + 0.0447068 \cdot \alpha \cdot Ar \end{aligned} \tag{11}$$

And the following equation is proposed for ER>1 considering the air limiting the combustion:

$$\begin{aligned} &CH_Y + \alpha(O_2 + 3.727587 \cdot N_2 + 0.0447068 \cdot Ar + \\ &\quad + 0.0015228 \cdot CO_2 + \beta \cdot H_2O) \rightarrow \\ &\left[\alpha \cdot \left(\frac{1}{1 + Y/4} + 0.0015228 \right) \right] \cdot CO_2 + \left[\alpha \cdot \left(\frac{Y}{4+Y} \cdot \beta \right) \right] \cdot H_2O + \\ &+ 3.727587 \cdot \alpha \cdot N_2 + \\ &+ \left[\alpha \cdot \left(1 - \frac{1}{(1 + Y/4)} - \frac{2 \cdot Y}{(4+Y)} \right) \right] \cdot O_2 + 0.0447068 \cdot \alpha \cdot Ar \end{aligned} \tag{12}$$

where: *Y* is the fuel hydrogen-carbon ratio; β is the water-air mass flow ratio; α is $(4+Y)/(4 \cdot ER)$.

The unburnt air is mixed to the combustion gases and the chemical composition of the gas leaving the burner is recalculated. Burner exit temperature can be either inputted or calculated based on the fuel flow. In both cases, the following equation is used to calculate the temperature from fuel flow or fuel flow from temperature:

$$WF \times LFHV = \frac{(\dot{m}_{OUT} \times h_{OUT}) - (\dot{m}_{IN} \times h_{IN})}{\eta_{cc}} \tag{13}$$

where *WF* is the fuel flow in kg/s; *LFHV* is the lower fuel heating value in J/kg; $\dot{m}_{IN}$ is the mass flow at burner inlet in kg/s; $\dot{m}_{OUT}$ is the mass flow at burnet outlet in kg/s; η_{cc} is the Ccombustion efficiency.

TURBINE

Turbine performance prediction is calculated as follows:

$$\frac{Tt_{OUT}}{Tt_{IN}} = 1 + \eta_t \times \left[\left(\frac{Pt_{OUT}}{Pt_{IN}} \right)^{\frac{(\gamma-1)}{\gamma}} - 1 \right] \tag{14}$$

where η_t is the turbine isentropic efficiency.

The expansion through the turbine generates the necessary power to drive the compressor mechanically linked to the turbine by a shaft. The turbine power can be calculated as follows:

$$W = \dot{m} \times (h_{out} - h_{in}) \tag{15}$$

where: $\dot{m}$ is the gas mass flow at turbine inlet in kg/s.

PROPELLING NOZZLE

In this model, 2 different nozzle geometries were implemented: convergent and convergent-divergent (con-di). For the con-di nozzle, 7 different flow configurations were implemented, as described by Devenport (2001) and Shapiro (1953). The pressure distribution in the nozzle for each configuration is shown in Fig. 3.

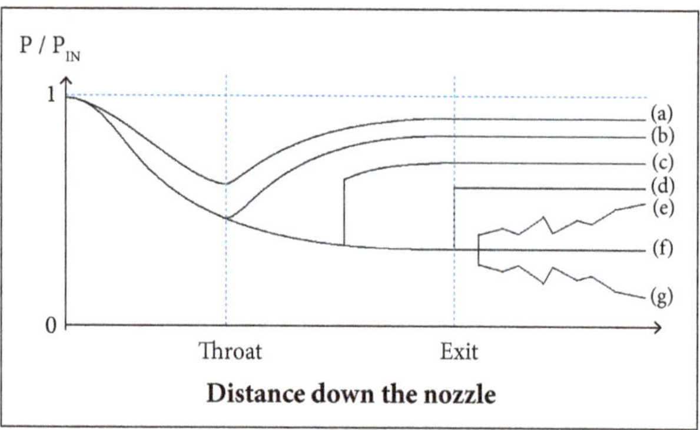

Figure 3. Pressure distribution through the nozzle (Devenport 2001). (a) Not choked at throat; (b) Just choked at throat; (c) Shock in nozzle; (d) Shock at exit; (e) Overexpanded; (f) Design condition; (g) Underexpanded.

GAS PROPERTIES

A good gas properties model is key for any thermodynamic cycle analysis. In order to keep the flexibility and accuracy of the engine performance simulations the gas properties model was developed with refined and detailed data from the Reference Fluid Thermodynamic and Transport Properties (REFPROP; Lemmon *et al.* 2013). All the main gases present in the air and combustion gases composition (N_2, O_2, CO_2, Ar and H_2O) were modeled separately. The gas property is so calculated depending on its chemical composition and the partial contributions of each specific gas enthalpy and molar mass. Enthalpy was modeled considering the effects of different temperatures and pressures.

OFF-DESIGN

The 3 major contributors who enabled the model to converge in few iterations and, therefore, short clock time were the powerful nonlinear system of equation solver, the maps interpolation method and the definition of the starting point of the iterative process.

NON-LINEAR SYSTEM OF EQUATION SOLVER

For the 3 shaft engine architectures the nonlinear system of equation is composed by 8 equations and 8 variables — equations: LP (low pressure) shaft work balance, LP shaft mass flow balance, IP (intermediate pressure) shaft work balance, IP shaft mass flow balance, HP (high pressure) shaft work balance, HP shaft mass flow balance, engine core mass flow balance, and fuel flow/Max cycle temperature constraint; variables: engine mass flow, fan pressure ratio, IP compressor pressure ratio, HP compressor pressure ratio, HP turbine pressure ratio, IP turbine pressure ratio, LP turbine pressure ratio and fuel flow).

The Broyden's Method (Broyden 1965) was selected from trade study that was conducted to define which system of equations solver would give the shortest clock time to find the solution.

The Broyden's method is a generalization of the secant method to nonlinear systems. The secant method replaces the Newton's method derivative by a finite difference:

$$f'(x_k) \approx \frac{f(x_k) - f(x_{k-1})}{x_k - x_{k-1}} \tag{16}$$

$$f'(x_k)(x_k - x_{k-1}) \approx f(x_k) - f(x_{k-1}) \tag{17}$$

where f is the function whose zeros are being searched; x is the free variable; k is the iteration number.

Broyden's gave a system of equation generalization:

$$J_F(x_k)(x_k - x_{k-1}) \approx f(x_k) - f(x_{k-1}) \tag{18}$$

where J_F is the Jacobian calculated for the system of equations; F is a matrix with the solution of each equation calculated for x_k.

Thus it is not necessary to calculate the Jacobian and all its derivatives of the Newton's method in every iteration, therefore this method is time saving at a cost of slightly lower convergence rate.

MAPS INTERPOLATION METHOD

The developed computer program make use of maps for compressors and turbines for off-design calculation. The implemented method to find the operating condition and interpolate within the map values is based on linear interpolation. However, in order to improve the interpolation time, the search for the nearest points for interpolation was enhanced. Usually the map would be read from the first line to the last looking for an interval that comprises the search point. It works fine if the interpolation point is close to the table head, usually close to the design point. However, the farthest the point is from the table head more data is necessary to be read and checked, which make the interpolation slow. In order to improve the searching for the nearest points it was implemented a procedure based on Point in Polygon (PIP) concept. The procedure consists in divide the map in four quadrants and check if the interpolation point is within one of the quadrants. The check is done by checking the sum of the angles between the interpolation point and the quadrant vertices. If the sum is 2π it means that it is in the quadrant and if it is 0, it is not. Once the quadrant that contains the interpolation point is found the same procedure is repeated reducing the quadrant size until the quadrant is formed only by the 4 nearest points, when it is ready for the interpolation. The closest three points defines a plane that comprises the interpolation point and therefore the interpolation within the plane can be calculated. The plane interpolation was implemented in order to avoid bilinear interpolation issues where the mass flow is constant and the interpolation in mass flow axis would lead to a division by 0.

Figure 4 shows a compressor map, as an example, and it is possible to observe how the map is divided into 4 quadrants successively until the quadrant is formed only by the 4 nearest points to the operating condition. Figure 5 shows how the angles between the operating point and the quadrant vertices shall be considered for PIP evaluation, whose possible values are described in Eqs. 19 and 20.

$$\text{If } \Sigma\theta_i = 2\pi \text{ then the point is within the polygon} \tag{19}$$

$$\text{If } \Sigma\theta_i = 0 \text{ then the point is out of the polygon} \tag{20}$$

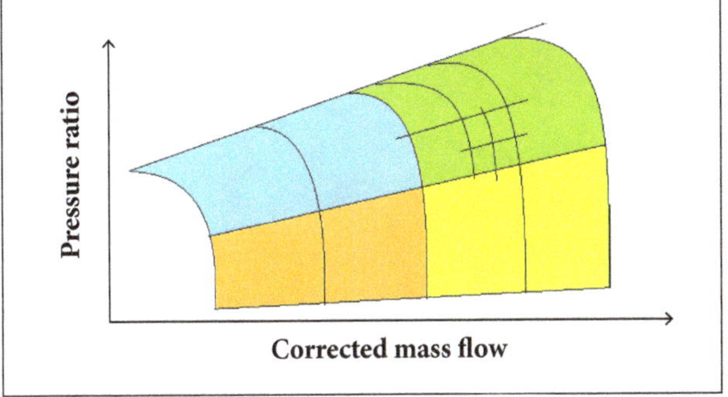

Figure 4. Quadrant division example in a compressor map.

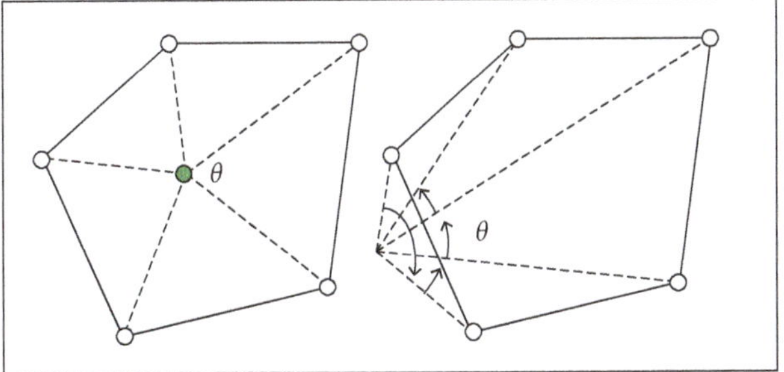

Figure 5. PIP graphic representation.

ITERATION STARTING POINT

Another extremely powerful feature of the model which improves both convergence success rate and time until the solution is the selection of the starting point close to the final solution. Obviously, the final solution is not known until the simulation is completed, but an approximation of the final solution can be estimated based on some engine parameters. In the model developed for this paper the engine parameters to start the iteration are set based on the flight condition and the a power setting parameter. The design point parameters are corrected to the off-design flight condition and then corrected to the input power setting. The power setting parameter defines the engine power such as fuel flow, burner exit temperature and shaft speed. All of them can be set as input to the model.

MODEL VERIFICATION

The developed model was compared in terms of thrust and fuel flow calculation with an existing commercial gas turbine performance model. The model for reference was GasTurb11® (Kurzke 2007) which is very known, reliable and flexible to receive the same kind of inputs necessary to set the model developed for this paper. The simulations, for all the 3 architectures, were based on a burner exit temperature sweep at ISA Sea Level Static condition and compared using the same compressors and turbine maps. Figures 6 to 11 show the comparison between the GasTurb11® and the developed model. The divergences found in thrust and fuel flow are due to differences in the combustion gas model. The gas model in GasTurb11® does not consider pressure in the enthalpy calculation while the developed model does. Also, the combustion gases composition calculation may lead to differences in the cycle calculation mainly downstream to the burner. The model could not be compared in terms of run time because no models were found in the literature with the ability to run in real time.

Three different engine architectures were simulated and compared in terms of thrust and fuel flow with the engines modeled in GasTurb11® with same configuration. The architectures are the most utilized in the aeronautic industry: single, 2 and 3 shafts direct drive engines with unmixed flows and convergent nozzles. The Design Point of the models is shown in Table 3.

Figures 6 and 7 show the thrust and fuel flow comparison with GasTurb11® for the turbojet architecture (one shaft direct

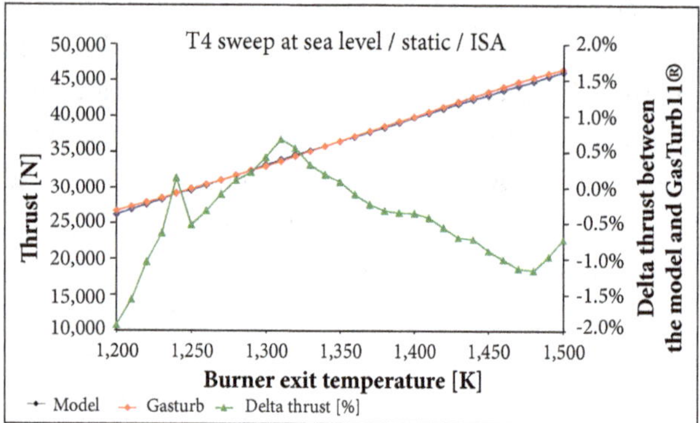

Figure 6. Single shaft engine thrust comparison.

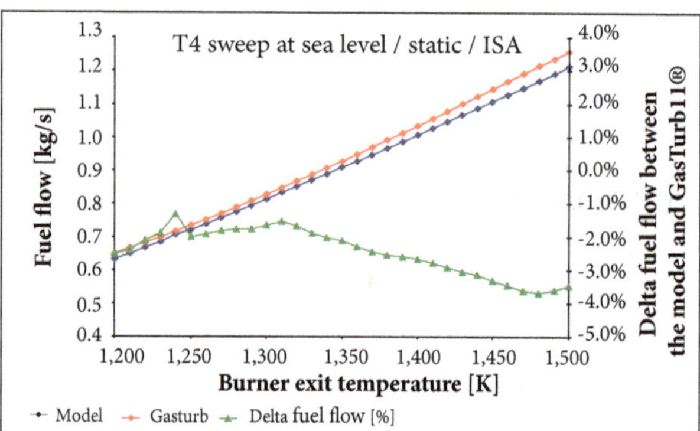

Figure 7. Single shaft simulated engine fuel flow comparison.

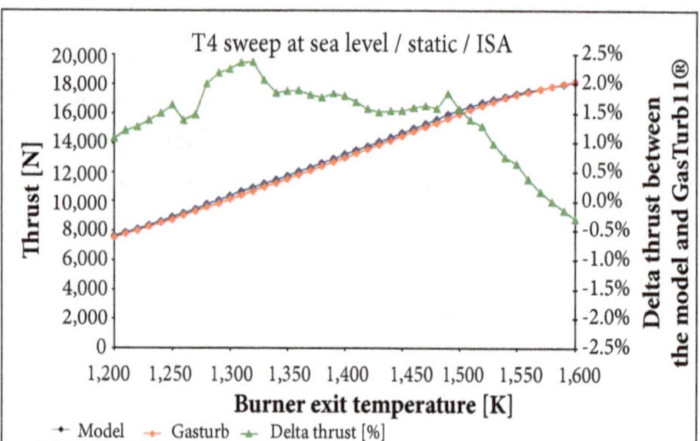

Figure 8. Two shaft simulated engine thrust comparison.

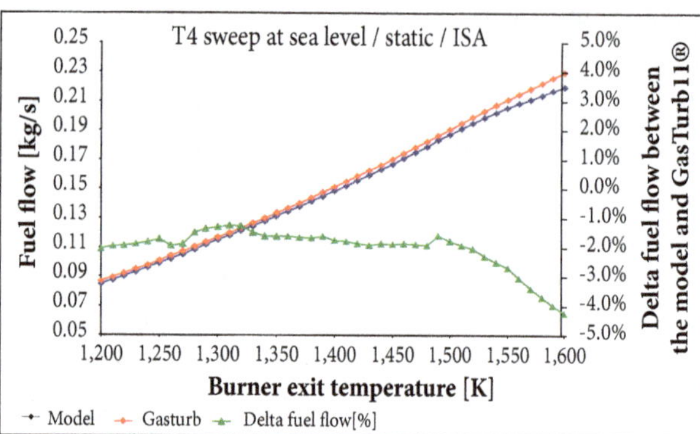

Figure 9. Two shaft simulated engine fuel flow comparison.

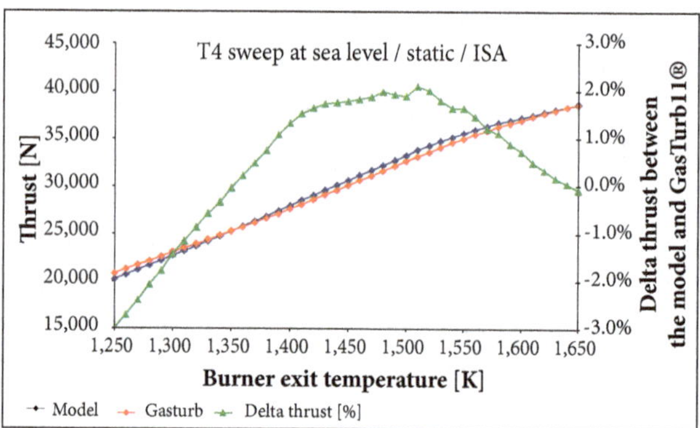

Figure 10. Three shaft simulated engine thrust comparison.

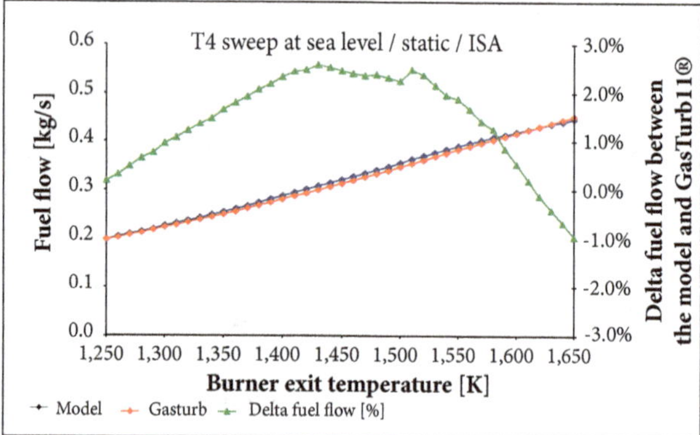

Figure 11. Three shaft simulated engine fuel flow comparison.

drive engine). Figures 8 and 9 show the same comparison for the 2 shaft direct drive turbofan. Finally, Figs. 10 and 11 are the comparison between the models for 3 shaft direct drive turbofan engine.

In Figs. 6 and 7 the blue and red curves refer to the calculated parameters, thrust or fuel flow, by this paper's model and GasTurb11®, respectively, and the values are in the left vertical axis. The difference between the values calculated by the model described in this paper and Gasturb11® are shown by the green curve whose values are in the right vertical axis. Differences are expected due to the gas properties model differences and premises in the 2 different simulation tools.

TEST MATRIX FOR RUN TIME EVALUATION

In order to test the convergence time, the model was run at different off-design conditions to explore different component map regions. The off-design conditions were set by inputting different altitudes, Mach Numbers, temperatures and 1 engine power set, burner exit temperature in this

Table 3. Simulated engines design point for output comparison with GasTurb11®.

	Single shaft	2shaft	3shaft
Altitude (m)	0	0	0
Temperature (K)	288.15	288.15	288.15
Flight Mach Number	0	0	0
Air inlet mass flow (kg/s)	50	50	100
Air intake pressure recovery	0.99	0.99	0.99
Bypass ratio	0	5	5
Inner LP compressor pressure ratio	-	2	1.8
Inner LP compressor isentropic efficiency	-	0.88	0.88
Outer LP compressor pressure ratio	-	1.8	2
Outer LP compressor isentropic efficiency	-	0.5	0.88
IP compressor pressure ratio	-	-	2
IP compressor isentropic efficiency	-	-	0.88
HP compressor pressure ratio	12	7	5
HP compressor isentropic efficiency	0.85	0.85	0.88
Fuel heating value (kJ/kg)	43,124	43,124	43,124
Burner exit temperature (K)	1,500	1,600	1,650
Burner pressure ratio	0.97	0.97	0.98
Burner isentropic efficiency	0.9999	0.9995	0.9995
HP turbine isentropic efficiency	0.89	0.9	0.9
HP shaft mechanical efficiency	0.99	0.99	0.98
IP turbine isentropic efficiency	-	-	0.9
IP shaft mechanical efficiency	-	-	0.99
LP turbine isentropic efficiency	-	0.9	0.9
LP shaft mechanical efficiency	-	0.99	0.99

assessment. Table 4 summarizes the chosen values used to simulate different engine operational conditions.

RESULTS

The run time distribution and the number of iterations until the convergence are shown in Figs. 12 and 13, respectively. The run times were achieved in a personal computer with Intel Core i7 920 at 2.67GHz and the solver convergence criteria was set to square root of the machine precision which was in the computer where the points were run, 10^{-8}. The results are disposed in a histogram chart where it is shown the distribution of the number of converged points, in the ordinates, by the elapsed time until convergence (Fig. 12) or number or iterations until the convergence (Fig. 13), in abscissas. The points and the operating conditions evaluated are described in Table 4.

An additional run time reducing opportunity was assessed in order to improve the model run time: iteration stopping criteria relaxing. In order to provide accuracy in the calculations the stopping criteria was chosen to be the square root of the machine precision. Figure 14 shows that the model converges very quickly to the solution and spends a lot of iterations refining

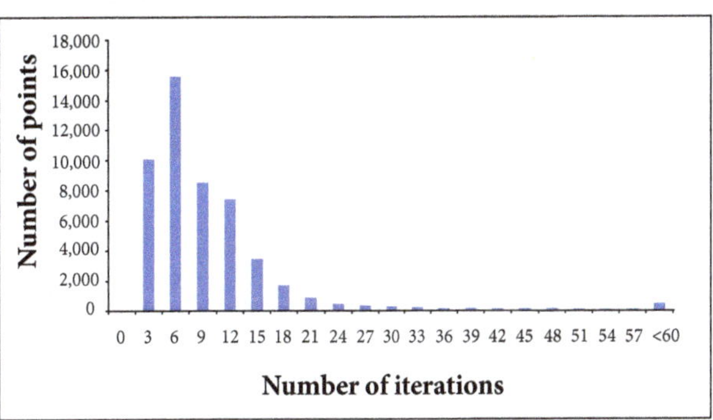

Figure 12. Run time histogram.

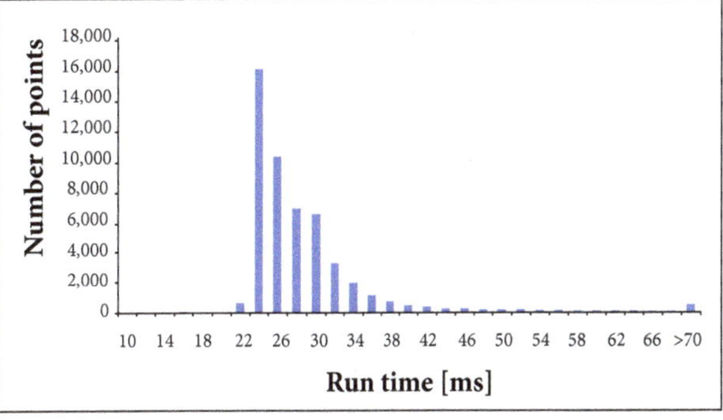

Figure 13. Number of iterations histogram.

Table 4. Simulation test matrix.

Altitude	Mach Number	Delta standard day	Burner exit temperature
Sea level - 15,000 m (steps of 500 m)	Static - 0.8 (steps of 0.05)	−30 °C +30 °C (steps of 5 °C)	1,800 K - 1,000 K (steps of 100 K)

the solution to meet the very tight stopping criteria. The chart shows the evolution of 3 of the equations in the nonlinear system of equations for off design calculation. When the equation goes to 0 it means it converged. It can be seen that the parameters converge very quickly to the solution, approximately 4 iterations in the example, and require another 5 iteration to refine the final solution to meet the excessively sharp stopping criteria.

The potential run time improvement due to the stopping criteria relaxation was assessed and the results are shown in Fig. 15. The chart shows the number of converged points in the ordinates and the run time in the abscissas. It can be seen that the peak and the average of the red columns, which represents the run time of the points with relaxed stopping criteria, are at lower run times when compared with the blue columns, which represents the points with original stopping criteria. It means that, by relaxing the stopping criteria, in general, the points converged faster, as expected.

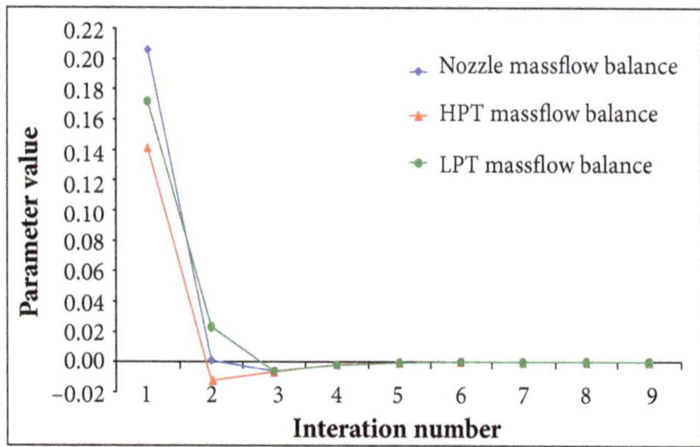

Figure 14. Iteration steps until the solution.

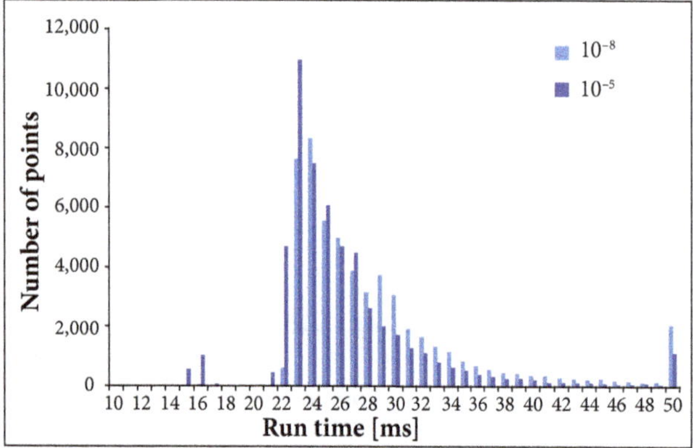

Figure 15. Stopping criteria relaxing benefit in run time.

Similar result can be verified in Fig. 16. The chart shows the benefit that the stopping criteria relaxing brought in terms of numbers of iterations. The peak and the average moved to the left, what means that more points converged at lower number of iterations.

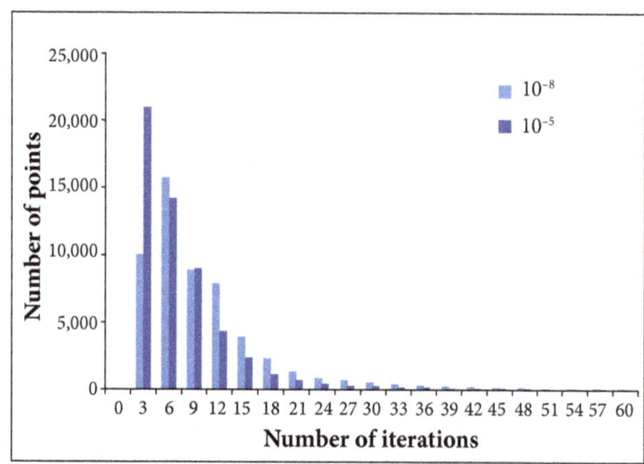

Figure 16. Stopping criteria relaxing benefit in the number of iterations.

CONCLUSIONS

A brand new engine performance prediction model was developed with the ability to run and reach the convergence in most of the times in less than 30ms, which is compatible with a high definition video format, whose refresh rate is 30 frames per second. The features implemented in the model to improve the run time were very effective and ensure good model performance, within the target run time. Additionally, the model did not lose accuracy and flexibility with those features. In fact, by setting the starting point close to the final solution, the convergence success rate was also improved. An additional feature which was also investigated, the relaxation in the iteration stopping criteria could improve even more the run time at a cost of some accuracy loss.

The aim of this study was to demonstrate a model with the ability to simulate the performance of single, 2 and 3 shaft gas turbines with run times compatible with real-time applications with high-fidelity accuracy. The developed model was verified using commercial gas turbine performance software.

ACKNOWLEDGEMENTS

The financial support from Empresa Brasileira de Aeronáutica (Embraer), Conselho Nacional de Desenvolvimento Científico e Tecnológico (CNPq), Centro de Pesquisa e Inovação Sueco-Brasileiro (CISB), and Svenska Aeroplan AB (SAAB) is acknowledged.

AUTHOR'S CONTRIBUTION

Gazzetta Junior H, Bringhenti C, Barbosa JR and Tomita JT conceived the idea; Gazzetta Junior H wrote the main text and prepared the figures. All authors discussed the results and commented on the manuscript.

REFERENCES

Antoine C (1888) Tensions des vapeurs; nouvelle relation entre les tensions et les températures. Comptes Rendus des Séances de l'Académie des Sciences (in French). Paris.

Baig MF, Saravanamuttoo HIH (1997) Off-design performance prediction of single-spool turbojets using gasdynamics. J Propul Power 13(6):808-810. doi: 10.2514/2.5240

Bringhenti C (1999) Análise de desempenho de turbinas a gás em regime permanente (Master's thesis). São José dos Campos: Instituto Tecnológico de Aeronáutica.

Bringhenti C (2003) Variable geometry gas turbine performance analysis (PhD thesis). São José dos Campos: Instituto Tecnológico de Aeronáutica. In Portuguese.

Broyden CG (1965) A class of methods for solving nonlinear simultaneous equations. Math Comp 19:577-593.

Devenport WJ (2001) Instructions; [accessed 2015 May 15]. http://www.engapplets.vt.edu/fluids/CDnozzle/cdinfo.html

Fishback LH, Koenig RW (1972) GENENG II: A program for calculating design and off-design performance of two and three spool turbofans with as many as three nozzles. (NASA TN D-6553).Washington: NASA.

Flack RD (1990) Analysis and matching of gas turbine components. International Journal of Turbo and Jet Engines 7(3-4):217-226. doi: 10.1515/TJJ.1990.7.3-4.217

Gordon S (1982) Thermodynamic and transport combustion properties of hydrocarbons with air. Lewis Research Center Cleveland.

Grönstedt T (2000) Development of methods for analysis and optimization of complex jet engine systems (PhD thesis). Göteborg: Chalmers University of Technology.

Ismail IH, Bhinder FS (1991) Simulation of aircraift gas turbine engines. J Eng Gas Turbines Power 113(1):95-99. doi: 10.1115/1.2906536

Koenig RW, Fishback LH (1972) GENENG: A Program for calculating design and off-design performance for turbojet and turbofan engines. (NASA TN D-6552). Washington: NASA.

Korakianitis T, Wilson DG (1994) Models for predicting the performance of Brayton-cycle engines. J Eng Gas Turbines Power 116(2):381-388. doi: 10.1115/1.2906831

Kurzke J (2007) GasTurb11: design and off-design performance of gas turbines. Aachen: GasTurb GmbH.

Lemmon EW, Huber ML, McLinden MO (2013) NIST Standard Reference Database 23: Reference Fluid Thermodynamic and Transport Properties-REFPROP, Version 9.1. Gaithersburg: National Institute of Standards and Technology.

Macmillan WL (1974) Development of a modular type computer computer program for the calculation of gas turbine off-design performance. Cranfield: Cranfield Institute of Technology.

Mckinney JS (1967) Simulation of turbofan engine - SMOTE: Description of method and balancing technique. AD-825197/AFAPL-TR-67-125. Air Force Aero Propulsion Lab. pt.1-2.

Palmer JR, Yang CZ (1974) TURBOTRANS: A programming language for the performance simulation of arbitrary gas turbine engines with arbitrary control systems. ASME Paper 82-GT-200.

Saravanamuttoo HIH, Rogers GFC, Cohen H (2001) Gas turbine theory. 5th edition. New York: Prentice Hall.

Sellers JD (1975) DYNGEN: a program for calculating steady-state and transient performance of turbojet and turbofan engines. NASA TN D-7901. Washington: NASA.

Shapiro AH (1953) The dynamics and thermodynamics of compressible fluid flow. Vol. I-II. New York: The Ronald Press Company.

Silva FJS (2011) Gas turbines performance study under the influence of variable geometry transients (PhD thesis). São José dos Campos: Instituto Tecnológico de Aeronáutica. In Portuguese.

Stamatis A, Mathioudakis K, Papailiou KD (1990) Adaptive simulation of gas turbine performance. J Eng Gas Turbines Power 112(2):168-175. doi: 10.1115/1.2906157

Szuk JR (1974) HYDES: a generalized hybrid computer program for studying turbojet or turbofan engine dynamics. NASA TM X-3014. Washington: NASA.

Walsh PP, Fletcher P (2004) Gas turbine performance. 2nd edition. Oxford: Blackwell Science.

Wittenberg H (1976) Prediction of off-design performance of turbojet and turbofan engines. CP-242-76, AGARD.

7

The Magnetic Tracker with Improved Properties for the Helmet-Mounted Cueing System

Michail Zhelamskij[1]

ABSTRACT: This article highlights both theoretical and experimental experiences in the field of helmet-mounted cueing systems. The current state of these systems is described as optical and hybrid. The adventures of the positioning under local magnetic field are considered, and the directions for further improvement of magnetic technology are identified. A new method is proposed for the local magnetic field creation to increase update rate, to reduce the influence of the Earth's magnetic field, and to reduce energy consumption of helmet-mounted cueing systems in relation to known prototypes. A mathematical model of positioning field is offered. The accuracy of the field mathematical description is studied for different shapes of windings. The transients are investigated in the source of positioning field and in the interior of the cockpit. In addition, a mathematical model of magnetic measurements is proposed, and the main sources of measurement and positioning errors are investigated. The calculation algorithm of the helmet's coordinates is considered based on the results of magnetic measurements. The results of physical models research are given, and the operation of a sample in the full range of angles is shown. The trial mapping is conducted for the field created by the source with a ferromagnetic core. Positioning of the helmet's movement on specified paths is performed, and the results make it possible to figure out the next generation of helmet-mounted cueing systems with extended angles range, higher angular and linear accuracy, increased update rate (200 Hz), and minimized influence of Earth's magnetic field.

KEYWORDS: Helmet, Cueing systems, Magnetic field.

INTRODUCTION

The helmet-mounted cueing system (HMCS) is a device used in modern combat aircraft, which allows the pilot to designate the on-board weapon and other equipment at the target in accordance with the direction of sight. Before the HMCS appeared, in close combat, the pilot had to align the aircraft to shoot at a target. Using the head angle as a pointer to direct the weapons, the pilot can point his head at the target to actuate a weapon. This enables making more attacks, without having to maneuver to the optimum firing position. These systems allow targets to be designated with minimal aircraft maneuvering, minimizing the time spent in the threat environment and allowing greater lethality, survivability, and pilot situational awareness. These devices were created first in South Africa (Mirage F1, mid-1970s), then in the Soviet Union (MIG-29, 1985), Israel (Python-4, 1990) and, finally, in the United States (AIX-9X missile, 1990) (Melzer, 1997). If the position of the helmet is used to point the missile, it thus must be calibrated and fit securely on the pilot's head. That is why HMCS should be considered from a scientific point of view.

Figure 1 shows the position and orientation of the helmet in the coordinate system of the aircraft. The task of targeting is the orientation calculing of the movable helmet coordinate system $X'Y'Z'$ in relation to the stationary coordinate system XYZ of aircraft. The line between the pilot's eye and the reticle on the visor is known as line of sight (LOS) between the aircraft and the intended target. The user's eye must stay aligned with the sight direction. To do it, the reticle R should be rigidly connected with the helmet and capture the view direction in

1.Polytechnical University – Department of Measurement and Technologies – Saint Petersburg – Russia.
Author for correspondence: Michail Zhelamskij | Polytechnical University – Department of Measurement and Technologies | Politekhnicheskaya ul., 29 | 195251 Saint Petersburg – Russia | E-mail: zhelamsk@rambler.ru

relation to it. Therefore, the pilot's eye always looks at the target through the reticle.

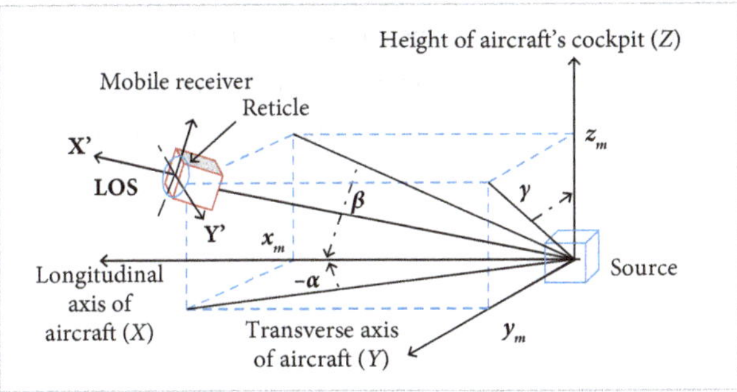

Figure 1. Linear position and orientation of mobile receiver relative to the source in the Cartesian coordinate system.

HMCS design must sense the elevation, azimuth and roll of the pilot's head in relation to the aircraft with sufficient precision even during maneuvering and rapid head movement. The *X'* axis indicates the direction of the target and corresponds to LOS. Azimuth (*a*) and elevation (*b*) determine the LOS direction. Linear coordinates *x*, *y*, and *z* of the helmet in the cockpit, calculated by any HMCS, can be used to monitor the pilot's status. Information about the roll is useful for stabilizing the image on the helmet-mounted display for the accounting of the pilot's vestibular.

The range of the helmet linear motion should not be lesser than 1 m. The precision of cueing, angular error between LOS and derived cue, is determined by the field of view (FOV) of the seeker of the air-to-air missile. The accuracy of the LOS positioning should be much lesser than the FOV of the missile, which is about 1.8° for an infrared heatseeker (Kopp 1982). The accuracy should be equal over all range of helmet angular motion, and the common field of view of HMCS, angular range over which the sight can still produce a suitably accurate measurement, should be maximum. The latency or slew rate, how much lag there is between the helmet and the cue, should be minimum. The weight of helmet-mounted part of HMCS should be minimum, as well as the power consumption. Otherwise, the update frequency should be maximized. It is enough the linear position coordinates accuracy around centimeters. The roll should be determined with accuracy at the level of units of angular degrees.

Any HMCS includes a movable part, located on the helmet, as well as another item, fixed in the cockpit. Both can be a receiver or a source of local physical fields. The computer, also included in HMCS, solves the mathematical positioning task. The helmet-mounted display will not be considered here.

The new theory is proposed with 6 degrees of freedom (DOF) of magnetic positioning at short distance. The way to organize the local magnetic field is described to improve accuracy of cueing. The mathematic models to create and measure the local positioning field are suggested. The proposal determines a concept of the new generation of magnetic trackers with improved properties. The test results of the first magnetic cueing sample are shown.

The results of numerical simulations of coordinates calculations in the positioning field are shown, besides the results of the mathematical and physical models investigations. The error estimation of the positioning field descriptions and measurements were done as well as positioning ranges concerning angular and linear displacement.

BASIC DEFINITIONS

It is known the HMCS based on the optical triangulation (Elbit Systems of America® 2016; Buganov 2016; Defencetalk.com 2007). The triad of one-by-one light emitting diodes (LEDs) is located on the helmet's surface. Two fixed on-board receivers are split in the cockpit room. The coordinates of each LED $[x_d, y_d]$ on the surface *XOY* can be obtained from the following system of equations:

$$\begin{cases} y_d = K_L(\alpha_L \cdot x_d + A_L(\alpha_L) \\ y_d = K_R(\alpha_R \cdot x_d + A_R(\alpha_R) \end{cases} \quad (1)$$

where: α_L and α_R are the bearing on each LED from 2 split receivers, obtained from the measurement; $K_L = -\text{tg}\alpha_L$; $K_R = -\text{tg}\alpha_R$; $A_L = x_2 \times \text{tg}\alpha_L + y_2$; $A_R = x_1 \times \text{tg}\alpha_R + y_1$; the coefficients; x_1, y_1, x_2, and y_2 are the coordinates of receivers in the cockpit.

The coordinates of the 3 LEDs together with the dimensions of emitting triad are enough to determine the spatial position of the helmet by the methods of analytic geometry, using the solution of Eq. 1 for each LED. The angular positioning accuracy is at the level of $\delta\varphi \geq 45'$. The optical HMCS operates within a cone not greater than ±45°. Accuracy depends on the helmet's orientation because the helmet itself closes the visibility between LED and receivers, and triangulation triangles are degenerated in the line.

A further approach appeared recently to extend the range of operation of the optical tracker, through the integration of helmet-mounted LEDs together with the gyroscopes and accelerometers (Thales Visionix, Inc. 2016; BAE Systems 2016). Hybrid inertial tracking systems employ a sensitive inertial measurement unit and optical sensor to provide reference to the aircraft. In the previous operating range, the same optical system is used. Beyond the range of optical tracker, the inertial sensors are used, which have a fundamentally permanent drift of output signals.

Micro-electro mechanical systems (MEMS) contain both gyroscopes and accelerometers and allow to measure full acceleration and orientation of the helmet together with the aircraft movements. It should be taken into account that the full acceleration (a) of the aircraft reaches 10 g during the maneuvers, whereas the head movement, just 0.01 g. An estimation showed that the modern MEMS like ADIS16448 (Analog Devices 2016) measures the helmet's orientation angles with error at level $\Delta\varphi^3 \pm 0.3$ angular degree per second for the aircraft acceleration at the level $a = 1$ g. In the same condition, the linear coordinates of the helmet are measured with the error at the level $\Delta x = \pm 0.22 \times t^2$ m. It is clear that the angular error exceeds requirements for HMCS during the first seconds, and the linear coordinates error reaches the level of the percentage meter. Thus, hybrid tracker allows expanding the ranges of positioning angles only by short-term use of inertial sensor out of operation range of optical HMCS. Common accuracy in the initial angular range is still determined by the worst element — optical tracker.

The magnetic tracking system includes the fixed source of local magnetic field, movable receiver on the helmet and on-board computer, which resolves 3 tasks simultaneously — source controlling, magnetic measurements and coordinates calculation. The procedure of active magnetic positioning intends to establish local non-uniform magnetic field with a known spatial distribution, in which the magnetic induction measurement is performed using sensors located at the helmet. The calculation of the position and orientation of a movable helmet-mounted receiver is associated with the solving of systems of non-linear equations, which contain the results of independent measurements of magnetic induction at the point of observation under specified parameters of the field source (the right parts) and unknown linear and angular coordinates of receiver in space of the positioning field source (the left parts of equations).

The known active magnetic trackers are based on alternating current (AC) with sinusoidal shape (Fig. 2; Raab 1977) or direct current (DC) with pulse shape (Fig. 3; Blood 1989) local magnetic field.

The tracker includes 3 orthogonal windings in the fixed source of local magnetic field and 3 orthogonal sensors in the mobile receiver. In the first case (AC), all windings of the source work simultaneously ($I_x + I_y + I_z$) on different frequencies. In the second case, they operate in the pulse mode, one by one, in sequence: $I_x \rightarrow I_y \rightarrow I_z$. The pause ($t_3 - t_4$) is designed to take into account the Earth's magnetic field (EMF), whose influence is subtracted from each of the measured signals, obtained from the windings.

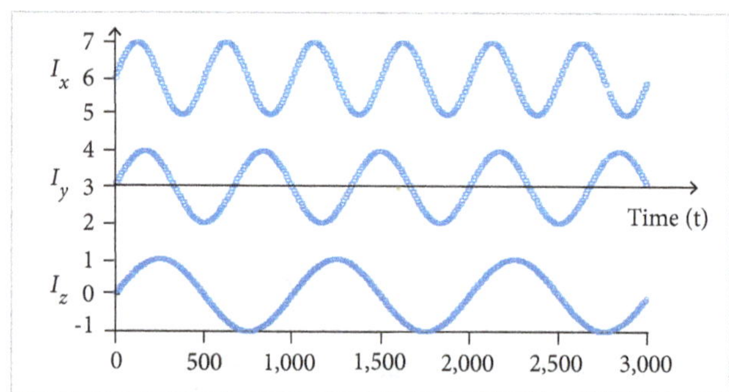

Figure 2. Dimensionless charts of the excitation currents through the windings of the sinusoidal magnetic positioning systems.

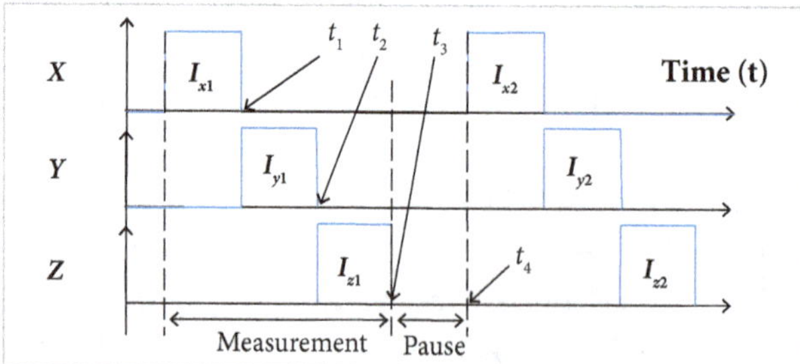

Figure 3. Dimensionless charts of the excitation currents in pulse magnetic positioning systems.

In both cases, the coordinates of the helmet are determined from the following equation system solution:

$$B_x \cdot \cos\varphi + B_y \cdot \cos\nu + B_z \cdot \cos\xi = B_{xx} \tag{2}$$

where: B_x, B_y and B_z are the computed values of axial components of magnetic fields induction vector, defined by mathematical models of positioning field, at coordinates of the initial approximation x_s, y_s, and z_s; cos φ, cos ν, and cos ξ are the values of guide cosines of Hall sensor in a fixed coordinate system; B_{xx} is the measured value of the induction of one sensor from one source winding. Designating $B_x \times \cos\varphi + B_y \times \cos\nu + B_z \times \cos\xi = F$ and $B_{xx} = \Theta$, we get:

$$F(\psi, X) = \Phi - \Theta = 0 \tag{3}$$

where: $\psi = [\mathrm{M}, t, X_c, T, C_0, I_{HG}, I_S, \Delta\mathrm{MS}, w]$ is the vector of parameters included in the mathematical model of establishment and measurement of positioning field, described in Eq. 9. The number of equations M in Eq. 3 shall not be less than the number of the desired coordinates $M \leq N \times K$, where N is the number of windings in the source and K is the number of sensors in the receiver. The calculation of the 6 coordinates is reduced to the solution of systems (Eq. 3), containing non-linear equations (Eq. 2). In this case, we have: $\mathbf{F}(\mathbf{X}, \psi) = 0$, where $\mathbf{F}$ is the vector function of $\mathbf{X}$ — vector of intended coordinates at the observation point $\mathbf{X} = (x, y, z, \alpha, \beta, \gamma)$:

$$\vec{F}(X,\Psi) = \begin{bmatrix} F_1(x,y,z,\alpha,\beta,\gamma,\Psi) \\ \dots\dots\dots\dots\dots\dots \\ F_6(x,y,z,\alpha,\beta,\gamma,\Psi) \end{bmatrix} = 0 \tag{4}$$

where: γ is the roll angle.

The desired linear coordinates of mobile receiver $[x_m, y_m, z_m]$, recorded in Eq. 2 as arguments to the mathematical description of the axial component of induction vector, $\mathbf{B}_m = [B_x(x_m, y_m, z_m), B_y(x_m, y_m, z_m), B_z(x_m, y_m, z_m)]^{\mathrm{T}}$, and orientation angles of the receiver, $[\alpha_m, \beta_m, \gamma_m]$, are present in guide cosines through the matrix of the movable receiver rotation: $A_{xyz} \equiv A_x(\alpha) \times A_y(\beta) \times A_z(\gamma)$. Thus, the system (Eq. 4), composed of 6 equations, has a strictly non-linear nature and can only be solved by numerical methods of iterative approximation.

Figure 4 shows a typical layout of magnetic HMCS in the cockpit through the example of the project Vista for the F-16D aircraft (Merryman 1994). A similar layout has another magnetic HMCS (Elbit Systems 2006; Kopp 1998; Thales Group 2016). Still in Fig. 4, there is a cubic source of the local magnetic positioning field, established behind the pilot's right shoulder. The movable receiver is hidden behind the external surfaces of the protective helmet. The main advantage of magnetic methods of positioning is that the LOS is not required between source and receiver. Therefore, the helmet cannot influence the ranges of operation due to transparency for the stationary magnetic field. This advantage sets out a broad range of the operation angles for magnetic HMCS. As a result, the operation distance — up to 1 m — is comparable with the cockpit size, and the angular range theoretically varies up to ±180°. Positioning accuracy of the magnetic HMCS is better than that of the optical one and declared at the level $\delta X \leq \pm 1$ mm and $\delta\varphi \leq \pm 0.5$ angular degree for up to 1-m distance for the stationary receiver with unchanged orientation.

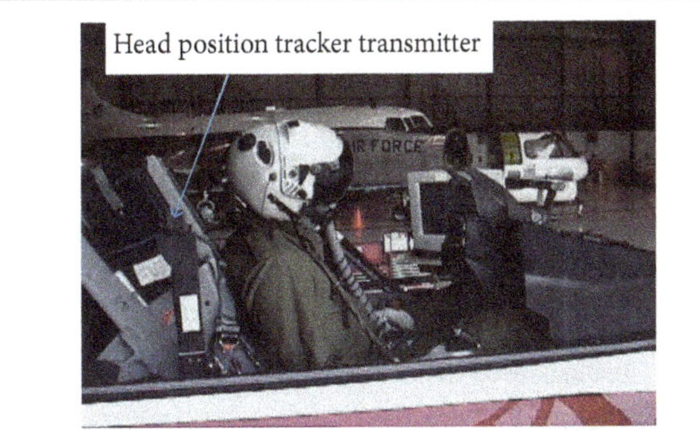

Figure 4. Magnetic positioning system of Vista layout aboard the F-16D.

Dielectric interior elements have no effect on the positioning accuracy for the magnetic method. Electrically conductive materials have effect for AC method only, depending on the size and distance between source, receiver and element. The mapping of the influence of eddy currents on the AC method is labor-intensive (Lescourret 1997) and not always yields the result, particularly for helicopters with cramped cockpits. The practice has shown that AC method cannot be used in helicopters (Egli *et al.* 1983). Conductive interior elements do not affect the DC method when a certain duration of the magnetic field pulse is selected (see next). The magnetic materials have effect on both AC and DC magnetic methods, but the magnetic elements in the cockpit interior are much lesser than the electrically conductive materials. Therefore, only DC method will be considered further.

The EMF vector is added to the field positioning. A contribution of EMF depends on the ratio between the object velocity and the update frequency and, for stationary object, it is zero. For DC method, the EMF is taken into account one time per full cycle of the pulse field switching. The magnetic field switches to 4 times faster than the output updated, as shown in Fig. 3. This is why the update frequency should be increased more over the current 100 Hz limit to reduce the EMF contribution.

The main parameters of AC and DC trackers, available in the literature, are given in Table 1. The main sources of measurement errors of the positioning field are shown in Zhelamskij (2014b), among which the influence of the sensors' size near the source is dominated as well as their spatial separation and the accuracy of the mathematical description of the positioning field. The sensors' size reaches 4 – 5 mm in modern aviation trackers (Kuipers 1975), which leads to increased measurement error of module induction vector near a source up to 10%. Therefore, the decreasing in the sensors' size is actual and accounts for their spatial separation, as described next. The already mentioned aircraft trackers prototypes use either inductive sensors in AC method or fluxgate sensor in DC one. In both cases, there are induction measuring windings with large area, which is sensitive to high-frequency interference. Therefore, when the sensitivity of measurements is at the level of 10 nTesla, the sustainability of mobile receiver to external disturbances, for example, radar's radiation, is relevant. A power consumption should be minimized for an on-board equipment in any case. The task is to find an opportunity to do it. Now we have about 100 W per sphere with 1-m radius. There is important to reduce the value of power consumption for on-board cueing system.

Table 1. Comparison of parameters of the prototypes (Polhemus 2016; Ascension Technology Corporation 2016).

Title	Source-receiver maximum distance (m)	Update frequency (Hz)	Static accuracy at the distance of 0.78 m
Polhemus "Patriot"	1.5	60	±1.5 mm; ±0.4°
Ascension "DriveBAY"	0.78	> 120	±1.4 mm; ±0.5°

The above review clearly showed that the pulse magnetic method of positioning is the most promising among others concerning the range of operation, precision, and interior influence. However, it can be seen that the existing magnetic tracking technology does not fully meet these requirements. Thus, this paper describes the results of an investigation of the possibility to improve the magnetic trackers for cueing task. It must be explored the possibility of reducing power consumption of the HMCS as well as the EMF's influence. It is necessary to increase the update frequency when possible and to increase the immunity HMCS against the external radar. Concerning the mapping, it can be supposed that this procedure is not needed, because the magnetic elements on-board the light modern airplane or helicopter are much smaller than the electrically conductive material. However, the movable ferromagnetics should not be situated inside the zone of the helmet movement. An effect of electroconductivity elements for DC method is determined by the ratio between magnetic field pulse duration and eddy current decay time constant. Therefore, the eddy current in the cockpit interior should be investigated. The main objective of this article is to declare the goals for a next generation of magnetic helmet-mounted cueing system with improved parameters as update frequency, EMF influence, power consumption, and outer noise immunity. The new mathematical models to create and measure the positioning magnetic field are presented. The advantages of the new method achieved in comparison with the prototype are described, the estimations of the eddy currents influence are given, and the algorithm of iterative calculation of coordinates of the mobile receiver is considered.

METHODOLOGY

An investigation of new magnetic cueing system was fulfilled with the methods described next, such as mathematic description and simulation, system analysis, comparison with known theories, physical modeling of non-standard elements of the system, and the actual movement of the mobile receiver at the final stage of positioning modeling.

To satisfy the formulated requirements, a new method is proposed to organize the positioning local magnetic field, and investigations are performed. The new method is theoretically considered as a means of reducing the impact of EMF and power consumption. A mathematical model of the positioning field is proposed. The investigation of the transition process is performed, as well as of the accuracy of the field descriptions in the presence of conductive elements. The mathematical model of movable receiver is also proposed and investigated regarding both measurement errors and eliminating the interference from on-board network and external sources. They are explored in different approaches for the iterative solving of the non-linear equations systems, linking the results of magnetic measurements with the desired coordinates. An investigation of the physical models allowed to estimate the mathematical description accuracy of the source with and without ferromagnetic core. Besides, the metrological research of movable receiver was performed. The 6-DOF tracking was fulfilled when the receiver was moving at the specified path.

A new (the 3rd in the world) method to organize the local magnetic positioning field is called alternating-direct current (ADC) method (Zhelamskij 2011). The bipolar current pulses are offered in the new method, running consecutively one by one without pause. Figure 5 shows the comparative chart of pulse currents in the source windings for DC prototype, separated by a pause, and for the new ADC method. The top graph is combined for 3 windings of DC method. Three lower graphics are separated for each of the windings of ADC method. In both

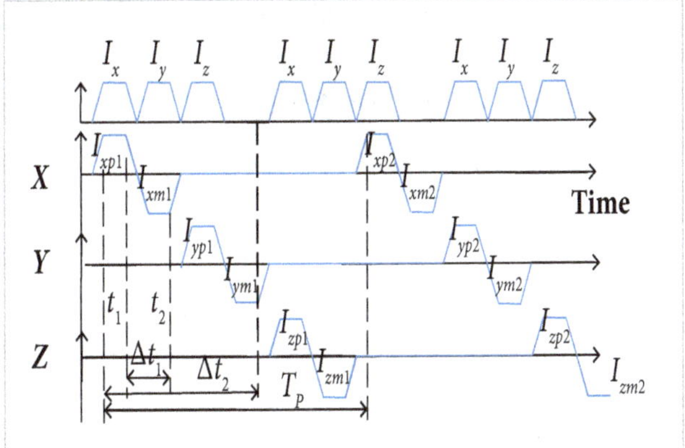

Figure 5. Schematic comparison of ADC and DC methods to organize the field of positioning.

cases, the graphics are post-poned for 3 orthogonal windings of local positioning systems (Raab 1977; Blood 1989), where the magnetic field pulses duration is equal. The graphics are schematically shown. Actually, the rise and fall of pulse current is much lesser than the pulse duration. At the top of each bipolar impulse, the positioning field vector is folded with the EMF at a mobile receiver:

$$\vec{B}_m = \vec{B}_{EMF} + \vec{B}_P \tag{5}$$

where: $\vec{B}_m$ is the total induction vector, measured by sensors of the mobile receiver; $\vec{B}_{EMF}$ is the EMF vector; $\vec{B}_P$ is the vector of positioning field.

One can write, from Eq. 5, the system of equations for time points t_1 and t_2, separated by an interval Δt_1, as shown in Fig. 5:

$$\begin{cases} \vec{B}_{EMF}(t_1) + \vec{B}_P(t_1) = \vec{B}_m(t_1) \\ \vec{B}_{EMF}(t_2) - \vec{B}_P(t_2) = \vec{B}_m(t_2) \end{cases} \tag{6}$$

This system can be resolved relatively to the vector of the positioning field $\vec{B}_P$. Subtracting the above equations from each other and assuming equality of currents through the source winding $|I(+)| = |I(-)|$ at the moments t_1 and t_2, one can write the common solution of the system (Eq. 6):

$$B_{meas} = 2\vec{B}_m = \left[\vec{B}_P(t_1) + \vec{B}_P(t_2)\right] + \left[\vec{B}_{EMF}(t_1) - \vec{B}_{EMF}(t_2)\right] \tag{7}$$

The residual value of the EMF, $\Delta B_{EMZ} = [\vec{B}_{EMF}(t_1) - \vec{B}_{EMF}(t_2)]$, from Eq. 7, depends on the angular velocity ω of the object and the time interval $\Delta t_1 = t_2 - t_1$. The value ΔB_{EMZ} is zero for the stationary object, when $\omega = 0$.

Hence, taking into account $\Delta t_2 \geq 2\Delta t_1$, the use of bipolar positioning field allows to double the amplitude of measured induction *versus* DC method and reduce twice the impact of EMF:

$$B_{meas} = 2B_p, \quad \Delta B_{EMFADC} \leq \frac{1}{2}\Delta B_{EMFDC}.$$

Besides, the condition $\Delta t_2 \geq 2\Delta t_1$ means twice-increased update rate *versus* the prototype.

POWER CONSUMPTION FOR ON-BOARD HMCS

The averaged power consumption for DC method (P_1) and ADC one (P_2) is compared as follows:

$$\frac{P_1}{P_2} = \frac{\frac{1}{t_p}\int_0^{t_p=4T} I_x^2 R\,dt}{\frac{1}{t_p}\int_0^{t_p=6T} I_{xpm}^2 R\,dt} \tag{8}$$

where: R is the full resistance of winding; t_p is the full period of the windings switching; I_x and I_{xpm} are the pulse current amplitudes for DC and ADC methods, respectively; T is the duration of one unipolar pulse of positioning field, which is equal for both DC and ADC methods.

It can be concluded from Eq. 8:

- If the amplitudes of pulsed current are equal ($I_x = I_{xp} = I_{xm}$), we have $(P_1/P_2)/(3/4)$ (Fig. 6, mode *A*).
- If the sweep of measured inductions is equal ($B_{DC} = 2B_{ADC}$), we have $(P_1/P_2)/3$ (Fig. 6, mode *B*).
- If the root mean square (RMS) of noise is given (σ_N = constant), then the signal-noise ratios for the prototype ($SNR_1 = B_1/\sigma_N$) and for the bipolar fields method ($SNR_2 = 2\cdot B_2/\sigma_N\cdot\sqrt{2}$) are identical ($SNR_1 = SNR_2$) for twice-reduced power consumption ($P_1/P_2 = 2$), as follows from the calculations (Fig. 6, mode *C*):

$$\frac{B_1}{\sigma_N} = \frac{2\cdot B_2}{\sigma_N\cdot\sqrt{2}}, \quad I_2 = I_1\cdot\frac{\sqrt{2}}{2}, \quad P_2 = P_1\cdot\frac{1}{2}.$$

The comparison of the parameters of DC and ADC methods is presented in Table 2. The module of induction vector at a great distance from the windings with current can be described by the following simplified formula: $B_M \approx \Psi/R^3$, where Ψ is the function that describes the effect of the winding sizes and the flowing current. In accordance with Fig. 6, for the same currents, the modules values used to calculate the coordinates for the prototype (B_1) and for the new method (B_2), which differ by half, will be read as $B_1 = \Psi/R_1^3$ and $B_2 = 2\Psi/R_2^3$. Equating $B_1 = B_2$, we get $(R_2/R_1)^3 = 2$ or $R_2 = 1.25 \times R_1$.

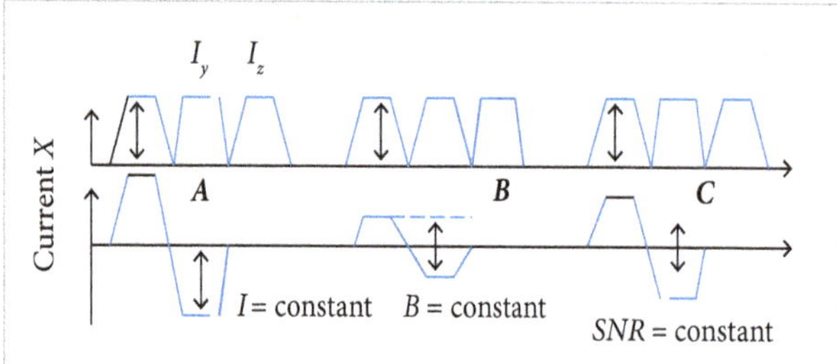

Figure 6. Three modes of ADC method operation.

Table 2 shows that the new proposed ADC method provides increased positioning zone radius on 25%, doubles the update rate and the magnitude of measured induction, at least half the impact of EMF, and, up to factor 3, reduces the power consumption. A slight increase in latency can be off-set by the application of Kalman filter, performing "forward extrapolation".

Table 2. Comparison of parameters of the prototype and the new method.

Parameter	DC	ADC
Update frequency for same pulse duration	F	2F
Measured inductance deviation	ΔB	$2\Delta B$
Coverage zone radius	R	1.25R
Latency	4T	6T
Power consumption	P	1.25P (I = constant) 0.33P (ΔB = constant) 0.5P (SNR = constant)
Earth's field influence	4T	2T

Due to the presented advantages, it can be verified a new way to organize the positioning field, because it better meets the conditions for HMCS as update rate, power consumption, and the impact of EMF.

MATHEMATICAL MODEL

To describe the spatial distribution of the positioning field, an expression for the induction vector projection can be written and is created at observation point with coordinates x_m, y_m, and z_m by one of the positioning field source windings on the normal to the sensor surface in the mobile receiver with orientation angles like α, β, and γ (Fig. 1):

$$B_m(x_m, y_m, z_m, \alpha, \beta, \gamma, t) = \Phi(X, M) = F_1(x_m, y_m, z_m) \quad (9)$$

where: $F_1(x_m, y_m, z_m) = [B_x(x_m, y_m, z_m, A, B, I), B_y(x_m, y_m, z_m, A, B, I), B_z(x_m, y_m, z_m, A, B, I)]^T$ is the function that defines the dependency between the induction vector and the linear coordinates; A and B are the side lengths of the rectangular winding; $F_2(\alpha, \beta, \gamma) = S_{1r} \times A_{xyz}$ is the function of the sensor orientation influence on the measured value, being $A_{xyz} \equiv A_x(\alpha) \times A_y(\beta) \times A_z(\gamma)$ the full rotation matrix on 3 angles in the sequence $\alpha \to \beta \to \gamma$ and $A_x(\alpha)$, $A_y(\beta)$, and $A_z(\gamma)$ 3 matrices with size 3 × 3, reflecting the coordinate transformation for the sequential object rotation around each axis of the stationary 3-D Euclidean space XYZ; $F_3(t)$ is the function that specifies the sequence of the independent fields over time, as illustrated in Fig. 3, respectively, for the winding $X \to Y \to Z$:

$$F_{3X}(t) = \begin{Bmatrix} +1 \\ -1 \\ 0 \end{Bmatrix} for \begin{cases} 0 < t < T \\ T < t < 2T \\ 2T < t < 6T \end{cases},$$

$$F_{3Y}(t) = \begin{Bmatrix} 0 \\ +1 \\ -1 \end{Bmatrix} for \begin{cases} 0 < t < 2T, 4T < t < 6T \\ 2T < t < 3T \\ 3T < t < 4T \end{cases}, \quad (10)$$

$$F_{3Z}(t) = \begin{Bmatrix} 0 \\ +1 \\ -1 \end{Bmatrix} for \begin{cases} 0 < t < 4T \\ 4T < t < 5T \\ 5T < t < 6T \end{cases}$$

where: T is the half period of bipolar pulse $T = \Delta t_1 = t_2 - t_1$ in Fig. 5.

The components of the induction vector included in factor F_1 of Eq. 9 are fully recorded in Zhelamskij (2014a, 2015) for flat rectangular coils, which have a magnetic moment oriented on each axis of the stationary coordinate system XYZ of aircraft from Fig. 1.

A given model (Eq. 9) allows to explore a positioning field in different conditions, as the shape and size of windings, their mutual location and orientation, orientation of mobile receiver, the frequency of the switching and transients, and influence of electrically conductive and ferromagnetic material on the tracking accuracy. The result of the simulation will allow verifying requirements for the elements of the distributed source.

To estimate the accuracy of the mathematical model, in any cases, the amendments $\Delta B = (B_C - B_M) \in [R]$ were investigated within the positioning zone R, where B_C and B_M are, respectively, the calculated and measured values of the induction vector components. At stage of numerical simulation for a coreless source as B_M, the values from Eq. 9 were used, written for single thin frame with current. In this case, the value of B_C described the result of the calculation for ultimate windings cross-section with different shapes. For the source with ferromagnetic core, the calculation was performed also in accordance with Eq. 9, but measured values B_M were taken from the mapping results for the helmet movement zone. As a result, the arrays of the amendments to the calculated values were obtained, which are used to modify Eq. 2:

$$(B_x + \Delta B_x)\cdot\cos\varphi + (B_y + \Delta B_y)\cdot\cos\nu + + (B_z + \Delta B_z)\cdot\cos\upsilon = B_{m1}, \quad (11)$$

where: B_x, B_y, and B_z are the calculated values of axial components of the induction vector from Eq. 9; ΔB_x, ΔB_y, and ΔB_z are the amendments from the mapping results; B_{m1} is the full measured value of induction; $\cos\varphi$, $\cos\nu$, and $\cos\upsilon$ are the guide cosines of the normal to the sensor surface from Eq. 2.

EDDY CURRENTS

Time dependence of the pulsed current $I_0(t)$ through the source windings from Eq. 10 may be written in detail in a mathematical model (Eq. 9) as follows:

$$F_3(t) \approx I_0(t) = -(-1)^N \cdot \begin{cases} I_0 \cdot \frac{t}{\tau_s}, if \ T_1 + \frac{T_n}{2}\cdot(N-1) < t < T_2 + \frac{T_n}{2}\cdot(N-1) \\ I_0, \ if \ T_2 + \frac{T_n}{2}\cdot(N-1) < t < T_4 + \frac{T_n}{2}\cdot(N-1) \\ I_0 \cdot e^{-t/\tau}, if \ T_4 + \frac{T_n}{2}\cdot(N-1) < t < T_5 + \frac{T_n}{2}\cdot(N-1) \end{cases}, \quad (12)$$

where: N is the number of half periods in contiguous sequence of bipolar current pulses in each measuring channel; τ_s and τ are the winding current charge and discharge time constants, respectively; $T_\text{п}$ is the full duration of bipolar current period; $(T_1 - T_2)$ represents the front duration; $(T_2 - T_4)$ is the plateau duration; $(T_4 - T_5)$ is the fall pulse duration.

The source design should provide the duration of front and fall of the pulse magnetic field much less than plateau duration $[(T_1 - T_2) \gg t] << (T_2 - T_4)$.

The mathematical model is created to investigate the influence of eddy currents on the accuracy of the positioning field measurement. Figure 7 shows a single winding W_1 of the positioning field source, oriented along the X axis of the fixed coordinate system XOY; helmet-mounded receiver R at the observation point with coordinates x_m, y_m, z_m; flat conductive element (EL) with radius r; linear coordinates of the center x_d, y_d, z_d; and direction of the normal to the surface of the element as a matrix of guide cosines NM = $[\cos\varphi, \cos\nu, \cos\upsilon]^T = [0, 1, 0]^T$, which reflects the properties of the helicopter cockpit covering ($\Delta = 2$ mm, $\rho = 2.7 \times 10^{-8}$ Ωm).

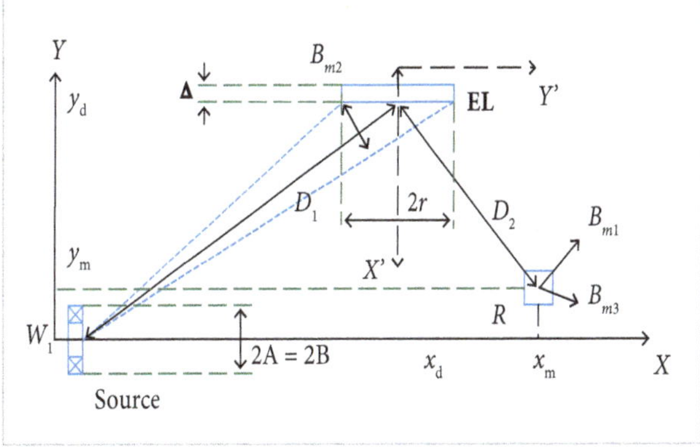

Figure 7. Design scheme of accounting effects of eddy currents on the interior.

Setting the current through the winding I_x and rectangular winding dimensions 2A and 2B, the mathematical models of the positioning field are: $B_{XX} = FXX(A, B, I_x, x, y, z)$, $B_{XY} = FXY(A, B, I_x, x, y, z)$, $B_{XZ} = FXZ(A, B, I_x, x, y, z)$.

The axial components and module of induction vector B_{m1}, generated by the winding at the movable receiver, as well as on the surface of the conductive element, B_{m2}, were calculated from Eq. 9 as shown in Fig. 7. Then, one can estimate the amount of additional field at the receiver B_{m3}, induced by eddy current in the interior EL. The model allows to estimate the ratio B_{m1}/B_{m3} depending on the coordinates of the observation point and the center of the EL disc for different r, Δ, ρ, and NM. The used formula (Khalfin 2004) for additional field on the receiver generated by eddy currents in single element of cockpit interior is:

$$Be \sim K_1 \cdot S/(D_1^3 \cdot D_2^3) \tag{13}$$

where: K_1 is the factor to take into account the conductive properties of the element EL; S is the projection of the element EL square, visible from the winding W_1 (see Fig. 7); D_1 is the distance between EL and the winding; D_2 is distance between EL and the receiver.

MAGNETIC INTERFERENCE

It is important to take into account the magnetic interference from an on-board network of 400 Hz, which can reach 10 – 15% of the amplitude of the positioning field. To implement a differencing method of compensation, it is necessary that the duration of half of the bipolar pulse positioning field ($t_2 - t_1$) be a multiple of the total duration of the periods $n \cdot T_N$ of the interference from the on-board network: $n \cdot T_N = (t_2 - t_1) = (t_3 - t_2) = (t_4 - t_3) = \ldots$, as shown in Fig. 8 for $n = 2$. In this case, $t_2 - t_1 = 5$ ms and $f = 100$ Hz. Correspondingly, for $n = 1$, $t_2 - t_1 = 2.5$ ms and $f = 200$ Hz; for $n = 3$, $t_2 - t_1 = 10$ ms and $f = 50$ Hz. Considering B_{EMF} constant and rewriting Eq. 2 in view of the contribution of interference from on-board network in case $t_2 - t_1 = 2n/2f_{400}$, it is possible to obtain:

$$\begin{cases} \vec{B}_{400}(t_1) + \vec{B}_P(t_1) = \vec{B}_m(t_1) \\ \vec{B}_{400}(t_2) - \vec{B}_P(t_2) = \vec{B}_m(t_2) \end{cases} \tag{14a}$$

After subtracting the equations, we have:

$$\begin{aligned} B_{meas} = (\vec{B}_m(t_1) - (\vec{B}_m(t_2)) = [\vec{B}_P(t_1) + \vec{B}_P(t_2)] + \\ + [\vec{B}_{400}(t_1) - \vec{B}_{400}(t_2)] \rightarrow 2\vec{B}_m \end{aligned} \tag{14b}$$

It follows from Eq. 14b that the difference result does not contain the influence of magnetic interference from the on-board network. The on-board power supply frequency $f = 400$ Hz is not as stable as in terrestrial networks of 50/60 Hz and depends on the aircraft speed engine. Therefore, the duration of the magnetic field pulse floats and strictly corresponds to the selected number of on-board network half period, using the hardware synchronization as shown in the lower graph of Fig. 8.

ACCURACY

New modern magnetometers with Hall sensors appeared recently, with the minimum weight of the units gram and the size of a millimeter level (Asahi Kasei Corp. 2016; Ivensense® 2016; Bosh Sensortec 2016). It is possible to evaluate the possibility of creating the helmet-mounted receiver on such magnetometers that would eliminate the problem of helmet's weight. The immunity of such a device against external interference should be considered as well.

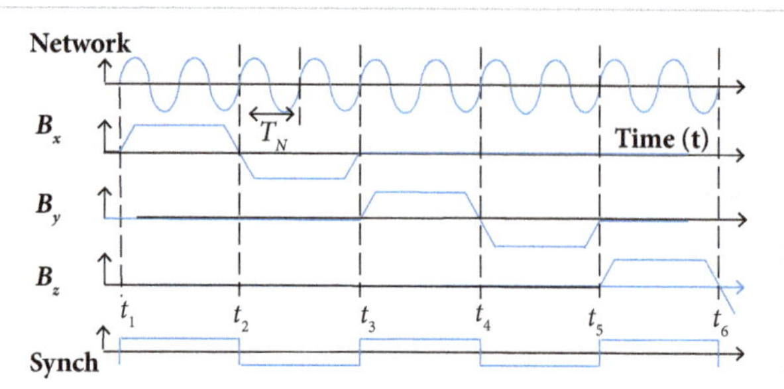

Figure 8. Compensation of magnetic interference influence from the on-board network.

These magnetometers are intended to measure homogeneous EMF only. The mathematical model proposed to apply a magnetometer to measure the inhomogeneous positioning field (Zhelamskij 2014b) is:

$$\begin{aligned} U_{out}(t) = \{K_M \cdot [B_{MT}(x,y,z,\alpha,\beta,\gamma,t) + \Delta B_{EMT}(t) + \\ + \Delta B_F(x,y,z)] \cdot (1 + \sum \varepsilon_{sk}(x,y,z))\} + \\ + U_{shift}(t, T) + U_{RAND}(t) \end{aligned} \tag{15}$$

where: $K_M = G(f) \times \mathrm{MS}(T = \text{constant} \pm \Delta T) \times I_{HG} \times I_S$ is the function which takes into account the sensitivity of the sensor MS, an amplification $G(f)$ of the measuring channel, the excitation currents through the sensors I_{HG} and field source I_S; $B_{MT} = F_1(x_m, y_m, z_m) \times F_2(\alpha, \beta, \gamma) \times F_3(t)$ is the projection of the induction vector of positioning field from Eq. 14b to be measured; $\sum_{k=1}^{8} \varepsilon_{sk}$ (X, Y, Z) is the total systematic measurement error, being ε_{s1} the effect of spatial diversity of the receiver's sensors, ε_{s2} the effect of temperature, ε_{s3} the mutual non-orthogonality of the sensors, ε_{s4} the effect of sensors' excitation current, ε_{s5} the effect of source excitation current, ε_{s6} the sensitivity uncertainty, ε_{s7} the error of ADC, ε_{s8} the influence of sensors' finite size; $\Delta B_{EMF}(t)$ is the residual impact of the EMF; $\Delta B_F(x, y, z)$ is the accuracy of the mathematical description of the positioning field; $U_{shift}(t, T = \text{constant} \pm \Delta T)$ is the off-set of sensor output voltage; $U_{RAND}(t)$ is a random process, modeling its own noise of measurement device.

It can be seen from Eq. 15 that the total induction is $B_S = B_{MT}(x, y, z, \alpha, \beta, \gamma, t) + \Delta B_{EMF}(t) + \Delta B_F(x, y, z)$, applied to the sensor, being converted to the measured voltage through K_M with an accuracy of $\sum_{k=1}^{8} \varepsilon_{sk}$ (X, Y, Z), thus being added the voltages $U_{shift}(t, T)$ and $U_{RAND}(t)$, which do not depend on the measured fields induction. As a result, the mathematical model of the systematic positioning error of the field measurement is proposed:

$$P(D) = \frac{\Delta B_x}{B_x}(D) = \sum_{k=1}^{8} f'_{xk} \cdot \Delta x_k + \varepsilon_{EMF}(D) = P_1(D) + P_2 \quad (16)$$

where: $P_1(D) = \varepsilon_{s1}(D) + \varepsilon_{s7}(D) + \varepsilon_{s8}(D) + \varepsilon_{EMF}(D)$; $P_2 = \varepsilon_{s3} + \varepsilon_{s2} + \varepsilon_{s4} + \varepsilon_{s5} + \varepsilon_{s6} \neq f(D)$; D is the distance between the receiver and the transmitter.

The specificity of the positioning field measurements comes from Eq. 16; the error depends not only on the magnitude, but also on the distance D. This fact allows us to consider Eq. 16 as a function of the distance $P(D)$, which has an extremum. The minimization of $P(D)$ was performed by the dichotomy method under the condition:

$$dP(D)/dD \rightarrow 0 \quad (17)$$

It is shown next that the condition (Eq. 17) runs at different distances D, depending also on the properties of the measuring system. A presented mathematical model of measurements of positioning field allowed to improve further the magnetic tracking theory as well as to explore the impact of the properties of receiver and analog-digital converters, the influence of distance, and the errors of mathematical description of the positioning field. The investigation of this model made it possible to create metrological ensuring of indoor-navigation theory and to formulate the requirements for mobile receiver.

NUMERICAL SIMULATION

Equation 3 can be solved by numerical methods, using the following iterative formula:

$$X_i = X_{i-1} - STEP \cdot grad(CF) \quad (18)$$

where: $\mathbf{X} = (x, y, z, \alpha, \beta, \gamma)$ is the vector of desired coordinates of movable receiver as already mentioned; i is the number of iterations; STEP is the step of the iterative procedure; CF ($\mathbf{X} \in$ R) $= \sum_{n=1}^{N} \sum_{k=1}^{K} ((_{knc}(X) - B_{knm})^2$ is the functionality to minimize, being n the number of windings in the source, k the number of sensors in the receiver, B_{knc} the calculated values of induction for the assumed coordinates, and B_{knm} the measured values of induction.

Different ways to determine the step in Eq. 18 were applied, like a constant step or the steepest descent method. In addition, it was researched the method of separation of variables into linear and angular, and, for each type of coordinates, it was solved the system of equations.

To connect both positioning and measurement errors, Monte Carlo method is applied to investigate the transformation of random measurement errors of positioning field induction (ΔB_N) into computation errors (ΔX) of coordinates. To simulate the noise of ΔB_N, centered signals from a random generator were added to induction measurements in each channel: $dB_x(j) = K_N \times 2 \times (Y(1,1) - 0.5)$, $dB_y(j) = K_N \times 2 \times (Y(1,2) - 0.5)$, and $dB_z(j) = K_N \times 2 \times (Y(1,3) - 0.5)$, where: $K_N = T \times j$ is the growing scale of the random process amplitude, the same for all channels; T is the rate of the amplitude growing; $Y = \text{rand}(1, 3)$ is the matrix of random uncorrelated numbers, with dimension 1×3 in scale 0 – 1.

Adding the same random process at all measuring channels models, the impact of external disturbances is $dB_x(j) = dB_y(j) = dB_z(j) = K_N \times 2 \times (Y(1,1) - 0.5)$. It was investigated the dependence of RMS of computed coordinates (σ_x) *versus* the RMS of input random process (σ_{Bmax}). The transformation of multiplicative measurement errors was explored by the Monte Carlo method for coincidence of the real coordinates and initial approximation. In this case, the measured signals were multiplied by $K_M = 1 + 0.01 \times j \times \varepsilon_S$, being ε_S the limit value of the multiplicative measurement errors, and j the point number at the interval 0 – 100 with step 1: $B_x(j) = B_x(0) \times K_M$; $B_y(j) = B_y(0) \times K_M$; $B_z(j) = B_z(0) \times K_M$.

The dependence was investigated on residual coordinate deviations

$$\delta x = \sqrt{\frac{1}{3} \cdot \left[(x - x_r)^2 + (y - y_r)^2 + (z - z_r)^2 \right]} \quad (19)$$

versus the increasing values of the multiplicative measurement error $\delta \mathbf{B}_m = f_1(j \times \varepsilon_S)$. For the study of transformation of the additive error component, it was suggested adding to the measured signals the following components: $B_x(j) = B_x(0) + 0.01 \times i \times \varepsilon_A \times B_x(0)$, $B_y(j) = B_y(0) + 0.01 \times i \times \varepsilon_A \times B_y(0)$, and $B_z(j) = B_z(0) + 0.01 \times i \times \varepsilon_A \times B_z(0)$, being ε_A the limit value of the additive measurement errors from Eq. 6, i the step number in the range of 0 – 100 with step 1, and $B_x(0)$ … $B_z(0)$ the values of the vector component of the induction at the observation point. Here, the dependence of dx from Eq. 19

was investigated in comparison with the increasing values of the additive measurement error $\delta \mathbf{B}_m = f_1(j \times \varepsilon_S)$.

PHYSICAL MODELS

At the final stage of investigation, in order to verify the created theory, the sample of active magnetic positioning system for targeting at a distance of up to 1 m was developed in the following composition: the movable receiver, consisted of 3 orthogonal Hall sensors (right part of Fig. 9); the source of the orthogonal fields with ferromagnetic core (left part of Fig. 9, Table 3); controller; computer with interface; and software to solve the task of active positioning in real time. The receiver's volume was -27×10^{-6} m^3, and the distance between receiver and computer was 10 m.

The objectives of the physical models study are: metrological research of movable receiver (the impact of temperature and the instability of the supply voltage on the accuracy of the measurement); calibration of receiver's sensitivity and orientation; metrological research of positioning field source (accuracy of the field description, eddy current decay time constant, amendments mapping, and amplitude stability of pulse field over time); trial positioning of movable receiver (stability "at the point", positioning on the specified path, influence of the interior, and operation ranges score).

Figure 9. Mobile receiver (right) and the source of positioning field.

RESULTS

The accuracy of positioning field description was estimated by numerical and physical simulation. The estimates were compared for different models — a single thin frame and winding with ultimate square and different shapes. Table 3 shows the main parameters of the source for modes with and without core. For the trial mapping of source with core, it was selected the area in 1 quadrant, delimited by the following coordinates: $0.13 < x < 0.61$, $-0.51 < y < -0.03$, and $-0.25 < z < -0.13$ m. The number of cubic element is $N = 15 \times 15 \times 3 = 675$, which contains $U = 16 \times 16 \times 4 = 1{,}024$ grid nodes. Each node was identified by discrepancy of 3 axial components of the field induction, created by each of the 3 orthogonal windings of the source. The total number of measurements is $M = 9{,}216$. The mapping of the source with ferromagnetic core showed that the amendments (Eq. 11) can be obtained as the difference between the measured and the calculated values for the corresponding nodes.

Table 3. Basic parameters of the positioning field source.

Parameter	Value
Weight	6.8 kG
Dimensions	~ 100 × 100 × 100 mm
Power consumption per winding	< 40 W
Amplitude of excitation current	8 A
Source-receiver maximum distance	0.75 m
Step of field mapping	30 mm
Induction value at maximum distance	0.1 Gauss
Accuracy of field description without core	≤ 0.02% (D > 0.5 m)
Amendments to analytical model due to core	≤ 25%
Own eddy current decay time constant	~ 70 mks
Magnetic field pulse duration	3 ms
Amplitude instability of pulse	≤ 0.15%
Factor of external field magnification by the core	3.5
Duration of current accelerated rise and discharging	(200 – 1,000) mks (adjustable)

The maximum voltage at the source winding in rise or discharge was < 400 V. Thus, to take into account individual features, inherent to the source with core, it is necessary the custom factory certification procedure by field mapping to get the array of amendments in respect to the model of "clean windings" from Eq. 9. The eddy current decay time constant in the source was experimentally estimated at the level of

55 ± 5 μks, which is much lesser than the discharge time constant τ from Eq. 12.

Numerical simulation of influence of eddy currents in the cockpit interior revealed that near the mobile receiver ($D_2 \to 0$) small items ($D < 20$ mm) of electroconductive materials do not affect positioning accuracy. The measurements show that small electroconductive elements (fastening to M_8), located directly at a movable receiver, affect the results of the coordinates computation at a level of ±0.1° and ±1 mm. Ferromagnetic items in this mode are not affected at the size up to M_3. In accordance with Eq. 13, the conductive plate with thickness of 3 mm and area up to $S \leq 0.1$ m^2 does not affect the moving off ($D \geq 0.5$ m) from the receiver. If the distance D_2 (see Fig. 7) is comparable with the size of the item $D_2 \leq 2r$, the influence of the electrically conductive element on the measured induction is at the level of $\Delta B/B \leq 1\%$. Large ferromagnetic elements at the level of mass $m \approx 1$ kg (like a gun) are affected on the distance $D \leq 0.3$ m, not being on the source-receiver line. Considering that the volume of ferromagnetic materials is much smaller than that of electrically conductive ones in the cockpit of a modern aircraft (and they are not situated between receiver and source), it can be assumed that the mapping of the cockpit may not be necessary. In addition, it is shown that, if not closer to the helmet's covering cabins, eddy currents do not affect the positioning accuracy.

The results of numerical simulation of measurement errors from Eq. 15 are shown in Table 4, where it can be seen that the total error of positioning field measurement has a minimal extremum at a distance $D = 0.5$ m which can match the original status of the helmet in the cockpit.

The numerical investigations of positioning error showed that, to obtain the error of linear coordinates calculation at a level $\Delta x = \pm 1$ mm and at a distance $L = 1$ m, the multiplicative error of magnetic induction measurement should be no worse than $\Delta K/K \approx \pm 0.5\%$. In the middle of the movement zone, where the value of the total additive error is $\Delta B_S + \Delta B_F/B = \pm 0.3\%$, it

Table 4. Total error of measurements at different distances from the source.

Distance/errors (%)	Multiplicative (ε_M)	Additive (ε_A)	Ramdom (ε_r)
D = 0.1 m	± 0.85	± 1.325	$\varepsilon_{r1} \to 0$
D = 0.5 m	± 0.85	± 1.1	$\varepsilon_{r1} \leq \pm 0.1\%$
D = 1 m	±0.85	±5.55	$\varepsilon_{r1} \leq \pm 0.3\%$

$\varepsilon_M = \varepsilon_{S2} + \varepsilon_{S4} + \varepsilon_{S5} + \varepsilon_{S6}$; $\varepsilon_A = \varepsilon_{S1} + \varepsilon_{S3} + \varepsilon_{S7} + \varepsilon_{EMF}$.

can be supposed the accuracy of linear positioning at the level $\Delta x \approx \pm 2$ mm or $\Delta x/x \approx 0.4\%$. The error of the angles calculation is no worse than ±0.3° throughout the full range of operation and is almost independent of the distance. To obtain the RMS of angles $\sigma_\varphi = 0.1°$, it is necessary to have the RMS of magnetic induction measurement at the level $\sigma_B = 10 - 20$ nTesla. In all cases, the maximum scores of errors are specified, so that they can be reduced to an exact task optimization.

According to the experimental research of the sample of mobile receiver, its own noise's RMS provides the signal-noise ratio no worse than 30 dB. To work in a wide range of ambient temperatures (inherent aviation), it is offered a micro-thermostat for the helmet's receiver, which reduces the temperature variation of the Hall sensor off-set in 80 times.

The algorithm is proposed to take into account the instability of the supply voltage, which allowed to reach the error of the mobile receiver coordinates calculation not worse than $\Delta x = \pm 2$ mm and $\Delta\varphi = \pm 0.5°$ for variations of the supply voltage of receiver and source windings within ±5%. The investigation showed that, to obtain a requested precision of positioning, the mutual orientation sensors in mobile receiver should be determined with an accuracy no worse than ±0.5°. The suggested algorithm for accounting of spatial sensors separation allowed to reduce the error of module calculation "near the source" from $\varepsilon_{s2} = (B_C - B_M)/B_C \times 100\% = 6\%$ to a value $< \pm 1.0\%$.

The RMS of calculated coordinates of the stationary receiver was not worse than $\sigma_X = 1$ mm and $\sigma_\varphi = 0.2°$, using the calculation algorithm from Eq. 18 For the modeling of helmet's linear motion within the cockpit, it was studied the movement of the receiver along the straight line $y = kx + b$ ($b = 0$, $k = 1$); z = constant on the surface XOY in the coreless source field. The orientation of the receiver remained unchanged during the movement. The investigation results are presented in Fig. 10, where linear coordinates are given in Fig. 10a and angles of orientation in Fig. 10b *versus* distance $D = \sqrt{x^2 + y^2 + z^2}$ from the source center.

It is seen the linear ramp-up of X and Y coordinates, which corresponds to the selected type of movement. The step between nearby points on the trajectory is not less than 50 mm, which is provided by the convergence of the computing procedure. In the coreless field source, the motion of the receiver onto the circle trajectory on a horizontal plane XOY of aircraft was also studied, which simulates the mode of the target selection by turning the pilot's head. Figure 11 shows the coordinates calculated

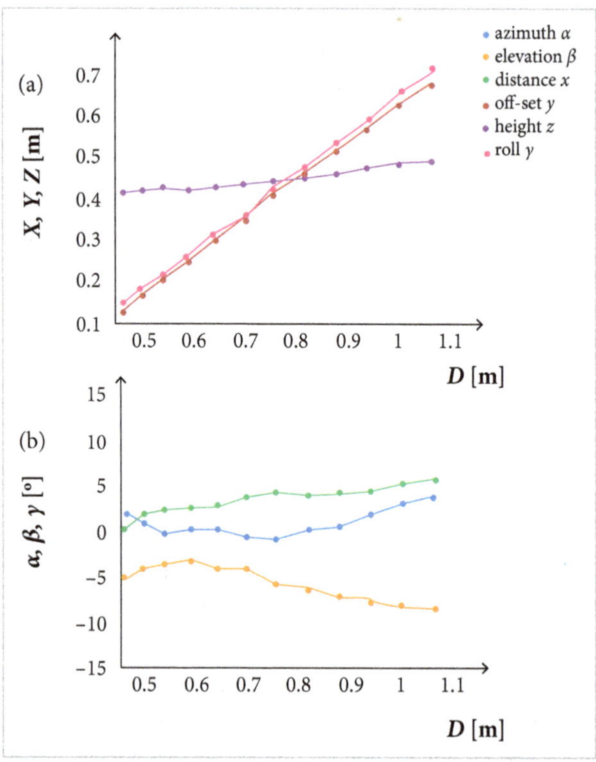

Figure 10. Motion along the straight line in field of coreless source. (a) Linear coordinates; (b) Angles of orientation.

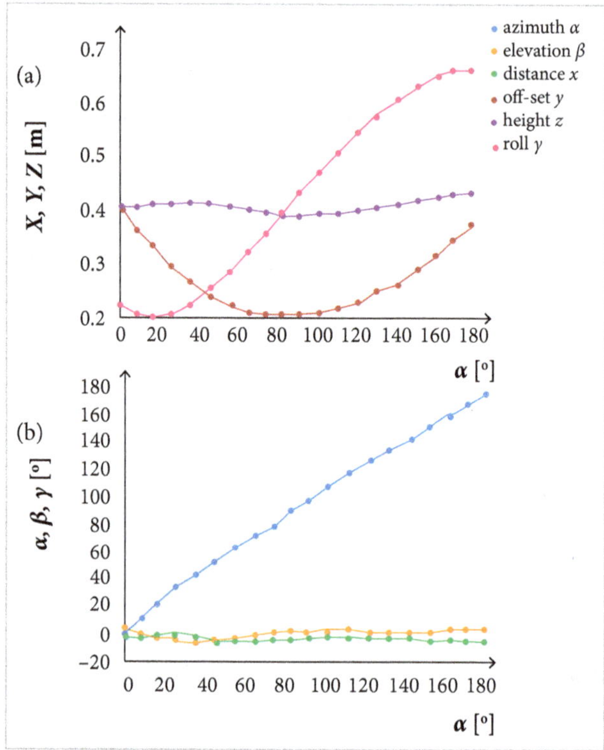

Figure 11. Traffic on the half-circle trajectory. (a) Linear coordinates; (b) Angles of orientation.

from Eq. 18 for the mobile receiver moving around the *Z* axis on the plane *XOY* along the circle trajectory in accordance with equation $(x - x_0)^2 + (y - y_0)^2 = R^2$, where $R = 0.2$ m and (x_0, y_0) are the radius and the central coordinates of rotation, respectively.

In both cases, a 6-DOF positioning is seen. The systematic errors of the calculated coordinates associated with insufficient precision of movable receiver model calibration at the stage of the 1st sample will be eliminated during the transition to the prototype through the R&D phase. To reduce energy consumption of on-board cueing system, it was researched the source of positioning fields with both ferromagnetic core and map of amendments. Figure 12 shows the results of a trial positioning of the mobile receiver during a movement along the full circle in *XOY* plane around the *Z* axis — in this case, we have $R = 0.16$ m. The map of amendments was used, obtained by mapping of the source field in free space, as already described.

The systematic errors in this case may be explained by a lack of accuracy in the amendments maps at the stage of the HMCS's 1st mockup sample. As a result of the

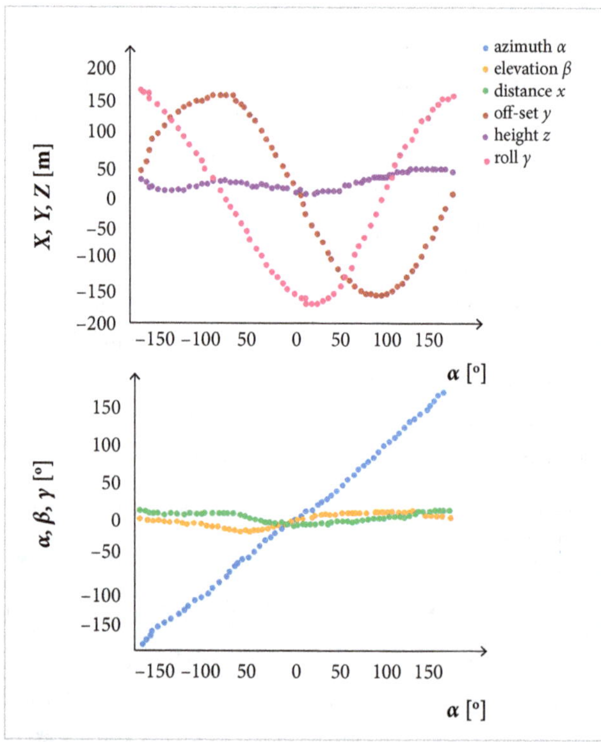

Figure 12. Movement on a full circle in the source field with core. (a) Linear coordinates relative to the center of rotation; (b) Orientation angles.

transition to the R&D phase, the accuracy in the amendments maps calculation will be increased. The investigations show that the created sample operates in full ranges of orientation angles of the receiver — $\alpha = \beta = \gamma = \pm 180°$. The core allowed a 3.5 times increase in the induction of an external field of positioning at the same amplitude of the excitation current — or at the same time reduction current at the same magnitude of the measured induction. Power consumption is reduced by an order of magnitude.

CONCLUSION

The proposed concept is to build a new generation of HMCS on magnetic principle, based on a new way of positioning field organization. The new approach suggests the measurement of the positioning field induction on the basis of modern magnetometers for smartphones, with minimum dimensions, which eliminates the problem of weight and dimensions of the helmet-mounted receiver. As a result, it is shown the possibility to increase the update rate and accuracy of measurements and positioning, in order to reduce power consumption and dimensions of the proposed HMCS in comparison with prototype.

The investigations have shown that mapping the cockpit is not necessary in the application of a new method for positioning field organization, due to negligible effect of the interior. The physical experiments show the functioning of HMCS sample in full ranges of orientation angles. Created mathematical models allow to simulate error of descriptions and measurements of positioning field induction as well as positioning errors to select the optimal configuration of HMCS for a specific aircraft.

Thus, it is possible to figure out the next generation of HMCS with enhanced features: angles range - unlimited; angular accuracy $\leq \pm 0.1°$; linear accuracy $\leq \pm 1$ mm; update rate - 200 Hz; and influence of EMF — twice reduced.

The enterprising researches in this area are being continued in Russia, where more than 30 scientific articles on this subject were already published.

REFERENCES

Analog Devices (2016) Inertial measurement units; [accessed 2016 Aug 19]. http://www.analog.com/en/products/mems/isensor-mems-inertial-measurement-units.html

Asahi Kasei Corp. (2016) AK8975/AK8975 3-axis electronic compass; [accessed 2016 Aug 26]. http://www.asahi-kasei.co.jp/asahi/en/

Ascension Technology Corporation (2016) 3D guidance; [accessed 2016 Aug 19]. http://www.ascension-tech.com/realtime/virtualreality.php

BAE Systems (2016) Striker® II digital helmet-mounted display; [accessed 2016 Aug 19]. http://www.baesystems.com/en-uk/product/striker-ii-digital-helmet-mounted-display

Bosh Sensortec (2016) 9-axis sensors; [accessed 2016 Aug 26]. http://www.bosch-sensortec.com/en/homepage/products_3/9_axis_sensors_5/9-axis_sensors

Blood EB, inventor; Device for quantitatively measuring the relative position and orientation of two bodies in the presence of metals utilizing direct current magnetic fields. 1989 Jul 18. Ascension. United States patent US 4849692 A.

Buganov VI (2016) KP "CDB" Arsenal — the improvement and creation of new opto-electronic aircraft equipment; [accessed 2016 Aug 19]. http://www.tinlib.ru/transport_i_aviacija/aviacija_i_vremja_2003_specvypusk/p12.php

Defencetalk.com (2007) Denel optronics head-tracker system for Eurofighter typhoon; [accessed 2016 Aug 19]. http://www.defencetalk.com/denel-optronics-head-tracker-system-for-eurofighter-typhoon-12206/

Egli WH, Jeffrey MS, Weir EJ, inventors; Helmet metal mass compensation for helmet-mounted sighting system. 1983 Jul 26. Honeywell Inc. United States patent US 4394831 A.

Elbit Systems of America® (2006) Helmet Mounted Display and Sight Systems; [accessed 2016 Aug 19]. http://defense-update.com/directory/elbit-hmd.htm

Elbit Systems of America® (2016) IHADSS; [accessed 2016 Aug 19]. http://www.elbitsystems-us.com/airborne-solutions/products-sub-systems/helmet-mounted-systems/helicopter-helmets/ihadss

Ivensense® (2016) 9 axis motion tracking; [accessed 2016 Aug 22]. http://www.invensense.com/products/motion-tracking/9-axis/

Khalfin I, inventor; Method and apparatus for electromagnetic position and orientation tracking with distortion compensation employing a modulated signal. 2004 Jul 13. Polhemus, Inc. United States patent US6762600 B2.

Kopp C (1982) Heat-seeking missile guidance; [accessed 2016 Aug 19]. http://www.ausairpower.net/TE-IR-Guidance.html

Kopp C (1998) Helmet mounted sights and displays; [accessed 2016 Aug 19]. http://www.ausairpower.net/hmd-technology.html

Kuipers J, inventor; Object tracking and orientation determination means, system and process. 1975 Feb 25. United States patent US 3868565 A.

Lescourret JL, inventor; Method of compensation of electromagnetic perturbations due to moving magnetic and conducting objects. 1997 Dec 2. Sextant Avionics. United States patent US 5694041 A.

Melzer JE, Moffitt KW (1997) Head-mounted displays: designing for the user. New York: McGraw-Hill.

Merryman RFK (1994) Vista Sabre II: integration of helmet-mounted tracker/display and high off-boresight missile seeker into F-15 aircraft. Proc SPIE 2218:1-12. doi: 10.1117/12.177361

Polhemus (2016) Motion tracking overview; [accessed 2016 Aug 19]. http://www.polhemus.com/motion-tracking/overview/

Raab FH, inventor; Remote object position locater. 1977 Oct 18. The Austin Company. United States patent US 4054881 A.

Thales Group (2016) TopOwl® helmet-mounted sight and display for helicopters; [accessed 2016 Aug 19]. https://www.thalesgroup.com/en/worldwide/defence/topowlr-helmet-mounted-sight-and-display-helicopters

Thales Visionix, Inc. (2016) Pilot; [accessed 2016 Aug 19]. http://www.thalesvisionix.com/pilot/

Zhelamskij MV, inventor; Method to define the linear position and orientation of movable object. 2011 Mar 10. Russia patent 2413957 RU.

Zhelamskij MV (2014a) Features of the construction of a positioning field for local navigation in enclosed spaces. Meas Tech 57(7):791-799. doi: 10.1007/s11018-014-0538-5

Zhelamskij MV (2014b) Features of the measurement of a local positioning magnetic field at short distances. Meas Tech 57(9):1032-1040. doi: 10.1007/s11018-014-0577-y

Zhelamskij MV (2015) The positioning of mobile objects in local magnetic field: theory and practice. Saarbrucken: Palmarium Academic Publishing. In Russian.

8

Anomalous Behavior of a Solid Rocket Motor Nozzle Insert During Static Firing Test

Ronald Izidoro Reis[1], Wilson Kiyoshi Shimote[2], Luiz Claudio Pardini[1]

ABSTRACT: This paper presents the study and development of a firing test used to evaluate the behavior of a solid rocket motor. The motivation for the development of a subscale solid rocket motor with end burning propellant grain geometry arose from the need to evaluate the nozzle inserts of graphite for the possible replacement with the carbon fiber-reinforced carbon composite. These subscale solid rocket motors, simulating full scale motor operating time, but with mass flow far below, aim to determine the ablative characteristics of composite materials as a function of operating time. The objective was to correlate the mass flow between subscale solid rocket motors and full scale using insert data materials such as graphite and carbon fiber-reinforced carbon composites, which have ablative characteristics determined in subscale solid rocket motors used at the Instituto de Aeronáutica e Espaço. The critical section to evaluate the test device is rocket nozzle throat region. Analysis of the materials of the subsonic and supersonic nozzle insert parts was performed after the burning tests. It was found the formation of a thin layer of material deposited after the test. The deposited coating layer was analyzed by electron dispersive x-ray analysis and scanning electron microscopy. The results analyzed by these methods showed that there were aluminum and carbon in the coating. Finally, the material was analyzed by x-ray diffraction, and the results showed the presence of aluminum oxide. It was also noticed that, because of the unexpected coating deposition forming material in exit conical and throat of the insert that the effect of ablation was not observed.

KEYWORDS: Propulsion, Subscale motor, Nozzle inserts, Graphite, Firing test.

INTRODUCTION

Subscales of solid rocket motors (SRM) are devices used for propellant development and for propellant ballistic parameters control. These subscale motor provide an estimation of full-scale motor performance at as small as possible motor size. Because of this, a reduced amount of propellant is needed, and reliable results related to the burning test can be achieved. Also, subscale motors are a classroom lesson for understanding the performance of solid rockets, since in general the delivered specific impulse is the main target (Geisler and Beckman 1998). The subscale SRM is also a useful tool to study problems related to the effect of multiple chamber lengths, submergence nozzle, end burners, nozzle inserts, insulation capability and ablation rate (Wermimont 1993; Cortopassi *et al.* 2009; Delaney *et al.* 1964). Thus, subscale rocket motors are those in a lab scale and a benchmark for real-size rocket launchers. Najjar *et al.* (2005) used a similar model of a SRM in their study.

Since the late 1960s, when the Instituto de Aeronáutica e Espaço (IAE) started the development of SRM for the Brazilian Space Program, subscale rocket motors have been used to perform testimony tests aiming the quality control of burning rate parameters and energetic characteristics for solid propellants. The burning rate determination method consists of burning a set of motors under different chamber pressures, which is accomplished basically by using different nozzle throat diameters.

Another important evaluation that can be done, besides burning time and ablation, which are related to the small amount of propellant burning characteristics, is the evaluation of the thermal protection materials, the behavior of the thermal protection materials and nozzle insert material. The thermal

1.Departamento de Ciência e Tecnologia Aeroespacial – Instituto de Aeronáutica e Espaço – Divisão de Materiais – São José dos Campos/SP – Brazil. **2.**Departamento de Ciência e Tecnologia Aeroespacial – Instituto de Aeronáutica e Espaço – Divisão de Propulsão Espacial – São José dos Campos/SP – Brazil.

Author for correspondence: Ronald Izidoro Reis | Departamento de Ciência e Tecnologia Aeroespacial – Instituto de Aeronáutica e Espaço – Divisão de Materiais | Praça Marechal Eduardo Gomes, 50 – Vila das Acácias | CEP: 12.228-015 – São José dos Campos/SP – Brazil | Email: izidororir@iae.cta.br

protections and the nozzle throat are adapted in order to test the material of the nozzle throat insert systems and allow definite burning times, from 10 up to 70 s, using a propellant grain size having appropriate end-burning configuration. This paper presents an investigation of the ablation phenomena in the nozzle throat region during firing tests using graphite inserts. Analysis of the materials from the subsonic and supersonic nozzle insert parts was performed after the end of burning tests. It has been found the formation of a deposited thin layer of burned material after the test. The deposited coating layer was analyzed by electron dispersive x-ray analysis (EDX) and scanning electron microscopy (SEM). The main changed components from the previous design of the SRM were the nozzle insert, steel tip and rear convergent thermal protection, as shown in Fig. 1, as well as the characteristics of propellant grain geometry, from start-burning to end-burning configuration, and the motor case thermal protection.

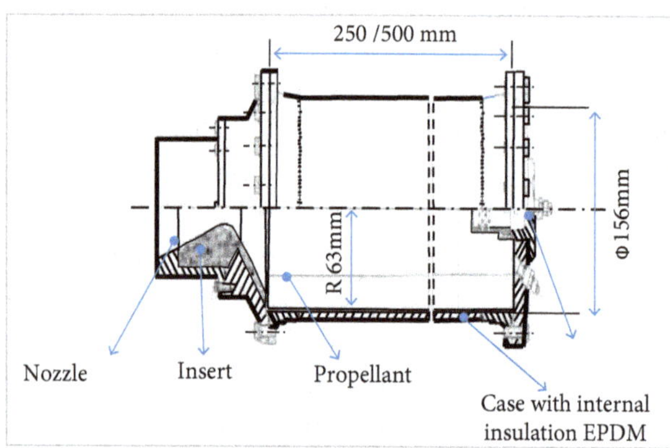

Figure 1. Subscale rocket motor used at IAE.

MATERIALS AND METHODOLOGY

A polycrystalline graphite HLM-85, from Sigri/SGL (Great Lakes Carbon Corporation), was used as a nozzle insert. Graphite and other carbon composites, such as carbon fiber-reinforced carbon composites, are materials that have a significant high emissivity (> 0.8). High emissivity is a performance requirement for rocket nozzle throats since they transfer and sink energy absorbed from the environment. Then, the metallic structure from the steel tip is isolated from high temperatures exerted from the propellant, during the lifetime of the firing test. Figure 2 shows a view of the rocket nozzle insert. The rocket nozzle exit used in the tests has a diameter of 18 mm and a critical diameter (smallest cross section) of 8 mm.

The rocket nozzle is assembled in a steel tip, as shown in Fig. 3, by a bond line between the insert ablative composite material/ graphite and the rear cover of the convergent subscale motor (Fig. 4). An appropriate bond line has been well-designed to avoid damage to the nozzle system during firing.

The rocket nozzle insert and the steel tip were integrated and further assembled in the rear cover convergent, as shown in Fig. 5a. The assembled nozzle is presented in Fig. 5b. The rear cover system has as a main function to provide the gas flow direction and structurally withstands the internal pressure of gases generated by burning of the propellant.

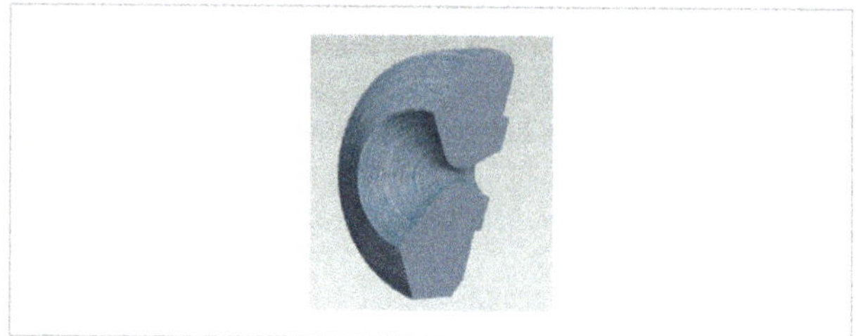

Figure 2. Cross section view and shape of a rocket nozzle insert of graphite.

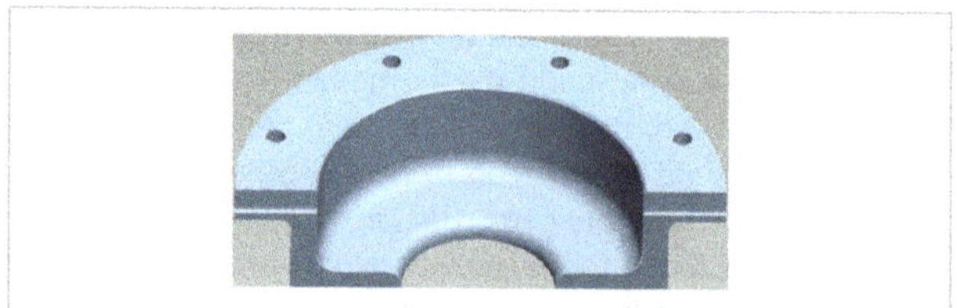

Figure 3. Steel tip from the nozzle system.

Figure 4. Rear cover convergent ablative material from the nozzle system.

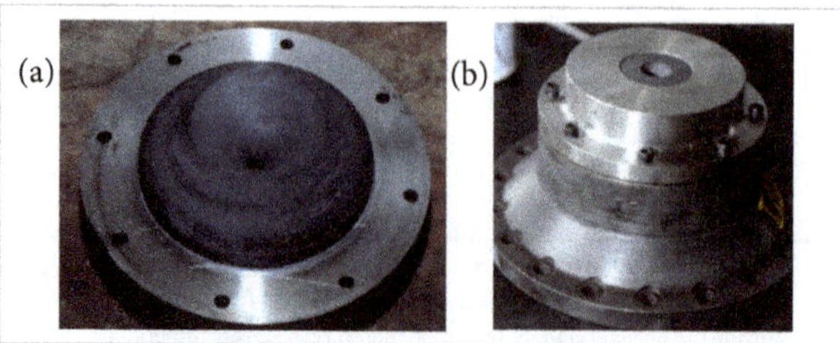

Figure 5. Rocket nozzle assembly. (a) Graphite nozzle insert; (b) Assembled nozzle.

The nozzle/steel tip is assembled in a cylindrical motor case, the rocket envelop, as schematically shown in Fig. 6, by fitting the component parts integrated by the steel tip. Figure 7 shows the subscale solid rocket motors e integrated with the exit nozzle.

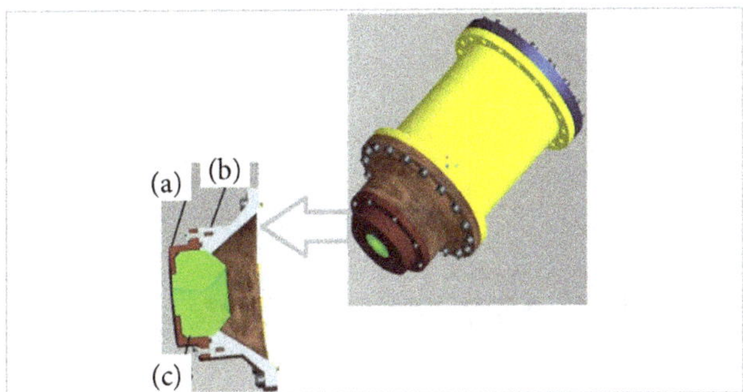

Figure 6. Subscale motor assembly. (a) Steel tip; (b) Rear convergent section; (c) Nozzle insert.

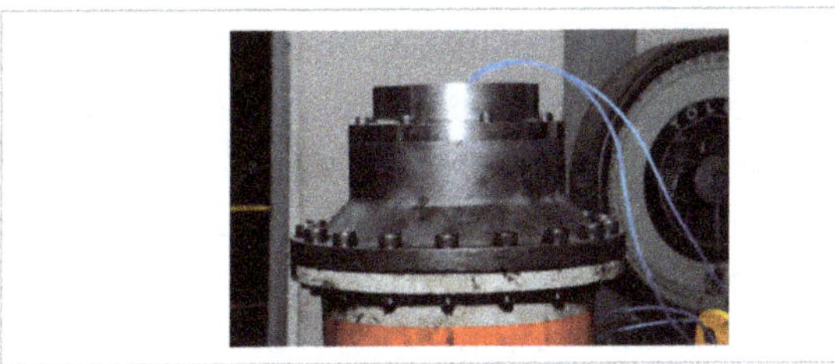

Figure 7. View of the rocket nozzle integrated to motor case.

During the burning test, the internal maximum chamber pressure in the motor case reaches a value of 9 MPa, and the thrust, a value of 250 N. The ablation phenomenon is strongly dependent on the operation time, chamber pressure and propellant combustion gases (Thakre and Yang 2007). The end-burning propellant grain geometry produces constant pressure and thrust curves in the motor case, suitable to study the ablation phenomenon because increased burning times are achieved by using the star-burning configuration. From the experimental thrust curves, the motor operation time is obtained directly, from the burning test, and chamber pressure is calculated by using Eqs. 3 to 7.

The solid composite propellants usually have a concentration of aluminum powder in the range of 1 to 15% by weight (Sciamareli *et al.* 2002). The propellant used in this study, during the static firing tests of the SRM, has 10% by weight of hydroxyl-terminated polybutadiene (HTPB) resin in liquid phase, 68% by weight of ammonium perchlorate (NH_4ClO_4) in solid phase and 16% of high powder aluminum concentration. The increase in the amount of aluminum power added to the propellant formulation increases the gas temperature, the propellant density and the propellant performance (Sutton 1986; Kubota 2007).

The firing tests were carried out at burning times from 9 to 56 s by changing the propellant mass: 3.35; 4.67 and 7.05 kg. The thrust was measured by using a load cell fitting the front lid of the motor case. This arrangement allows the assessment of the behavior of ablation from both graphite nozzle insert and the thermal protections.

Measurements of the surface internal profiles of the nozzle insert and from the convergent section were done before and after the firing test by a coordinate measuring machine Zeis Accura. These measurements allow to estimate the insulation efficiency of the thermal protection of the convergent section and nozzle insert material during the motor static firing test. An anomalous increase in the area of the critical section (nozzle throat) is detrimental to the operation of the motor due to the reduction in the nozzle expansion ratio and the value of rocket motor specific impulse.

According to Wani *et al.* (2012), most composite propellants contain typically ammonium perchlorate at proportions ranging from 65 to 70% by weight, a metallic fuel like aluminum powder at proportions ranging from 15 to 20% by weight and a rubber-like binder, such as HTPB, at proportions ranging from 10 to 15% by weight. These 3 constituents can correspond approximately to the amount of 95% by weight of the propellant weight (Glotov 2006; Susuki and Chiba 1989). A similar propellant formulation has been applied in the rocket motors used in the Brazilian Space Program. Particularly, in this research, a propellant formulation based on a composition of ammonium perchlorate, aluminum powder and HTPB resin was also used according to Sciamareli *et al.* (2002).

After the firing tests, the analysis of the internal surface of nozzle inserts was carried out by EDS through an OXFORD equipment, model 7059, as well as software INCA and SEM, through LEO model 435VPi. X-ray diffraction was done in a PANalytical equipment, model XPert PRO, and the analysis conditions were: CuKα radiation (λ = 1.54056 A); voltage 40 kV and 45 mA of current; the scan ranged from 5 to 90, and the step was $0{,}016^{o}\cdot min^{-1}$.

The subscale motor has end burning grain geometry. In this case, it is expected that the thrust curve remains constant throughout the firing test. The dependency of the critical section area on the rocket nozzle as a function of pressure and thrust values is given by Eqs. 1 to 8. The nozzle thrust coefficient, C_F (Eq. 3), is proportional to the chamber pressure and to the critical section area.

$$I_{sp} = C^{*}.C_F \tag{1}$$

where: I_{sp} is the solid rocket motor specific impulse ($m \cdot s^{-1}$); C^* represents propellant characteristic velocity ($m \cdot s^{-1}$); C_F means nozzle thrust coefficient.

C^* is given by Eq. 2, and C_F, by Eq. 3:

$$C^* = \frac{\sqrt{R.T_0}}{\sqrt{\gamma}.\left(\frac{2}{\gamma+1}\right)^{\frac{(\gamma+1)}{2.(\gamma-1)}}} \quad (2)$$

where: R represents the gas constant (312 $J \cdot kg^{-1} \cdot K^{-1}$); T_0 is the adiabatic flame temperature (K); γ represents specific heat ratio of the burning gas.

$$C_F = \sqrt{\frac{2.\gamma^2}{\gamma-1}.\left(\frac{2}{\gamma+1}\right)^{\frac{\gamma+1}{\gamma-1}}.\left[1-\left(\frac{p_e}{p_c}\right)^{\frac{\gamma-1}{\gamma}}\right]} + \frac{A_e}{A_{cr}}.\left(\frac{p_e}{p_c}-\frac{p_a}{p_c}\right) \quad (3)$$

where: p_e is the nozzle exit pressure (Pa); p_c is the combustion chamber pressure (Pa); p_a is the atmospheric pressure (Pa); A_e is the nozzle exit section area (m^2); A_{cr} is the nozzle critical section area (m^2).

The nozzle thrust coefficient (C_F) is a function of gas specific heat (γ), the nozzle area ratio (A_e/A_{cr}), and the pressure ratio across the nozzle (p_e/p_c). The propellant mass flow rate ($kg \cdot s^{-1}$) is given by:

$$\dot{m} = \frac{p_c.A_{cr}}{C^*} \quad (4)$$

The combustion chamber pressure is given by:

$$p_c = (a.C^*.\rho_p)^{\frac{1}{1-n}}.\left(\frac{S_b}{A_{cr}}\right)^{\frac{1}{1-n}} \quad (5)$$

where: S_b is the propellant burning surface area (m^2); ρ_p is the propellant density ($kg \cdot m^{-3}$); a is an empirical constant influenced by propellant temperature and is known as burning rate coefficient variable, being dimensionless; n is the burning rate pressure exponent (dimensionless).

According to Davenas (1993), the coefficient a and the pressure exponent n are given by the equation of Saint Robert and Vieille:

$$V_b = a.P_c^n \quad (6)$$

where: V_b is the solid propellant burning rate ($mm \cdot s^{-1}$); p_c is the combustion chamber pressure (MPa); a and n are obtained from the burning of subscale test motor at different chamber pressures.

The solid rocket motor thrust (in N) is given by:

$$F = p_c.A_{cr}.C_F \quad (7)$$

or

$$F = \dot{m}.I_{sp} \quad (8)$$

RESULTS AND DISCUSSION

Figure 8 shows a typical theoretical behavior of a solid rocket motor firing test where thrust and pressure are given as a function of time. The curve gives a theoretical idea of the influence of nozzle insert ablation on the pressure and thrust. Results show theoretical calculations from the chamber pressure as a function of the propellant burning rate characteristics.The graph of Fig. 8 was calculated by using Eqs. 5 and 7, defined by the coefficients a = 3.42 $mm \cdot s^{-1}$and n = 0.25 from Eq. 6, the propellant characteristic velocity (C^*= 1,551 m/s) was calculated from Eq.2 where the adiabatic flame temperature (To = 3,200 K) and specific heat ratio (γ = 1.2); the nozzle thrust coefficient (C_F= 1.4) was calculated from Eq.3. In this equation atmospheric pressure (Pa = 98.4 kPa), the nozzle exit pressure (P_e = 9.4 ×104 Pa), nozzle exit section area (A_e = 2.54 × 10^{-4} m^2) and the nozzle critical section area (A_{cr} = 5 × 10^{-5} m^2). Finaly the combustion chamber pressure was calculated from Eq.5 where propellant density (ρ_p = 1,700 $kg \cdot m^{-3}$) and the propellant grain burning surface (S_b = 7.24 × 10^{-3} m^2).

The values of thermodynamic properties, T_0, γ and R, are obtained from CETPC 273 (1994) thermochemical equilibrium calculation software, which takes into consideration the propellant chemical composition, the reactants mass percentage, the reactants heat of formation, the expansion nozzle ratio and the operating stagnation pressure in the SRM combustion chamber (Zucrow and Hoffman 1976).

The coefficient a and pressure exponent n are given by Eq. 6, which shows the dependence of the solid propellant regression rate on the operating pressure, and were obtained by burning subscale SRM at different chamber pressures. With fine ammonium perchlorate (3 μm), the pressure exponent n is around 0.50 – 0.55, while, for somewhat large perchlorate (10 μm), n may drop to as low as 0.45 (Davenas 1993). In this study, the ammonium perchlorate grain size is between 30 and 400 μm, then the pressure exponent is near 0,25.

According to Davenas (1993), the range of composite solid propellant burning rate varies from 5 to 50 mm·s^{-1}. Therefore, by knowing the value of n, it is possible to calculate the value of a through Eq. 6. In the present study, the propellant burning rate value is 5.35 mm·s^{-1} for operating pressure of 6.0 MPa.

For end-burning grain configuration, the burning surface S_b is kept constant during the motor operation time. On the other hand, if the nozzle critical section increases, the chamber pressure and thrust decrease marginally, as shown in Fig. 8.

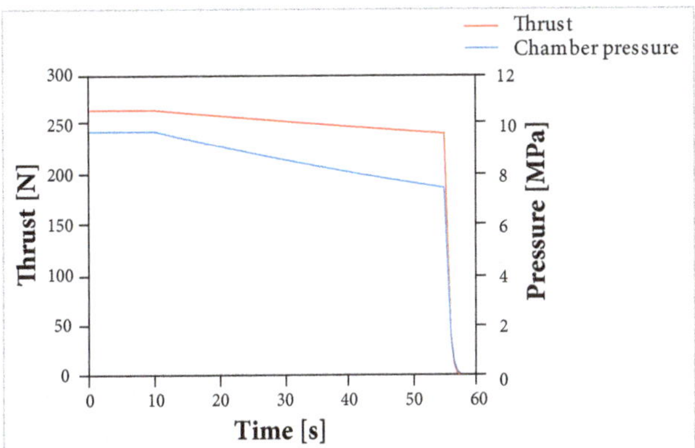

Figure 8. Theoretical behavior of a SRM showing the pressure and thrust as a function of time during a firing test.

The pressure chamber reduces significantly as the propellant is consumed.

In a real firing test, the critical diameter increases and the thrust coefficient reduces. Thus, the pressure in the rocket motor chamber and the propellant burning velocity reduce and the burning time increases. The thrust losses can be calculated in terms of I_{sp}, which is defined in Eq. 8.

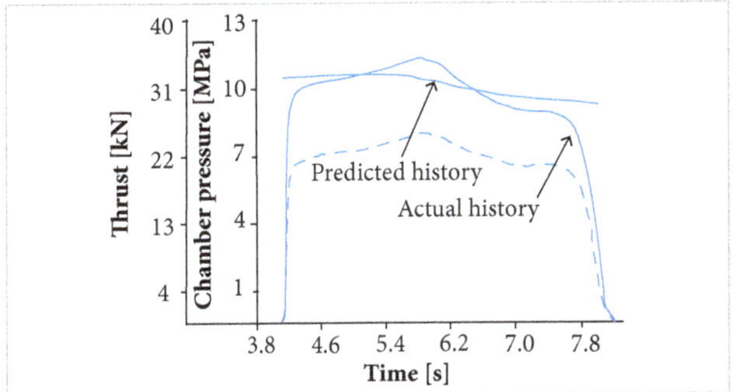

Figure 9. Profiles of pressure (continuous line) and thrust (dotted line) of a 70-pound motor as a function of time, during a firing test. Predicted behavior is also depicted. Source: Geisler and Beckman (1982).

Geisler and Beckman (1982) reported the behavior of a solid motor test firing, as shown in Fig. 9. A similar performance can be found when a comparison is made with the results presented in Fig. 8. Figure 9 shows that pressure and thrust reduce as the firing test goes to the end.

Figure 10 shows the results of thrust as a function of time history during static firing tests. All the tests showed a sharp increase in pressure chamber immediately after the beginning of the firing test. For the 9-s test (Fig. 10a), the thrust remains constant throughout the test, as a function of time at a thrust pressure near 325 N. For the 37-s test (Fig. 10b), the thrust reduces just after reaching the peak at 350 N, stabilizing the thrust at 250 N. This behavior is similar to the one found in the test conducted for 55 s (Fig. 10c), where the thrust also reached a peak at 350 N, stabilizing the thrust after 30 s of firing.

The results of these 3 firing tests were analyzed in this study to evaluate the ablation of EPDM rubber as a thermal protection

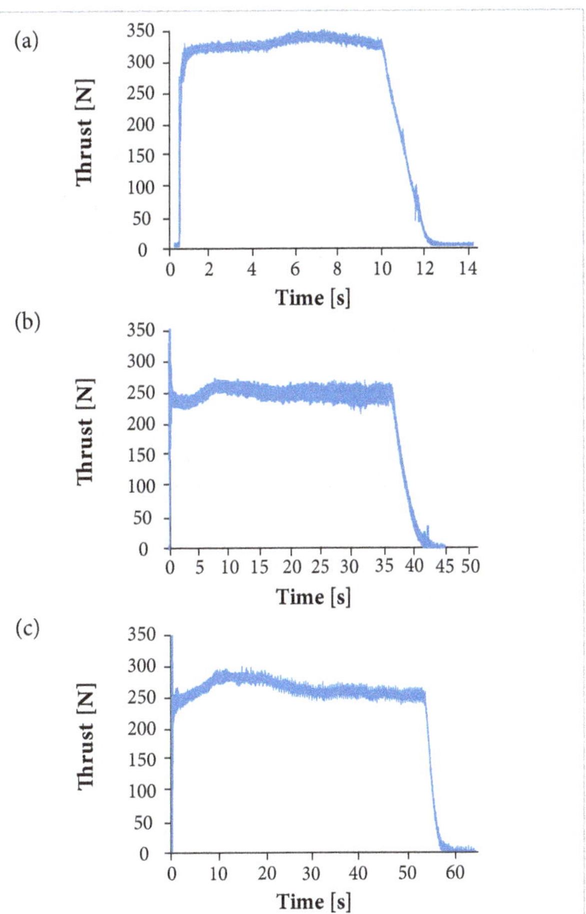

Figure 10. Experimental profile curves of the solid motor test showing the thrust at various throat diameters. (a) 9 s, 6 mm throat; (b) 37 s, 8 mm throat; (c) 55 s, 8 mm throat.

from the convergent exit cone section of the nozzle. The initial tests were performed considering different burning times in order to compare the firing behavior. The operation time of the first firing test was 9 s (Fig. 11a), and the second one, 37 s (Fig. 11b). The last firing test was performed in 56 s (Fig. 11c).

In all firing tests, the formation of a dense coating deposit over the surface of the nozzle throat insert (convergent and divergent region) was observed. The condensed coating material was taken from the surface of the nozzle insert, as shown in Fig. 11, for analysis. Figure 12 shows close-up views of the extracted deposited coating at the nozzle entrance.

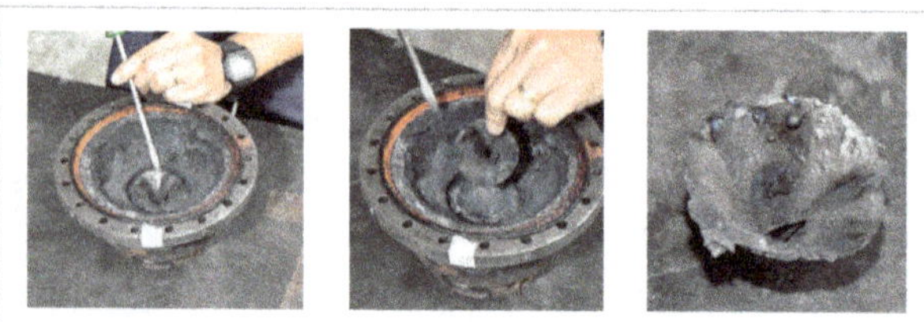

Figure 11. Removal of condensed material coating from the nozzle entrance after the firing test.

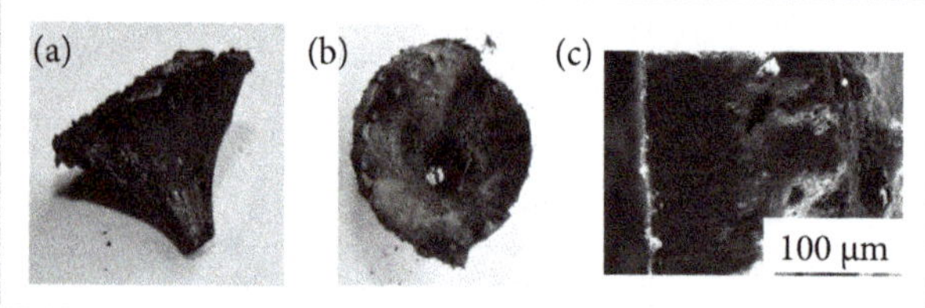

Figure 12. Images of the material deposited over the nozzle throat insert. (a) Side view; (b) Top view; (c) Cross section (38X magnification).

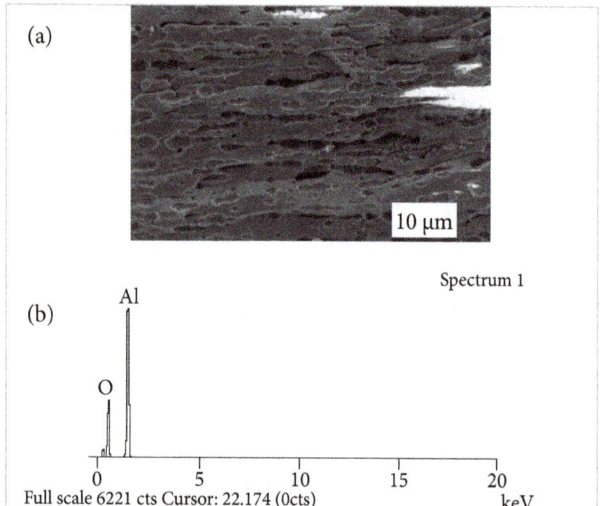

Figure 13. View of the cross section of the material. (a) SEM with magnitude of 1,000X; (b) Spectrum of EDX in this section: 19.23% C, 44.36% O_2 and 36.41% Al.

Figure 13 shows images taken from the SEM performed at the edge of the deposited coating. In such a way, the effect of ablation at the nozzle throat was not observed due to the formation of the condensed material. The condensed material exhibits a laminar and fuzzy deposited pattern, as can be seen in Fig. 13a. The deposit coating was analyzed by EDS, according to Fig. 13b, where the presence of aluminum was mainly detected, which is the main component of the propellant. Carbon was also found due to the residues of the firing test. It was also observed, after the firing test, that the condensed aluminum oxide formed at the graphite nozzle insert decreased the nozzle throat area.

In order to identify the chemical species present in the structure of the coating material deposited over the rocket nozzle surface, XRD analysis was conducted. The sample was ground and analyzed in the form of a powder. The results are shown in Fig. 14. It can be seen from the XRD analysis (Fig. 14) that the coating material removed from the nozzle essentially consists of aluminum oxide.

The coordinate (MMC) measurements were done using a CNC Zeis Accura. The erosion that occurs in the diameter of the insert was evaluated. The regions investigated were the internal diameter of the converging entrance, where the exit of gases occurs during firing, and the inner diameter from the exit cone. The evaluation was done by setting a z axis across the center of the insert, considering a pitch distance of 2 mm. Thirty-two measurements throughout the diameter were performed at the exit cone divergent surface region and another 26 measurements, at the convergent surface. It can be seen from Fig. 15 that the measurements were largely unchanged. Then, no ablation effect was observed in the graphite insert. This fact is attributed to the formation of the coating material over the surface of the rocket nozzle, as can be seen in Fig. 11.

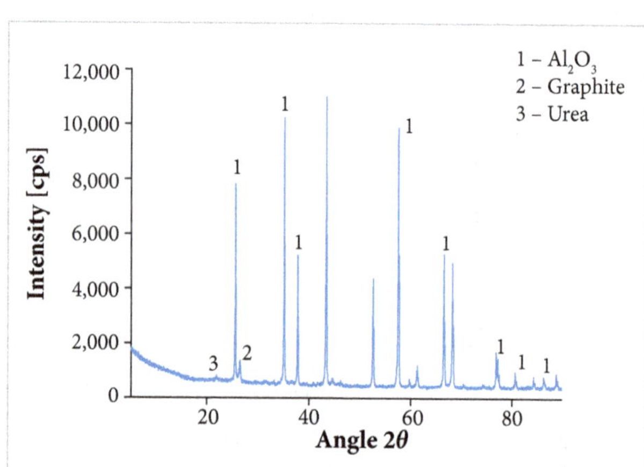

Figure 14. Diffractogram of the coating material removed from the nozzle.

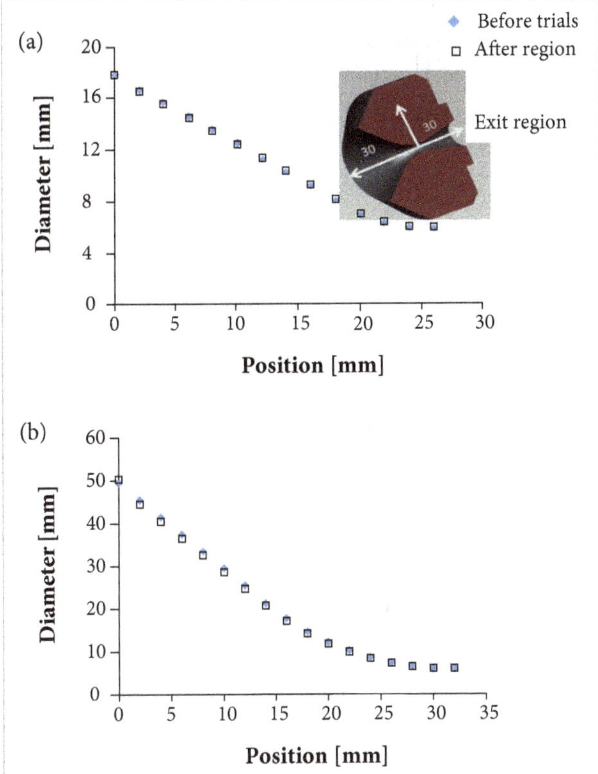

Figure 15. Measurements of the cross section diameter of the convergent and divergent sections from the rocket nozzle. (a) Entrance region; (b) Exit region.

CONCLUSIONS

A new SRM test was designed, based on a similar device described in the literature, to evaluate the ablation, thrust and the thermal protection system. The results of thrust as a function of firing time, which is the main performance parameter of the SRM, are constant and independent of the throat diameter.

The firing test of the SRM showed that the measured thrust was similar to the results calculated theoretically. This suggests that the formation of the coating may happen during the firing test continuously, *i.e.* there was no effect on the erosion of the nozzle. The coating deposited in the nozzle area consists basically on aluminum oxide, which is due to the aluminum from propellant formulation.

The formation of deposit in the throat and conical exit of the insert prevented evaluating the effect ablation of the insert material. Therefore, investigations will be done on the phenomena involved in the formation of the anomalous coating materials to overcome SRM malfunction. These studies involve a detailed analysis of the propellant formulation, conducting further tests to evaluate if the deposit coating is eliminated.

ACKNOWLEDGEMENTS

The authors are grateful to the Space Systems Division (ASE), Chemical Division (AQI), Space Propulsion Division (APE), Mechanical Division (AME), Materials Division (AMR) and Integration and Test Division (AIE) of the Instituto de Aeronáutica e Espaço (IAE) and the CENIC Company for cooperation and technical assistance in the preparation and execution of this research.

AUTHOR'S CONTRIBUTION

Reis RI did the design of the SRM, write the paper, the experiments and discuss results. Shimote WK also designs the SRM testing device and writes the paper and discuss the results. Pardini LC did the experiments, writing also and discusses the results. All authors conceived the idea, wrote the main text and performed the experiments.

REFERENCES

CETPC 273 (1994) Revision 2.0. Thermochemical Equilibrium Calculation Software.

Cortopassi AC, Boyer E, Kenneth KK (2009) Update: a subscale solid rocket motor for characterization of submerged nozzle erosion. Proceedings of the 45th AIAA/ASME/SAE/ASEE Joint Propulsion Conference and Exhibit; Denver, USA.

Davenas A (1993) Solid rocket propulsion technology. Oxford; New York: Pergamon Press.

Delaney LJ, Eagleton LC, Jones WH (1964) A semi-quantitative prediction of the erosion of graphite nozzle inserts. Proceedings of the Solid Propellant Rocket Conference; Palo Alto, USA.

Geisler R, Beckman C (1982) Ballistic anomaly trends in subscale solid rocket motors. Proceedings of the 18th AIAA/ASME/SAE/ASEE Joint Propulsion Conference and Exhibit; Cleveland, USA.

Geisler R, Beckman C (1998) The history of the BATES motors at the Air Force Rocket Propulsion Laboratory. Proceedings of the 34th AIAA/ASME/SAE/ASEE Joint Propulsion Conference and Exhibit; Cleveland, USA.

Glotov OG (2006) Condensed combustion products of aluminized propellants. IV. Effect of the nature of nitramines on aluminum agglomeration and combustion efficiency. Combust Explo Shock+ 42(4):436-449. doi: 10.1007/s10573-006-0073-z

Kubota N (2007) Propellants and explosives: thermochemical aspects of combustion. 2nd edition. Weinheim: Wiley-VCH Verlag.

Najjar FM, Massa L, Fiedler R, Haselbacher A, Wasistho B, Balachandar S (2005) Effects of aluminum propellant loading and size distribution in BATES motors: a multiphysics computational analysis. Proceedings of the 41th AIAA/ASME/SAE/ASEE Joint Propulsion Conference and Exhibit; Tucson, USA.

Sciamareli J, Takahashi MFK, Teixeira JM, Iha K (2002) Propelente sólido compósito polibutadiênico: I- influência do agente de ligação. Quím Nova 25(1):107-110. doi: 10.1590/S0100-40422002000100018

Susuki S, Chiba M (1989) Combustion efficiency of aluminized propellant. Proceedings of the 25th AIAA/ASME/SAE/ASEE Joint Propulsion Conference and Exhibit; Monterey, USA.

Sutton GP (1986) Rocket propulsion elements: an introduction to the engineering of rockets. 5th edition. New York: Wiley.

Thakre P, Yang V (2007) Graphite nozzle material erosion in solid propellant rocket motors. Proceedings of the 45th AIAA Aerospace Sciences Meeting and Exhibit; Reno, USA.

Wani V, Mehihal M, Jain S, Singh PP, Bhattacharya B (2012) Studies on the influence of testing parameters on dynamic and transient properties of composite solid rocket propellants using a dynamic mechanical analyzer. J Aerosp Technol Manag 4(4):443-452. doi: 10.5028/jatm.2012.04044012

Wermimont EJ (1993) 48-inch subscale motor material testing of Space Shuttle advanced solid rocket motor nozzle carbon cloth phenolic ablatives. Proceedings of the 29th AIAA/ASME/SAE/ASEE Joint Propulsion Conference and Exhibit; Monterey, USA.

Zucrow MJ, Hoffman JD (1976) Gas dynamics. Vol. I. New York: Wiley.

the spatial discretization and numerical methods involved. The geometric discretization typically consists of a pre-processing step in which a mesh is generated around the object of interest. Such mesh must be loaded in memory during computation to provide data on model geometry and domain boundaries.

The present study uses a 2-D, finite volume, unstructured CFD solver as a starting point for the development of a high-order method (Breviglieri *et al.* 2010a). The primary objective is to use the SFV numerical method to solve compressible flows at various speed regimes. The SFV method in 2-D allows one to assess the difficulties and advantages of high-order reconstruction in representative test cases, with proper discontinuity and boundary treatment techniques, in a manageable framework. Such techniques are, in principle, extendable to other high-order methods and also for 3-dimensional (3-D) problems.

Two complementary techniques for the high-order method are explored here, namely, the high-order boundary representation and limited reconstruction. In order to keep the high-order method competitive with the lower-order ones, a high-order representation of the geometric boundaries is necessary to reduce the total number of mesh cells. Such high-order boundary representation improves the numerical resolution and convergence aspects of the method, as indicated in Wang and Liu (2006). The use of limiters is necessary when the flow solution contains discontinuities, in order to remove spurious oscillations that may eventually lead to divergence of the numerical solution. Previous study on limiter implementations for high-order methods (Breviglieri *et al.* 2010b, 2008) was based on problem-dependent parameters to find out which cells need limiting, which can limit too many or too few elements of the solution. In the first case, the order of the method is seriously reduced and, in the second one, divergence can occur. To circumvent this drawback, the study here discussed uses a parameter-free generalized moment limiter (Yang and Wang 2009) based reconstruction to deal with discontinuities. The new limiter does not require input constants from the user, rendering the code more robust.

The numerical solver is implemented for the solution of the 2-D Euler equations in a cell-centered finite volume context for triangular meshes. The reported findings and tools are relevant for the long-term goal of numerical analysis over complex 3-D configurations. The study is also motivated by the authors' institute mission, which is to design and build satellite launchers and probes. As a consequence, the research addresses high Mach number flows in the context of high-order schemes.

GOVERNING EQUATIONS

The 2-D Euler equations are solved in their integral form as:

$$\frac{\partial}{\partial t}\int_V Q dV + \int_V (\nabla \cdot \vec{P}) dV = 0 \ , \tag{1}$$

where $\vec{P} = E\hat{\imath} + F\hat{\jmath}$. The application of the divergence theorem to Eq. 1 yields:

$$\frac{\partial}{\partial t}\int_V Q dV + \int_S (\vec{P} \cdot \vec{n}) dS = 0 \ . \tag{2}$$

The vector of conserved variables, Q, and the convective flux vectors, E and F, are given by:

$$Q = \left\{ \begin{array}{c} \rho \\ \rho u \\ \rho v \\ e_t \end{array} \right\}, E = \left\{ \begin{array}{c} \rho u \\ \rho u^2 + p \\ \rho u v \\ (e_t + p)u \end{array} \right\}, F = \left\{ \begin{array}{c} \rho v \\ \rho u v \\ \rho v^2 + p \\ (e_t + p)v \end{array} \right\} . \tag{3}$$

The standard CFD nomenclature is being used here. Hence, ρ is the density, u and v are the Cartesian velocity components in the x and y directions, respectively, p is the pressure, and e_t is the total energy per unit volume. The system is closed by the equation of state for a perfect gas:

$$p = (\gamma - 1)\left[e_t - \frac{1}{2}\rho(u^2 + v^2)\right] , \tag{4}$$

where the ratio of specific heats, γ, is set as 1.4 for all computations in this study. In the finite volume context, for stationary meshes, Eq. 2 can be rewritten for the i-th mesh cell as:

$$\frac{\partial Q_i}{\partial t} = -\frac{1}{V_i}\int_{S_i} (\vec{P} \cdot \vec{n}) dS \ , \tag{5}$$

where Q_i is the cell averaged value of Q at time t; V_i is the volume, or area in 2-D, of the i-th mesh element; S_i is the surface that surrounds the V_i volume.

NUMERICAL FORMULATION

SPATIAL DISCRETIZATION

In order to solve Eq. 5, a k-th-order approximation of the integral is computed. The computational domain, Ω, is discretized into N non-overlapping triangles such that:

$$\Omega = \bigcup_{i=1}^{N} SV_i \ . \tag{6}$$

These triangles are referred to as the spectral volumes (SVs).

9

Further Development and Application of High-Order Spectral Volume Methods for Compressible Flows

Carlos Breviglieri[1], João Luiz F Azevedo[2]

ABSTRACT: The present paper investigates the high-order spectral finite volume method with emphasis on applicability aspects for compressible flows. The intent is to improve the understanding and implementation of numerical techniques related to high-order unstructured grid schemes. In that regard, a hierarchical moment limiter and high-order mesh capability are developed for a 2-dimensional Euler spectral finite volume solver. The limiter formulation and geometry interpreter for high-order mesh generation are new contributions for the spectral finite volume method. Literature test cases are evaluated to assess the interaction of curved mesh, limiter and spatial reconstruction features of the spectral finite volume scheme. An order-of-accuracy study is presented along with steady and unsteady problems with strong shock waves and other discontinuities typical of compressible flows. Moreover, second, third and fourth-order spatial resolution analyses are explored and the spectral finite volume results are compared with those from different numerical methods.

KEYWORDS: Spectral volume method, Compressible flows, High-order reconstruction, Unstructured grids.

INTRODUCTION

High-order numerical schemes represent the natural extension of current Computational Fluid Dynamics (CFD) methods, which were developed over the past 30 years for aerospace simulations. The current generation methods are mostly 2nd-order accurate in space and have achieved a level of maturity and robustness desirable for everyday use in aeronautical engineering scenarios. Likewise, several complementary methods were developed for time integration, convergence acceleration, shock capturing and geometry flexibility. However, there are many problems that cannot be fully simulated using low-order methods, such as vortex dominated flows. This observation has motivated the CFD community to consider high-order methods for unstructured meshes. Application of discretization orders larger than 2nd-order has been an area of ongoing research for the last decades (Abgrall 1994; Barth and Frederickson 1990; Ollivier-Gooch 1997; Wang *et al.* 2013). The present paper is aligned with such effort.

The spectral finite volume (SFV) scheme, as proposed by Wang and co-workers (Liu *et al.* 2006; Sun *et al.* 2006; Wang 2002; Wang and Liu 2002, 2004; Wang *et al.* 2004), shares the functionality of a finite volume solver, copes with unstructured meshes and is an efficient alternative to other classes of 2-dimensional (2-D) high-order methods, for instance, the essentially non-oscillatory (ENO) and weighted ENO (WENO) families of schemes (Wolf and Azevedo 2006, 2007). The CFD solution resolution, or quality, is directly related to

1.Departamento de Ciência e Tecnologia Aeroespacial – Instituto Tecnológico de Aeronáutica – Divisão de Ciências da Computação – São José dos Campos/SP – Brazil. **2.**Departamento de Ciência e Tecnologia Aeroespacial – Instituto de Aeronáutica e Espaço – Divisão de Aerodinâmica – São José dos Campos/SP – Brazil.

Author for correspondence: João Luiz F Azevedo | Departamento de Ciência e Tecnologia Aeroespacial | Instituto de Aeronáutica e Espaço – Divisão de Aerodinâmica | Praça Marechal Eduardo Gomes, 50 – Vila das Acácias | CEP: 12.228-904 – São José dos Campos/SP – Brazil | E-mail: joaoluiz.azevedo@gmail.com

The solution process involves the definition of proper initial and boundary conditions to the computational domain.

For a given order of spatial accuracy, k, using the SFV method, each SV_i cell must be further divided in sub-cells or control volumes (CVs).

$$N_k = \frac{(k+d-1)!}{(k-1)!\,d!} \tag{7}$$

where d is the physical dimension of the problem of interest. If one denotes by $CV_{i,j}$ the jth control volume of SV_i, the cell-averaged conserved variables, q, for $CV_{i,j}$ at time t are computed as:

$$q_{i,j} = \frac{1}{V_{i,j}} \int_{CV_{i,j}} q(x,y)dV, \tag{8}$$

where $V_{i,j}$ is the volume of $CV_{i,j}$.

Once the CV cell-averaged conserved variables are available for all CVs within SV_i, a polynomial $p_i(x, y) \in P^{k-1}$ of degree $k-1$ can be reconstructed to approximate each component of q as:

$$p_i(x,y) = q(x,y) + O(h^{k-1}),\ (x,y) \in SV_i, \tag{9}$$

where h represents the maximum edge length of all CVs within SV_i. The polynomial reconstruction process is discussed in detail in the following section. For now, it is sufficient to say that this high-order reconstruction is used to update the cell-averaged state variables at the next time step for all the CVs within the computational domain. Note that this polynomial approximation is valid within SV_i and the use of numerical fluxes are necessary across SV boundaries.

Integrating Eq. 5 in $CV_{i,j}$, one can obtain the integral form for the CV mean state variable:

$$\frac{dq_{i,j}}{dt} + \frac{1}{V_{i,j}} \sum_{r=1}^{nf} \int_{A_r} (\vec{f}\cdot\vec{n})dS = 0, \tag{10}$$

where $\vec{f} = E\hat{\imath} + F\hat{\jmath}$ at the CV level; A_r represents the length of the r-th edge of $CV_{i,j}$; nf is the number of edges of $CV_{i,j}$. The boundary integral in Eq. (10) can be further discretized into the convective operator form:

$$C(q_{i,j}) \equiv \int_{S_{i,j}} (\vec{f}\cdot\vec{n})dS = \sum_{r=1}^{nf} \int_{A_r} (\vec{f}\cdot\vec{n})dS. \tag{11}$$

The integration on the right side of Eq. 11 can be performed numerically with a k-th order accurate Gaussian quadrature formula as:

$$C(q_{i,j}) = \sum_{r=1}^{nf}\sum_{q=1}^{nq} w_{rq}\vec{f}(q(x_{rq},y_{rq}))\cdot\vec{n}_r A_r + O(A_r h^k)\ , \tag{12}$$

where (x_{rq}, y_{rq}) and w_{rq} are, respectively, the Gaussian quadrature point coordinates and the weights on the r-th edge of $CV_{i,j}$; nq = integer$[(k+1)/2]$ is the number of quadrature points required on the r-th edge for k-th order accuracy.

Due to the discontinuity of the reconstructed values of the conserved variables over SV boundaries, one must use a numerical flux function to approximate the flux values along the spectral volume boundaries. This means that $\vec{f}\,(q(x_{rq}, y_{rq}))$, which appears in Eq. 12, must come from an appropriate numerical flux if the r-th edge of $CV_{i,j}$ is also an external edge of SV_i. Moreover, at any moment during the simulation, one can compute the SV-averaged conserved variable vector, Q_i, for the i-th spectral volume, as:

$$Q_i = \frac{1}{V_i}\sum_{j=1}^{N_k} q_{i,j}V_{i,j}\ . \tag{13}$$

The calculation of the SV-averaged values is important at the end of the computation in order to analyze the high-order numerical solution at the original grid level. The average is also used to recover the conserved variable vectors for the SVs, which are required for the limited reconstruction process as discussed in the section "Limited Reconstruction".

The flux integration across CV boundaries that lie on the SV edges involves 2 discontinuous states, to the left and to the right of the edge. A numerical flux is used to solve for this discontinuous state. Note that the normal flux component needs to be continuous in order to maintain conservation. In the present study, the Roe flux difference splitting method (Roe 1981) with entropy fix is used to compute the numerical flux. As the method is based on one of the forms of a Riemann solver, it introduces the upwind effects and, hence, the artificial dissipation terms into the SFV method.

The semi-discrete SFV scheme for updating the values of conserved properties for the CVs can be written as:

$$\begin{aligned}\frac{dq_{i,j}}{dt} &= -\frac{1}{V_{i,j}}C(q_{i,j}) \\ &= -\frac{1}{V_{i,j}}\sum_{r=1}^{nf}\sum_{q=1}^{nq} w_{rq} f_{Roe}(q_L(x_{rq},y_{rq}), q_R(x_{rq},y_{rq}), \vec{n}_r) A_r\ ,\end{aligned} \tag{14}$$

where the summations in the right hand side of Eq. 14 are

the equivalent convective operator, $C(q_{i,j})$, for the j-th control volume of SV_i. The numerical flux can be expressed as:

$$\vec{f} \cdot \vec{n} = f_{roe}(q_L, q_R, \vec{n}) = \frac{1}{2}\left[\vec{f}(q_L) + \vec{f}(q_R)\right] \cdot \vec{n} - \\ - \frac{1}{2}|B|(q_R - q_L) \ . \tag{15}$$

Here, B is the Roe matrix (Roe 1981) in the edge-normal direction, which has 4 real eigenvalues, namely, $\lambda_1 = v_n - a$, $\lambda_2 = \lambda_3 = v_n$, $\lambda_4 = v_n + a$ where v_n is the velocity component normal to the edge and a is the speed of sound. Let T be the matrix composed of the right eigenvectors of B. Then, this matrix can be diagonalized as:

$$T^{-1}BT = \Lambda \ , \tag{16}$$

where Λ is the diagonal matrix composed of the eigenvalues of B, which can be written as:

$$\Lambda = diag\left[v_n - a, v_n, v_n, v_n + a\right] \ . \tag{17}$$

Hence, the $|B|$ matrix is formed as:

$$|B| = T\,|\Lambda|\,T^{-1}, \tag{18}$$

where Λ and T are calculated as a function of the Roe averaged properties (Roe 1981). Furthermore, $|\Lambda|$ is formed with the magnitude of the eigenvalues.

It is important to emphasize that some edges of the CVs, resulting from the partition of the SVs, lie inside the original SV in the region where the polynomial is continuous. For such edges, there is no need to compute numerical fluxes, as previously described. Instead, one uses analytical formulas for the flux computation and, hence, no artificial dissipation is required for such edges and the flux computation is extremely fast.

TEMPORAL DISCRETIZATION

In order to advance Eq. 14 in time, an explicit, multi-stage, Runge-Kutta time-stepping method is considered, unless otherwise stated. In particular, the 3-stage total variation diminishing (TVD) Runge-Kutta scheme is used, *i.e.*,

$$\begin{aligned} q_{i,j}^{(1)} &= q_{i,j}^{n} - \frac{\Delta t}{V_{i,j}} C(q_{i,j}^{n}) \ , \\ q_{i,j}^{(2)} &= \alpha_1 q_{i,j}^{n} + \alpha_2 \left[q_{i,j}^{(1)} - \frac{\Delta t}{V_{i,j}} C(q_{i,j}^{(1)})\right] \ , \\ q_{i,j}^{n+1} &= \alpha_3 q_{i,j}^{n} + \alpha_4 \left[q_{i,j}^{(2)} - \frac{\Delta t}{V_{i,j}} C(q_{i,j}^{(2)})\right] \ . \end{aligned} \tag{19}$$

as discussed in Wolf and Azevedo (2006), where the n and $n+1$ superscripts denote, respectively, the values of the properties at the beginning and at the end of the n-th time step. The α coefficients are $\alpha_1 = 3/4$, $\alpha_2 = 1/4$, $\alpha_3 = 1/3$, and $\alpha_4 = 2/3$. The C operator represents the discretized convective operator as indicated in Eq. 14.

GENERAL FORMULATION FOR HIGH-ORDER RECONSTRUCTION

The evaluation of the conserved variables at the quadrature points is necessary in order to perform the flux integration over the mesh cell faces or mesh cell edges, in the 2-D case. These evaluations can be achieved by reconstructing conserved variables in terms of some base functions using the degrees-of-freedom (DOFs) within a SV. The present paper has carried out such reconstructions using polynomial base functions. Hence, let P^{k-1} denote the space of $(k-1)$-th degree polynomials in 2 dimensions. Then, the minimum dimension of the approximation space that allows P^{k-1} to be complete is defined by Eq. 7. In order to reconstruct q in P^{k-1}, it is necessary to partition the SV into N_k non-overlapping CVs, such that:

$$SV_i = \bigcup_{j=1}^{N_k} CV_{i,j} \ . \tag{20}$$

The reconstruction problem, for a given continuous function in SV_i and a suitable partition, can be stated as finding $p_{k-1} \in P^{k-1}$ such that:

$$\int_{CV_{i,j}} p_{k-1}(x,y)dS = \int_{CV_{i,j}} q(x,y)dS. \tag{21}$$

With a complete polynomial basis, $e_l(x, y) \in P^{k-1}$, it is possible to satisfy Eq. 21. Hence, p_{k-1} can be expressed as:

$$p_{k-1} = \sum_{\ell=1}^{N_k} b_\ell e_\ell(x,y), \tag{22}$$

where e is the basis function vector, $[e_1, \ldots, e_{Nk}]$; b is the reconstruction coefficient vector, $[b_1, \ldots, b_{Nk}]^T$. It should be further observed that p_{k-1} here denotes the $(k-1)$-th order polynomial for the standard SV cell, *i.e.*, mapped into the standard domain. Hence, the same polynomial can be used for similarly partitioned SVs. The substitution of Eq. 22 into Eq. 21 yields:

$$\frac{1}{V_{i,j}} \sum_{\ell=1}^{N_k} b_\ell \int_{CV_{i,j}} e_\ell(x,y) dS = q_{i,j} \; . \tag{23}$$

If $\bar{q}$ denotes the $[q_{i,1}, \ldots, q_{i,Nk}]^T$ column vector, Eq. 23 can be rewritten in matrix form as:

$$S b = \bar{q} \; , \tag{24}$$

where the S reconstruction matrix is given by

$$S = \begin{bmatrix} \frac{1}{V_{i,1}} \int_{CV_{i,1}} e_1(x,y) dS & \cdots & \cdots & \frac{1}{V_{i,1}} \int_{CV_{i,1}} e_{N_k}(x,y) dS \\ \vdots & \cdots & \cdots & \vdots \\ \frac{1}{V_{i,N_k}} \int_{CV_{i,N_k}} e_1(x,y) dS & \cdots & \cdots & \frac{1}{V_{i,N_k}} \int_{CV_{i,N_k}} e_{N_k}(x,y) dS \end{bmatrix} . \tag{25}$$

Hence, the reconstruction coefficients, b, can be obtained as:

$$b = S^{-1} \bar{q} \; , \tag{26}$$

provided that S is non-singular. Substituting Eq. 26 into Eq. 22, p_{k-1} is, then, expressed in terms of shape functions, $L = [L_1, \ldots, L_{Nk}]$, defined as $L = eS^{-1}$, such that one could write:

$$q(x,y) = p_{k-1}(x,y) = \sum_{j=1}^{N_k} L_j(x,y) q_{i,j} = L\bar{q} \; . \tag{27}$$

Equation 27 gives the value of the conserved state variable, $q(x, y)$, at any point within the SV and its boundaries, including the quadrature points. Note that $\bar{q}$ in the equation takes the form as a column vector, as presented in Eq. 24. The previous equation can be interpreted as an interpolation of a property at a point using a set of cell averaged values and the respective weights, which are set equal to the corresponding cardinal base value evaluated at that point.

Once the polynomial base functions, e_ℓ, are chosen, the L shape functions are uniquely defined by the partition of the spectral volume. The shape and partition of the SV can be arbitrary, as long as the S matrix is non-singular. A detailed discussion of quality and stability analysis of the SFV method partitions can be found in Breviglieri *et al.* (2008). The partitions used in the present paper follow the orientations given in van den Abeele and Lacor (2007) and they are shown in Fig. 1.

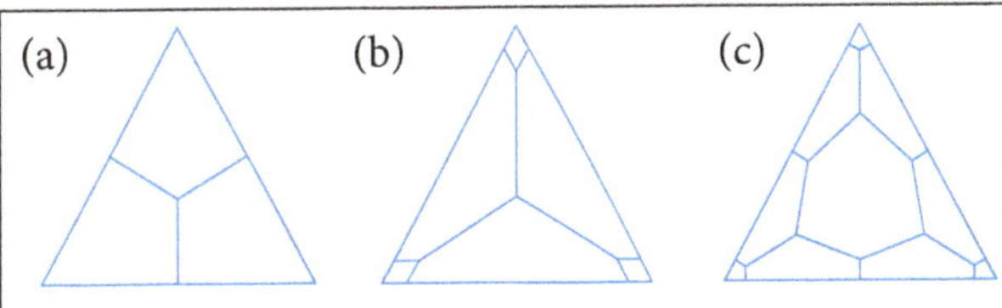

Figure 1. Triangular spectral volume partitions for (a) linear, (b) quadratic and (c) cubic reconstructions.

HIGH-ORDER BOUNDARY REPRESENTATION

From the formulation described so far, it is clear that any input mesh will be divided into a finer mesh and, in principle, render the computation more costly. In the standard 2nd-order finite volume scheme, the mesh boundaries are represented as line segments. This coarse approximation of the geometry results in a cluster of mesh nodes into highly-curved boundaries simply to represent the curved nature of it, for instance, in regions such as the leading edge of an airfoil.

If such approach is carried over to the SFV method, there is no gain in computational performance. The solution is to use high-order geometric elements in the mesh, effectively curving the edges of cells along the domain boundaries. For the present study, quadratic and cubic boundary representations are addressed, respectively, for the 3rd- and 4th-order SFV schemes and only for the elements located at wall boundaries.

In order to perform this representation, one can adopt isoparametric SV cells and map them to the boundary data. However, this particular SV will differ in the partition design from the other SVs. Thus, it will require a dedicated reconstruction and shape function values for property interpolations.

For a quadratic SV, one needs to specify 6 nodes, while, for a cubic SV, 10 nodes, as shown in Fig. 2. In order to perform the transformation, one can use the following relation to map a particular triangle of the mesh to the standard element, partition it there, and map it back to the physical domain to get the new nodes of the CV faces,

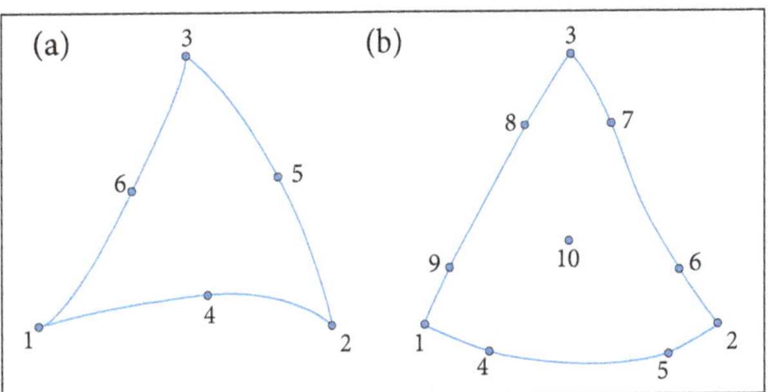

Figure 2. Quadratic (a) and cubic (b) isoparametric SV elements.

$$\vec{r} = \sum_{j=1}^{m} M_j(\xi,\eta)\vec{r}_j \quad (28)$$

where $\vec{r} = (x, y)$ and M_j are the shape functions for the transformation.

Once the "curved" SV is partitioned, the interpolation shape functions and the CV edge normals must be recalculated. Note that, typically, only 1 edge of the SV stands at a boundary, as depicted in Fig. 3.

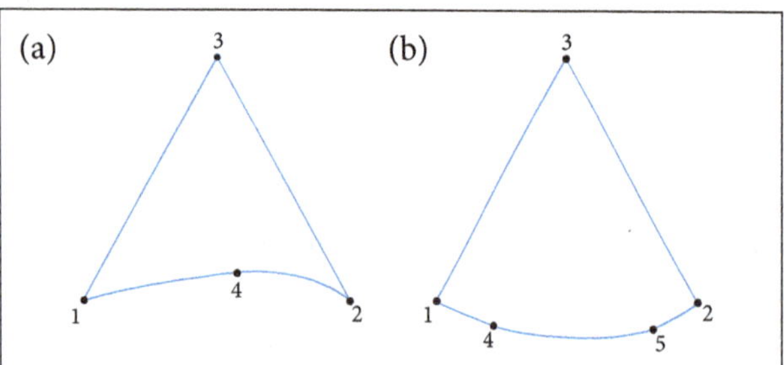

Figure 3. Simplified quadratic (a) and cubic (b) SVs with one curved boundary edge.

Therefore, one could use a simplified formulation for this specific edge. In such case, the mapping becomes

$$\begin{aligned} M_1(\xi,\eta) &= 1 - 3\xi + 2\xi(\xi+\eta) - \eta \,, \\ M_2(\xi,\eta) &= -\xi + 2\xi(\xi+\eta) \,, \\ M_3(\xi,\eta) &= \eta \,, \\ M_4(\xi,\eta) &= 4\xi(1-\xi-\eta) \,, \end{aligned} \quad (29)$$

for the quadratic SV, and

$$\begin{aligned} M_1(\xi,\eta) &= 1 - \frac{9}{2}\xi - \eta + 9\xi(\xi+\eta) - \frac{9}{2}\xi(\xi+\eta)^2 \,, \\ M_2(\xi,\eta) &= \xi\left[1 - \frac{9}{2}(\xi+\eta) + \frac{9}{2}(\xi+\eta)^2\right] , \\ M_3(\xi,\eta) &= \eta \,, \\ M_4(\xi,\eta) &= 9\xi\left[1 - \frac{5}{2}(\xi+\eta) + \frac{3}{2}(\xi+\eta)^2\right] , \\ M_5(\xi,\eta) &= \xi\left[-\frac{9}{2} + 18(\xi+\eta) - \frac{27}{2}(\xi+\eta)^2\right] , \end{aligned} \quad (30)$$

for the cubic SV. The simplified formulation is utilized throughout this study.

One particular challenge is to precisely obtain the mid-face node at the curved edge. The authors previously tried to work with neighborhood interpolation to determine its position, but such approach is not generally applied to "difficult" geometries. The interpolation could render wrong values if one of the neighbors is close to a sharp corner or if mirrored meshes are used, with the same x-coordinates for some mesh nodes.

A solution to this problem was to implement a geometry interpreter inside the solver. Within every simulation, the user provides the geometry prescribed by splines in the IGES format (Gruttke 1995). No matter how complicated the geometric construct is for a particular configuration, the present approach always obtains the correct node correspondence, as illustrated in Fig. 4.

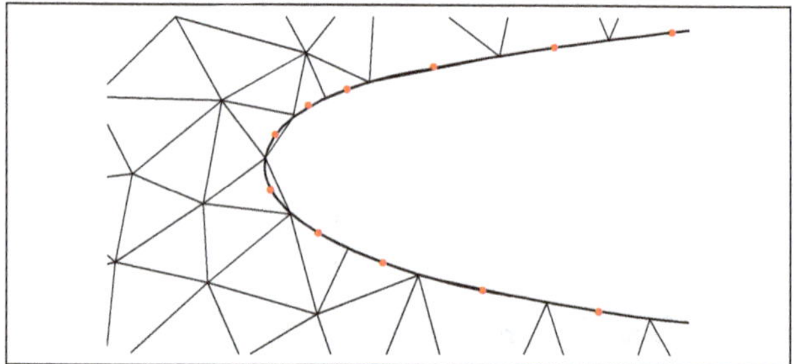

Figure 4. Mid-face nodes, in red, for quadratic boundary reconstruction computed from actual geometry definition.

The surface integral in the physical domain, Eq. 12, is also transformed to a surface integral for the standard element in the computational domain. Let the equation of the r-th face of $C_{i,j}$ in the standard SV be

$$\eta = \eta_r(\xi) \,, \quad \xi_{r1} < \xi < \xi_{r2} \,. \quad (31)$$

Then, along this face,

$$d\eta = \eta_r' d\xi \,. \quad (32)$$

Since $\eta = \eta_r(\xi)$ is a line segment in the standard element, η_r' is a constant, evaluated as $(\eta_2 - \eta_1)/(\xi_2 - \xi_1)$. Furthermore, along such a face,

$$\begin{aligned} dx &= \frac{\partial x}{\partial \xi}d\xi + \frac{\partial x}{\partial \eta}d\eta = \left(\frac{\partial x}{\partial \xi} + \frac{\partial x}{\partial \eta}\eta_r'\right)d\xi \\ dy &= \frac{\partial y}{\partial \xi}d\xi + \frac{\partial y}{\partial \eta}d\eta = \left(\frac{\partial y}{\partial \xi} + \frac{\partial y}{\partial \eta}\eta_r'\right)d\xi \\ dS &= (dx^2 + dy^2)^{\frac{1}{2}} = \left[\left(\frac{dx}{d\xi}\right)^2 + \left(\frac{dy}{d\xi}\right)^2\right]^{\frac{1}{2}} d\xi \,, \end{aligned} \quad (33)$$

where

$$\begin{aligned} \frac{dx}{d\xi} &= \frac{\partial x}{\partial \xi} + \frac{\partial x}{\partial \eta}\eta_r' \\ \frac{dy}{d\xi} &= \frac{\partial y}{\partial \xi} + \frac{\partial y}{\partial \eta}\eta_r' \; . \end{aligned} \tag{34}$$

The face unit normal vector, $\vec{n} = (n_x, n_y)$, is defined as:

$$\begin{aligned} n_x &= \frac{dy}{dS} \\ n_y &= -\frac{dx}{dS} \; . \end{aligned} \tag{35}$$

Hence, it is recomputed for the curved faces. It easily follows from Eq. 28, for instance, that

$$\begin{aligned} \frac{\partial x}{\partial \xi} &= [-3 + 2(\xi + \eta) + 2\xi]x_1 + [-1 + \\ &\quad + 2(\xi + \eta) + 2\xi]x_2 + [4(1 - \xi - \eta) - 4\xi]x_4 \\ \frac{\partial x}{\partial \eta} &= (-1 + 2\xi)x_1 + (2\xi)x_2 + x_3 - (4\xi)x_4 \end{aligned} \tag{36}$$

for the simplified quadratic SV. Analogous formulation is applied for the y derivatives and, also, for the simplified cubic SV. The surface integral, then, becomes

$$\begin{aligned} \int_S (\vec{P} \cdot \vec{n}) dS &= \int_S \vec{P} \cdot \left(\frac{dy}{dS}, -\frac{dx}{dS} \right) dS = \\ &= \int_{\xi_{r1}}^{\xi_{r2}} \vec{P} \cdot \left(\frac{dy}{d\xi}, -\frac{dx}{d\xi} \right) d\xi \; . \end{aligned} \tag{37}$$

This line integral in the standard element can be evaluated using the standard Gauss quadrature formulation:

$$\begin{aligned} &\int_{\xi_{r1}}^{\xi_{r2}} \vec{P} \cdot \left(\frac{dy}{d\xi}, -\frac{dx}{d\xi} \right) d\xi = \\ &= (\xi_{r2} - \xi_{r1}) \sum_{q=1}^{nq} w_q \, f_{Roe}(\xi_{rq}) \; , \end{aligned} \tag{38}$$

where w_q represents the Gauss quadrature weights; f_{Roe} is the numerical flux function.

The numerical integration, then, follows the discussion presented in section "Spatial Discretization".

LIMITED RECONSTRUCTION

For the Euler equations, in the presence of flow discontinuities, it is necessary to limit reconstructed properties at flux integration points to produce a converged simulation. The present limiter technique involves 2 stages. First, the solver must find out and mark "troubled cells" which are, in the 2nd stage, limited.

For the detection and limiting process, the limiter employs a Taylor series expansion for the reconstruction (van Altena 1999) with regard to the cell-averaged derivatives. The troubled cells are, then, limited in a hierarchical manner, *i.e.*, from the highest-order derivative to the lowest-order one. If the highest derivative is not limited, the original polynomial is preserved and so is the order of the method at the element level.

This limiter technique is capable of suppressing spurious oscillations near solution discontinuities without loss of accuracy at local extrema in smooth regions. Originally, this limiter methodology was developed for the spectral difference method in Yang and Wang (2009). In the present study, the formulation is extended for the SFV method. The extension here reported applies some ideas from previous research on ENO and WENO schemes (Wolf and Azevedo 2006) in the calculation of the limited properties. It should be emphasized that there are other approaches which implement a similar hierarchical moment limiter formulation for the SFV scheme (Xu *et al.* 2009), but with slight differences, in comparison to the present approach, on the calculation of the derivatives for the limited reconstruction.

Several markers (or sensors) were developed and employed for unstructured meshes over the past decades. For an in-depth review, the interested reader is referred to Qiu and Shu (2006). The limiter marker used in the present paper is termed Accuracy-Preserving TVD (AP-TVD) marker (Yang and Wang 2009). One important aspect is that the troubled-cell and limited properties are inherent to the SV element, and not to the CVs in which the flux calculations are performed. Once the SV element is confirmed as a troubled cell, its polynomial, based on CV cell-averaged variables, can no longer be used in any flux integration point, because the property function is no longer smooth within such SV. Hence, it is of utmost importance to limit as few SVs as possible.

To that end, the marker is designed to first check for the flux integration points in each SV and mark those that do not satisfy the monotonicity criterion. However, if an extremum is smooth, the first derivative of the solution, at such point, should be locally monotonic. Hence, in a second moment, and using derivative information, as described in the forthcoming discussion, the limiter sensor unmarks those SVs that have smooth local extrema and were unnecessarily marked as troubled cells. Therefore, for instance, for a quadratic reconstruction, the limiting process can be summarized in the following stages:

1. For a given spectral volume, SV_i, compute the minimum and maximum cell averages using a local stencil which includes its immediate face neighbors, *i.e.*,

$$Q_{min,i} = \min\left(Q_i\ ,\ \min_{1\leqslant r\leqslant nf} Q_r\right)$$
$$Q_{max,i} = \max\left(Q_i\ ,\ \max_{1\leqslant r\leqslant nf} Q_r\right) \tag{39}$$

2. The i-th cell is considered as a possible troubled cell if

$$p_i(x_{rq}, y_{rq}) > 1.001\, Q_{max,i} \quad \text{or} \quad p_i(x_{rq}, y_{rq}) < 0.999\, Q_{min,i} \tag{40}$$

where (x_{rq}, y_{rq}) identifies a quadrature point in the outer edges of SV_i. Note that p_i here denotes the reconstruction polynomial of order $k-1$ of the i-th SV cell, in the sense of Eq. 22. The 1.001 and 0.999 constants are not problem-dependent. They are simply used to overcome machine error, when comparing 2 real numbers, and to avoid the trivial case of when the solution is constant in the neighborhood of the spectral volume considered. This step is performed in order to check the monotonicity criterion over the SV cells for which a troubled reconstruction might exist.

3. Since the previous steps may flag more SVs than strictly necessary, the next operations attempt to unmark SVs in smooth regions of the flow. Hence, for a given marked spectral volume, a minmod TVD function is applied to verify whether the cell-averaged 2nd derivative is bounded by an estimate of the 2nd derivative obtained using cell-averaged 1st derivatives of the neighboring spectral volumes. Such test is performed as:
   - If the unit vector in the direction connecting the centroids of the i-th and nb-th cells is denoted $\vec{l} = l_x\hat{i} + l_y\hat{j}$, where nb indicates the face-neighbor of a marked SV_i, the 2nd derivative in such direction is defined as:

$$Q_{ll,i} = Q_{xx,i}l_x^2 + 2Q_{xy,i}l_xl_y + Q_{yy,i}l_y^2 \tag{41}$$

   - In a similar fashion, the first derivatives in the same $\vec{l}$ direction, for both i-th and nb-th cells, can be computed as:

$$Q_{l,i} = Q_{x,i}l_x + Q_{y,i}l_y\ ,$$
$$Q_{l,nb} = Q_{x,nb}l_x + Q_{y,nb}l_y\ ; \tag{42}$$

   - Another estimate of the second derivative, in the $\vec{l}$ direction, can be obtained as:

$$\tilde{Q}_{ll,i} = \frac{Q_{l,nb} - Q_{l,i}}{|\vec{r}_i - \vec{r}_{nb}|}\ , \tag{43}$$

   where $\vec{r}$ is the centroid coordinate vector.
   - A scalar limiter for this face is computed according to

$$\phi_{i,nb}^{(2)} = \text{minmod}\left(1, \frac{\tilde{Q}_{ll,i}}{Q_{ll,i}}\right)\ ; \tag{44}$$

   - The process is repeated for the other faces of SV_i, and the scalar limiter for the SV is the minimum among those computed for the faces, *i.e.*,

$$\phi_i^{(2)} = \min_{nb}\left(\phi_{i,nb}^{(2)}\right)\ ; \tag{45}$$

   - If $\phi_i^{(2)} = 1$, the second derivatives are bounded, as previously defined, and, hence, SV_i is actually in a smooth region of the flow. Therefore, SV_i is unmarked.

4. If the previous test is not satisfied, this means that the particular SV_i spectral volume should indeed be limited. In this case, the limiter for the first derivative reconstruction must also be computed. The calculation procedure follows the same approach as for the second derivatives, and it can be summarized as:
   - An estimate of the first derivative in the $\vec{l}$ direction is calculated as:

$$\tilde{Q}_{l,i} = \frac{Q_{nb} - Q_i}{|\vec{r}_i - \vec{r}_{nb}|}\ ; \tag{46}$$

   - Such estimate is compared to the cell-average first derivative in the $\vec{l}$ direction, computed according to Eq. 42, in order to obtain the scalar limiter for the face as:

$$\phi_{i,nb}^{(1)} = \text{minmod}\left(1, \frac{\tilde{Q}_{l,i}}{Q_{l,i}}\right)\ ; \tag{47}$$

   - As before, the scalar limiter for the cell is the minimum of those limiters computed for the faces, *i.e.*,

$$\phi_i^{(1)} = \min_{nb}\left(\phi_{i,nb}^{(1)}\right)\ . \tag{48}$$

The cell-averaged derivatives for the i-th cell, necessary to perform the previous calculations, are obtained by solving a quadratic least-squares reconstruction problem, for a 3rd-order scheme, or a cubic least-squares reconstruction problem, for a 4th-order scheme. For the quadratic reconstruction presented here, one would impose the mean conservation constraint in the first row of the least-square system and solve the following linear system:

$$\begin{bmatrix} M_{x^1y^0}|_{nb1} & M_{x^0y^1}|_{nb1} & M_{x^2y^0}|_{nb1} & M_{x^1y^1}|_{nb1} & M_{x^0y^2}|_{nb1} \\ M_{x^1y^0}|_{nb2} & M_{x^0y^1}|_{nb2} & M_{x^2y^0}|_{nb2} & M_{x^1y^1}|_{nb2} & M_{x^0y^2}|_{nb2} \\ M_{x^1y^0}|_{nb3} & M_{x^0y^1}|_{nb3} & M_{x^2y^0}|_{nb3} & M_{x^1y^1}|_{nb3} & M_{x^0y^2}|_{nb3} \\ M_{x^1y^0}|_{nb4} & M_{x^0y^1}|_{nb4} & M_{x^2y^0}|_{nb4} & M_{x^1y^1}|_{nb4} & M_{x^0y^2}|_{nb4} \\ M_{x^1y^0}|_{nb5} & M_{x^0y^1}|_{nb5} & M_{x^2y^0}|_{nb5} & M_{x^1y^1}|_{nb5} & M_{x^0y^2}|_{nb5} \end{bmatrix} \begin{Bmatrix} Q_x \\ Q_y \\ Q_{xx} \\ Q_{xy} \\ Q_{yy} \end{Bmatrix} = \begin{Bmatrix} Q_{nb1} - Q_i \\ Q_{nb2} - Q_i \\ Q_{nb3} - Q_i \\ Q_{nb4} - Q_i \\ Q_{nb5} - Q_i \end{Bmatrix} . \quad (49)$$

The matrix terms are the SV area moments and they can be computed, up to the desired order of accuracy, by numerical integration as:

$$M_{x^m y^n}|_i = \int_{SV} (x_{rq} - x_i)^m (y_{rq} - y_i)^n dV \ . \quad (50)$$

The SV area moments are computed during an initial stage of the numerical solver and kept in memory for efficiency. The $nb1$ to $nb5$ subscripts represent the neighbors of SV_i that form the computational stencil to compute the averaged derivatives. This stencil is determined by an ENO-based search, as in Wolf and Azevedo (2006), for the smoothest SV set. Since only a small number of SVs is selected for limited reconstruction, the overhead of this search does not adversely affect the overall performance of the scheme.

It is also important to observe that, considering all the information already available in a SFV method implementation, there are other possible approaches to compute the averaged derivatives. Actually, such approaches can be more computationally efficient than the one here adopted. The interested reader is referred, for instance, to the study presented in Yang and Wang (2009) for further details of one of such alternative approaches.

Finally, the quadratic limited polynomial, which is used in order to obtain property values at the quadrature points for a troubled SV_i spectral volume, is given by:

$$p_i^{\text{limited}}(x_{rq}, y_{rq}) = Q_i + \phi_i^{(1)} \left[\frac{1}{V_i} (Q_x M_x + Q_y M_y)_i \right] + \phi_i^{(2)} \left[\frac{1}{V_i} \left(\frac{1}{2} Q_{xx} M_{x^2} + Q_{xy} M_{xy} + \frac{1}{2} Q_{yy} M_{y^2} \right)_i \right] . \quad (51)$$

The area moments terms, $M\ldots$ in the previous equation, are computed as in Eq. 50 by replacing the m and n exponents with the appropriate order. The limited reconstruction is based on primitive variables $\{\rho, u, v, p\}^T$, instead of conserved variables, as one can readily check for non-physical values as, for instance, negative pressures (Bigarella and Azevedo 2007, 2012). Once these properties are available from the limited reconstruction, the vector of conserved variables is easily obtained to resume the numerical flux integration.

RESULTS

The results presented here attempt to validate the implementation of the data structure, temporal integration, numerical convergence stability and resolution of the SFV method. The overall performance of the method is compared with that of a 2nd-order monotonic upstream-centered scheme for conservation laws (MUSCL) scheme and a 3rd-order WENO scheme implementations. The latter uses an oscillation indicator proposed by Jiang and Shu (1996), with the modification of Friedrich (1998). The Roe numerical flux is also used with both schemes. Moreover, the geometric coefficients for the WENO reconstructions are computed in a pre-processing step and kept in memory during the computation. For more details on the MUSCL and WENO scheme formulations, the interested reader is referred to the study in Barth and Jespersen (1989) as well as Wolf and Azevedo (2006, 2007).

For the results here reported, density is made dimensionless with respect to the freestream condition, and pressure is made dimensionless with respect to the freestream density times the freestream speed of sound squared. All numerical simulations are carried out on a 16-core 3.2 GHz PC Intel64 architecture, with Linux OS. The code is written in Fortran 95 language, and the Intel Fortran Compiler® with optimization flags is used. For the performance comparisons, which are presented in this section, all residuals are normalized by the first iteration residue.

SCALAR CONSERVATION LAWS

The accuracy of the SFV method is tested for the linear scalar advection equation:

$$\frac{\partial u}{\partial t} + \frac{\partial u}{\partial x} = 0 \ , \quad (52)$$

for $-1 \le x \le 1$ and $u(x, 0) = u_0(x)$with periodic boundary condition at the domain extremes. For this setup, the initial condition is $u_0(x) = \sin(\pi x)$. The previously described 3rd-order TVD Runge-Kutta method is employed for time integration with a Δt value of 10^{-4}, in order to make the discretization error time-step independent. Furthermore, the hierarchical limiter is considered in this test to verify if its marker formulation is able to ignore smooth solutions.

Table 1 shows the L_1 and L_∞ error norms produced using the SFV method with equidistant CVs for $t = 1$. One can observe that the 2nd-, 3rd- and 4th-order schemes are capable of achieving the expected order of accuracy even on coarse grids. However, the performance of the 5th-order method is not as good.

This behavior is related to the oscillatory pattern of the polynomial interpolation, due to the equidistant distribution, as previously observed by Wang and Liu (2004). As the grid is refined, the errors actually increase in both norms, which give negative orders of accuracy.

The same problem is simulated with the Gauss-Lobatto point distributions and the results are presented in Table 2. For the 2nd-order scheme, the Gauss-Lobatto point is actually in the middle of the domain, resulting in an equidistant CV partition, and therefore the corresponding results are not shown in Table 2.

One can note that all schemes are, now, able to achieve their expected order of accuracy. A comparison of the data in the tables indicates that the Gauss-Lobatto partitions yield smaller errors in both norms, when compared to the equidistant CV distributions. In the present study, nDOF represents the number of degrees of freedom for the problem, which, in this case, is given by $n\text{DOF} = N_k \times N$, where N_k is given by Eq. 7 and N is the number of SVs. The results shown in Tables 1 and 2 can

Table 1. Accuracy assessment of the 1-D SFV method for the linear scalar advection equation. Equidistant CV partition.

Order	N	$1/n$ DOF	L_∞ error	L_∞ order	L_1 error	L_1 order
2	40	1.25×10^{-2}	5.34×10^{-2}	–	3.29×10^{-2}	–
	80	6.25×10^{-3}	1.41×10^{-2}	1.92	8.72×10^{-3}	1.92
	160	3.13×10^{-3}	3.56×10^{-3}	1.98	2.24×10^{-3}	1.96
	320	1.56×10^{-3}	8.94×10^{-4}	1.99	5.65×10^{-4}	1.99
	640	7.81×10^{-4}	2.24×10^{-4}	2.00	1.42×10^{-4}	1.99
	1,280	3.91×10^{-4}	5.59×10^{-5}	2.00	3.55×10^{-5}	2.00
3	60	5.56×10^{-3}	4.12×10^{-3}	–	2.40×10^{-3}	–
	120	2.78×10^{-3}	5.31×10^{-4}	2.96	3.08×10^{-4}	2.96
	240	1.39×10^{-3}	6.67×10^{-5}	2.99	3.90×10^{-5}	2.98
	480	6.94×10^{-4}	8.34×10^{-6}	3.00	4.90×10^{-6}	2.99
	960	3.47×10^{-4}	1.04×10^{-6}	3.00	6.14×10^{-7}	3.00
	1,920	1.74×10^{-4}	1.30×10^{-7}	3.00	7.69×10^{-8}	3.00
4	40	6.25×10^{-3}	3.24×10^{-3}	–	1.64×10^{-3}	–
	80	3.13×10^{-3}	1.91×10^{-4}	4.08	1.13×10^{-4}	3.86
	160	1.56×10^{-3}	1.47×10^{-5}	3.70	7.04×10^{-6}	4.01
	320	7.81×10^{-4}	8.44×10^{-7}	4.12	4.30×10^{-7}	4.03
	640	3.91×10^{-4}	5.39×10^{-8}	3.97	2.71×10^{-8}	3.99
	1,280	1.95×10^{-4}	3.50×10^{-9}	3.95	1.69×10^{-9}	4.00
5	40	5.00×10^{-3}	1.27×10^{-3}	–	4.42×10^{-4}	–
	80	2.50×10^{-3}	7.55×10^{-5}	4.07	2.19×10^{-5}	4.33
	160	1.25×10^{-3}	8.15×10^{-6}	3.21	3.98×10^{-6}	2.46
	320	6.25×10^{-4}	1.06×10^{-5}	–0.38	4.58×10^{-6}	–0.20
	640	3.13×10^{-4}	4.61×10^{-4}	–5.44	1.80×10^{-4}	–5.29

be visualized in graphical form in Fig. 5. The results shown in this figure refer to error measured in the L_∞ norm.

The next test case addressed considers the linear advection of a Gaussian pulse. Hence, a pulse with a half width equal to 0.05 dimensionless units is used as initial condition:

$$u(x,0) = \exp\left[-\left(\frac{x-0.5}{0.05}\right)^2\right]. \tag{53}$$

The linear advection problem again assumes periodic boundary conditions. The simulation is carried out for various orders of accuracy, $k = 1, 2, 3, 4$ and 6, and up to a final time $t = 2$ dimensionless units. The Gauss-Lobatto point distribution is used for CV partitioning.

Figure 6 shows the analytical and numerical solution profiles for a grid with 100 SVs. The 1st-order simulation smeared the pulse so much that it is hardly recognizable. This is due to the extra amount of dissipation associated with such low-order scheme. The 2nd-order simulation retained the shape of the initial condition, but also smeared the pulse and produced an oscillatory behavior, which, in turn, is associated with the dispersion properties of the method. The 3rd-,

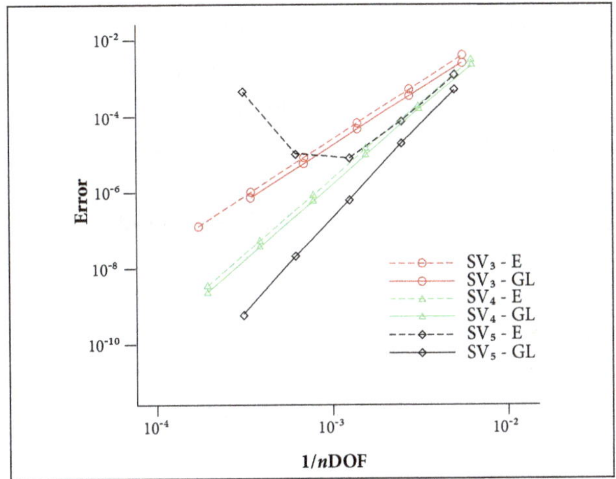

Figure 5. Accuracy assessment of the 1-D SFV method for the linear scalar advection equation. Effect of equidistant (E) and Gauss-Lobatto (GL) CV partition. Error shown in the L_∞ norm.

Table 2. Accuracy assessment of the 1-D SFV method for the linear scalar advection equation. Gauss-Lobatto CV partition.

Order	N	$1/n$ DOF	L_∞ error	L_∞ order	L_1 error	L_1 order
3	60	5.56×10^{-3}	2.67×10^{-3}	–	1.24×10^{-3}	–
	120	2.78×10^{-3}	3.65×10^{-4}	2.87	1.61×10^{-4}	2.95
	240	1.39×10^{-3}	4.67×10^{-5}	2.97	2.05×10^{-5}	2.97
	480	6.94×10^{-4}	5.90×10^{-6}	2.98	2.59×10^{-6}	2.98
	960	3.47×10^{-4}	7.41×10^{-7}	2.99	3.23×10^{-7}	3.00
4	40	6.25×10^{-3}	2.26×10^{-3}	–	7.30×10^{-4}	–
	80	3.13×10^{-3}	1.60×10^{-4}	3.82	5.07×10^{-5}	3.85
	160	1.56×10^{-3}	9.72×10^{-6}	4.04	3.18×10^{-6}	3.99
	320	7.81×10^{-4}	6.15×10^{-7}	3.98	2.00×10^{-7}	3.99
	640	3.91×10^{-4}	3.85×10^{-8}	4.00	1.26×10^{-8}	3.99
	1,280	1.95×10^{-4}	2.41×10^{-9}	4.00	7.87×10^{-10}	4.00
5	40	5.00×10^{-3}	5.30×10^{-4}	–	1.46×10^{-4}	–
	80	2.50×10^{-3}	1.96×10^{-5}	4.76	4.58×10^{-6}	4.99
	160	1.25×10^{-3}	6.50×10^{-7}	4.91	1.49×10^{-7}	4.94
	320	6.25×10^{-4}	2.13×10^{-8}	4.93	4.91×10^{-9}	4.92
	640	3.13×10^{-4}	6.13×10^{-10}	5.12	1.57×10^{-10}	4.97
6	60	2.78×10^{-3}	1.28×10^{-5}	–	2.57×10^{-6}	–
	120	1.39×10^{-3}	1.88×10^{-7}	6.09	4.08×10^{-8}	5.98
	240	6.94×10^{-4}	2.98×10^{-9}	5.98	6.49×10^{-10}	5.97
	480	3.47×10^{-4}	4.51×10^{-11}	6.05	1.04×10^{-11}	5.96

4th- and 6th-order simulations yield good results, which are very similar to the analytical solution shown in the figure.

Table 3 shows the error and order measured from the numerical experiment. Figure 7 shows graphically the table data. One can observe that the design order is obtained considering the Gauss-Lobatto CV partition scheme. As the order of the method increases, the error is consistently reduced to machine zero. It is worth mentioning that the limiter is enabled for these simulations and it can be observed that it is not activated for such smooth solutions.

Another qualitative test case, which addresses a combination of smooth and discontinuous profiles, as in Krivodonova (2007), is also performed to check the SFV method capabilities. The advected signal consists of a smooth Gaussian pulse, a square pulse, a triangle and half an ellipse, which are defined in the domain $-1 \leq x \leq 1$.

Hence, the problem initial condition is defined as:

$$u(x,0) = \begin{cases} [G(x,\beta,z-\delta) + G(x,\beta,z+\delta) \\ \quad +4G(x,\beta,z)]/6, & -0.8 \leq x \leq -0.6\,, \\ 1, & -0.4 \leq x \leq -0.2\,. \\ 1 - |10(x-0.1)|, & 0 \leq x \leq 0.2\,. \\ [F(x,\alpha,a-\delta) + F(x,\alpha,a+\delta) \\ \quad +4F(x,\alpha,a)]/6, & 0.4 \leq x \leq 0.6\,. \\ 0, & \text{otherwise}\,. \end{cases} \quad (54)$$

with

$$G(x,\beta,z) = e^{-\beta(x-z)^2}\,,$$
$$F(x,\alpha,a) = \sqrt{\max(1-\alpha^2(x-a)^2, 0)}\,.$$

where $a = 0.5$, $z = 0.7$, $\delta = 0.005$, $\alpha = 10$ and $\beta = \log 2/36\,\delta^2$.

Once more, the Gauss-Lobatto distribution is used for CV partitioning and, for these tests, there are 100 SVs in the mesh. The same time discretization algorithm and Δt are used as in the previous case. The simulation is carried out with various orders of accuracy, for $k = 2, 3, 4$ and 6, up to a final time $t = 2$ dimensionless time units. The limiter

Table 3. Accuracy assessment of the 1-D SFV method for the Gaussian pulse problem.

Order	N	1/n DOF	L_∞ error	L_∞ order
2	100	5.00×10^{-3}	3.44×10^{-1}	-
	200	2.50×10^{-3}	1.45×10^{-1}	1.25
	400	1.25×10^{-3}	4.42×10^{-2}	1.71
	800	6.25×10^{-4}	9.81×10^{-3}	2.17
3	100	3.33×10^{-3}	1.08×10^{-2}	-
	200	1.67×10^{-3}	9.17×10^{-4}	3.56
	400	8.33×10^{-4}	6.50×10^{-5}	3.82
	800	4.71×10^{-4}	5.31×10^{-6}	3.61
4	100	2.50×10^{-3}	2.48×10^{-4}	-
	200	1.25×10^{-3}	9.24×10^{-6}	4.74
	400	6.25×10^{-4}	5.29×10^{-7}	4.13
	800	3.13×10^{-4}	3.25×10^{-8}	4.02
5	100	2.00×10^{-3}	6.39×10^{-5}	-
	200	1.00×10^{-3}	1.93×10^{-6}	5.05
	400	5.00×10^{-4}	6.23×10^{-8}	4.95
	800	2.50×10^{-4}	1.88×10^{-9}	5.05
6	100	1.43×10^{-3}	3.14×10^{-7}	-
	200	7.14×10^{-4}	2.32×10^{-9}	7.08
	400	3.57×10^{-4}	5.44×10^{-11}	5.41
	800	1.79×10^{-4}	1.99×10^{-11}	1.45

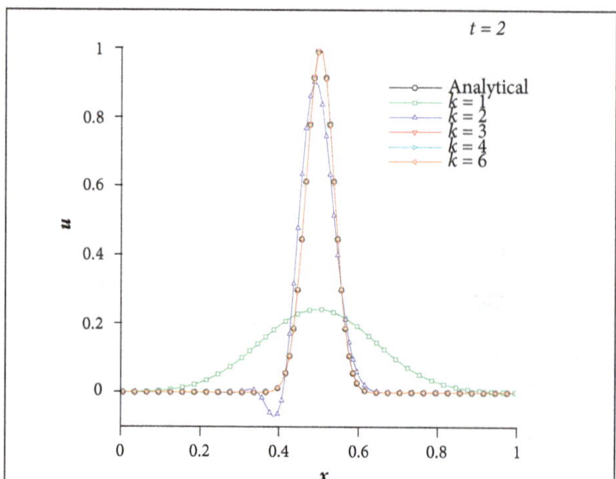

Figure 6. Simulation of a travelling Gaussian pulse with SFV schemes of various orders at $t = 2$ dimensionless units.

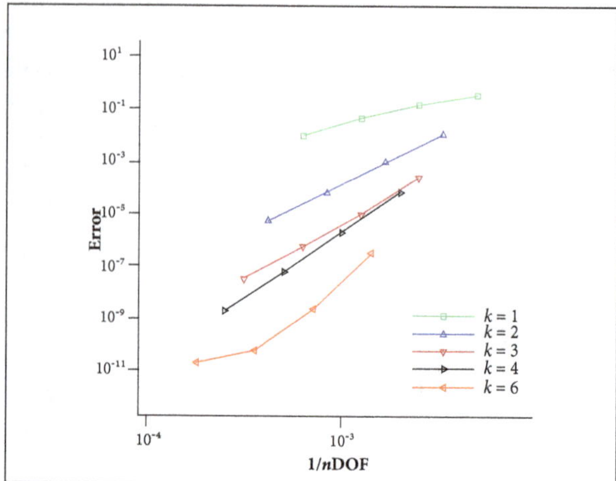

Figure 7. Accuracy assessment of the 1-D SFV Gaussian pulse. Error shown in the L_∞ norm.

is enabled to test its ability to mark non-smooth states of the solution and reduce the oscillations observed in high-order interpolations. The results can be seen in Figs. 8 and 9, plotted for the CV mesh.

The 2nd-order case retains some of the initial features, but it significantly smears the profiles. No oscillations are noticeable due to the use of the limiter formulation. The 3rd-order solution resolves the Gaussian pulse with improved accuracy but there is a noticeable difference in the base of the discontinuous profiles. The same behavior is observed for the 4th-order solution with a tendency to better approach the peak numerical values. However, the result distances itself from the analytical one near steep gradients in the solution. Once again, no oscillations are observed in the numerical solution.

The 6th-order result achieves the better approximation with the analytical data. The smooth profiles are hardly distinguishable from the real solution. The reconstruction is not able to reproduce the triangle and square wave profiles exactly

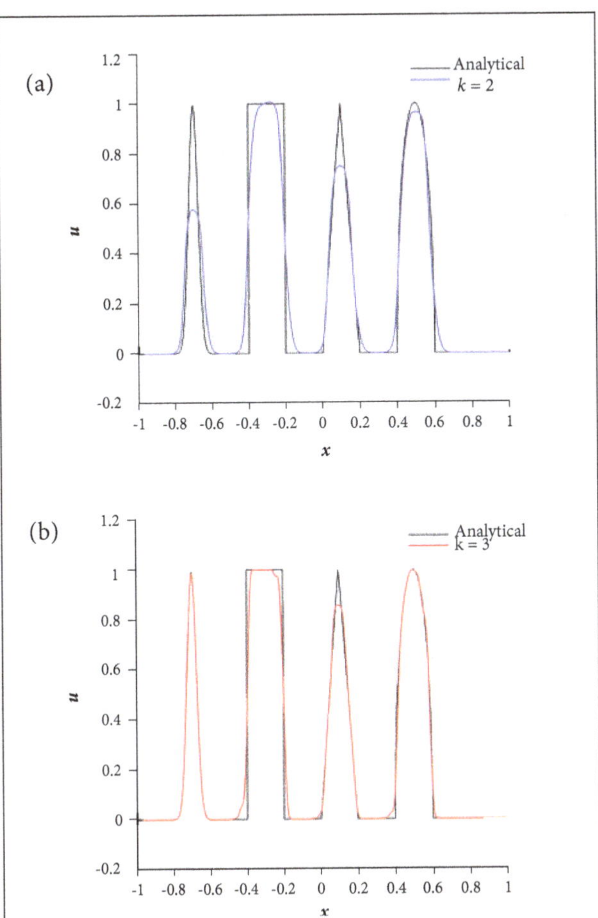

Figure 8. Simulation of travelling discontinuous profiles with (a) 2nd- and (b) 3rd-order SFV schemes at $t = 2$ dimensionless time units.

but yields an approximation with no apparent oscillations and a better resolution before and after such profiles. These results indicate that high-order schemes tend to better resolve the analytical profile without numerical instabilities.

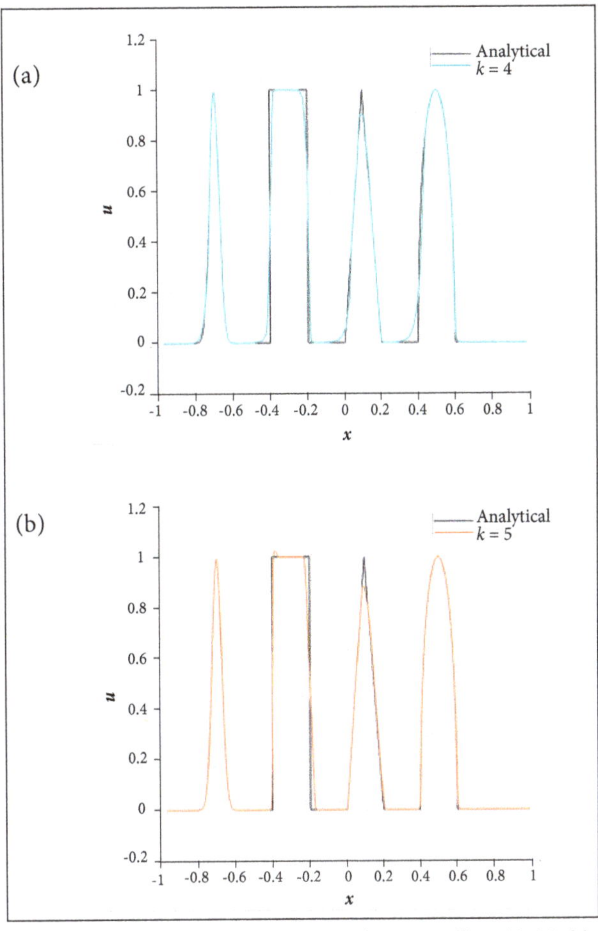

Figure 9. Simulation of travelling discontinuous profiles with (a) 4th- and (b) 6th-order SFV schemes at $t = 2$ dimensionless time units.

RINGLEB FLOW

The Ringleb flow simulation consists of an external flow, which has an analytical solution for the Euler equations derived with the hodograph transformation (Shapiro 1953). The analytical solution is used as initial condition for all simulations here discussed. The flow depends on the inverse of the stream function, κ, and the velocity magnitude, v_t. In the present simulations, these parameters are chosen as $\kappa = 0.4$ and 0.6, in order to define the lateral boundaries of the domain, and $v_t = 0.35$ to define the inlet and outlet boundaries. For such configuration, the test case represents an irrotational and isentropic flow around a symmetric blunt obstacle. An interesting property of the Ringleb test case is that transition of flow regime, from supersonic to subsonic, for example, is shockless (Wang and Liu 2006).

In order to measure the order of the implemented SFV method, 4 meshes are considered for the grid refinement study, corresponding to 128; 512; 2,048 and 8,192 spectral volume cells. The analytical solution is computed for all meshes in order to measure how close the numerical results are to the exact solution. The error with respect to the analytical solution is computed using the L_1 and L_∞ norms of the density. Figure 10 shows the 2,048-cell grid and the Mach number contours computed in this grid with the 4th-order SFV method, using the corresponding high-order boundary representation.

The L_∞ norms of the error for the density values obtained in the converged solutions with the 3rd- and 4th-order SFV method are shown in Fig. 11 for the 4 meshes considered. The figure indicates that the theoretical orders of accuracy are actually recovered by the simulations with the high-order boundary representation. It should be pointed out that the same numerical test case was studied in Breviglieri *et al.* (2008), considering only a linear boundary representation. It was observed in this effort that the low-order boundary treatment causes a shock wave to develop close to the inner boundary, which, then, makes the limiter active. Eventually, the shock wave propagates and it causes the simulation to diverge.

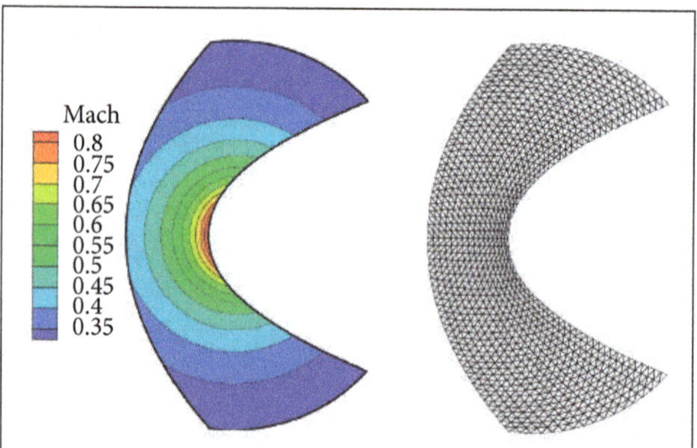

Figure 10. Computational mesh and Mach number contours calculated with the 4th-order SFV method for the 2,048-cell grid.

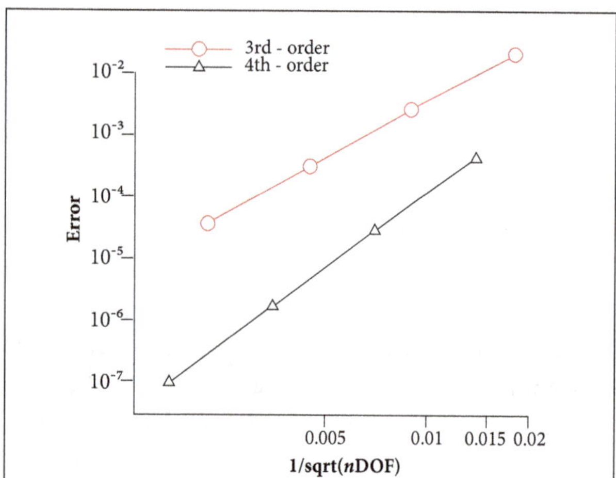

Figure 11. Ringleb flow error measurement with the 3rd- and 4th-order SFV method. Density error shown in the L_∞ norm.

In the present paper, however, which considers the higher-order boundary representation, reasonable results are always obtained for this test case, including the simulations with the 4th-order SFV method. As previously discussed, for the 3rd-order scheme, a quadratic polynomial is used to represent the SV edges which lie along the domain boundaries. In a similar fashion, for the 4th-order scheme, a cubic polynomial is employed instead, which is compatible with the internal polynomial order of each SV. Table 4 presents the L_1 and L_∞ error norms of the density for the present calculations with the high-order boundary representation. The table also shows the actual measured order of accuracy for the 3rd- and 4th-order SFV methods. The orders of accuracy in the results shown in Table 4 are calculated as indicated in Wang and Liu (2006). The actual orders of accuracy here obtained are in good agreement with those shown in the cited reference.

Table 4. Accuracy assessment of SFV method for the Ringleb flow test case.

Order	SV cells	L_1 error	L_1 order	L_∞ error	L_∞ order
3	128	2.41×10^{-2}	-	2.11×10^{-2}	-
	512	4.14×10^{-3}	2.54	2.65×10^{-3}	2.99
	2,048	6.27×10^{-4}	2.72	3.13×10^{-4}	3.08
	8,192	8.67×10^{-5}	2.85	3.60×10^{-5}	3.12
4	128	5.77×10^{-4}	-	4.37×10^{-4}	-
	512	6.48×10^{-5}	3.16	2.82×10^{-5}	3.95
	2,048	6.15×10^{-6}	3.39	1.68×10^{-6}	4.07
	8,192	6.87×10^{-7}	3.16	9.75×10^{-8}	4.11

NACA 0012 AIRFOIL

For the NACA 0012 airfoil simulations, 2 meshes are considered. The 1st simulations are performed on a mesh with 716 cells and 358 nodes, of which 40 nodes define the airfoil wall. This mesh is denoted as the coarse grid. On the other end, there is a fine mesh with 7,117 cells and 3,555 nodes, of which 116 nodes represent the airfoil surface. Both of these meshes are *O*-type grids and they are presented in Fig. 12. The airfoil geometry is collapsed at the trailing edge. The far field boundary radius is 50 chord units. Differently from all other simulations in the present paper, the LU-SGS implicit time marching scheme, as discussed in Breviglieri *et al.* (2010b) and Parsani *et al.* (2010), is used for this test case. A CFL value of 1.0×10^6 is considered. The main objectives of the test case are to assess the SFV method accuracy and convergence for a transonic steady-state flow regime, as well as to provide further insight into the effects of a high-order boundary treatment.

The freestream flow replicates the conditions of the experimental data (McDevitt and Okuno 1985), that is, freestream Mach number of $M_\infty = 0.8$ and 0 deg. angle-of-attack. Simulations with the 2nd-, 3rd- and 4th-order SFV schemes are performed, along with the 1st-order Roe scheme. Figure 13 presents density contours for the coarse mesh, for the 3rd-order SFV method solution and its comparison to the 1st-order Roe scheme. The 3rd-order flow solution considers quadratic curved boundary reconstruction. The figure shows 30 evenly-spaced density contours, with values ranging from 0.8 to 1.6 in dimensionless density. A high-order post-processor tool would be necessary in order to accurately see and analyze these results. Nevertheless, one can already see that the 1st-order solution has clearly smeared out the shock wave that occurs in this flow.

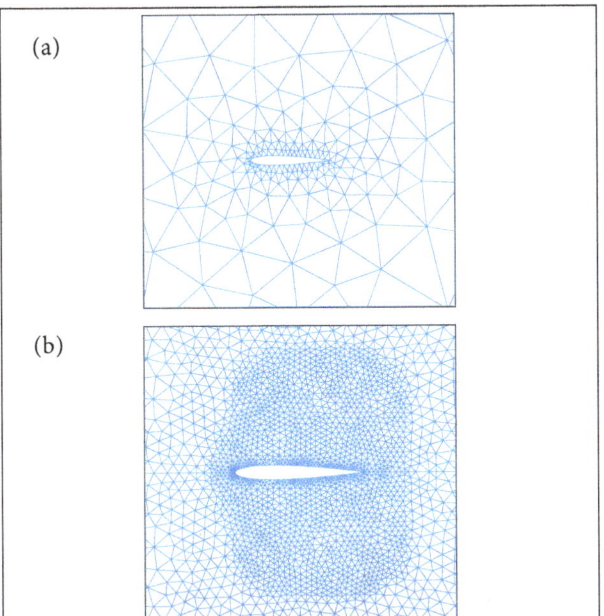

Figure 12. Computational meshes used in the NACA 0012 simulations (a) Coarse mesh and (b) Fine mesh.

Figure 14 shows the corresponding pressure coefficient (*Cp*) distributions for the same mesh and for the same methods, as compared to the experimental data for this flight condition. It is clear from the *Cp* distributions that the 1st-order scheme introduces too much dissipation and, essentially, smears out the shock wave, as already pointed out. The high-order *Cp* distribution, on the other hand, is remarkably close to the experimental results, particularly for the shock position and considering the crude geometric discretization.

These results illustrate the potential of high-order methods to ease the burden on the mesh generation process for

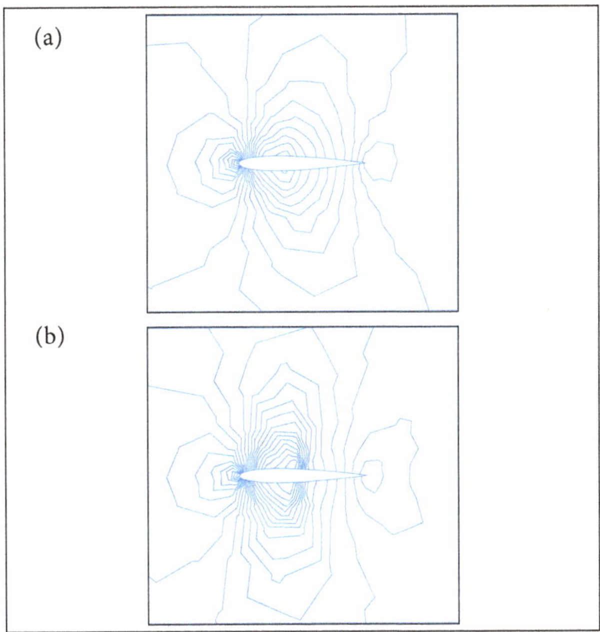

Figure 13. Evenly-spaced density contours on the coarse mesh: 30 contour lines are shown for dimensionless density values ranging from 0.8 to 1.6. (a) 1st-order Roe scheme; (b) 3rd-order SFV scheme

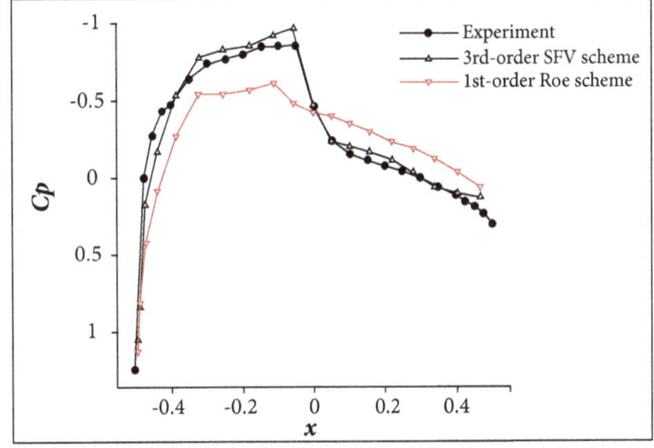

Figure 14. *Cp* distributions for the coarse mesh solutions.

aeronautical applications. One should observe, however, that the experimental results consider the presence of the boundary layer and the consequent shock-boundary layer interaction that necessarily occurs in the experiment. For the numerical solution, the shock is typically captured as a sharper discontinuity, as one should expect for an Euler simulation. In this case, however, the high-order solution seems to follow exactly the experimental data due to the extremely coarse mesh used.

The other set of results considers the fine mesh. Figure 15 presents density contours for the same range of dimensionless density variation, as in the coarse grid simulations, for the 1st-order Roe method and for the 2nd- and 3rd-order SFV schemes. One can observe that the high-order solutions present much more features and sharper flow gradients when compared to the 1st-order method, mainly in the vicinity of the shock wave. These results confirm the higher accuracy and resolution of the SFV methods, even though a limiter is used.

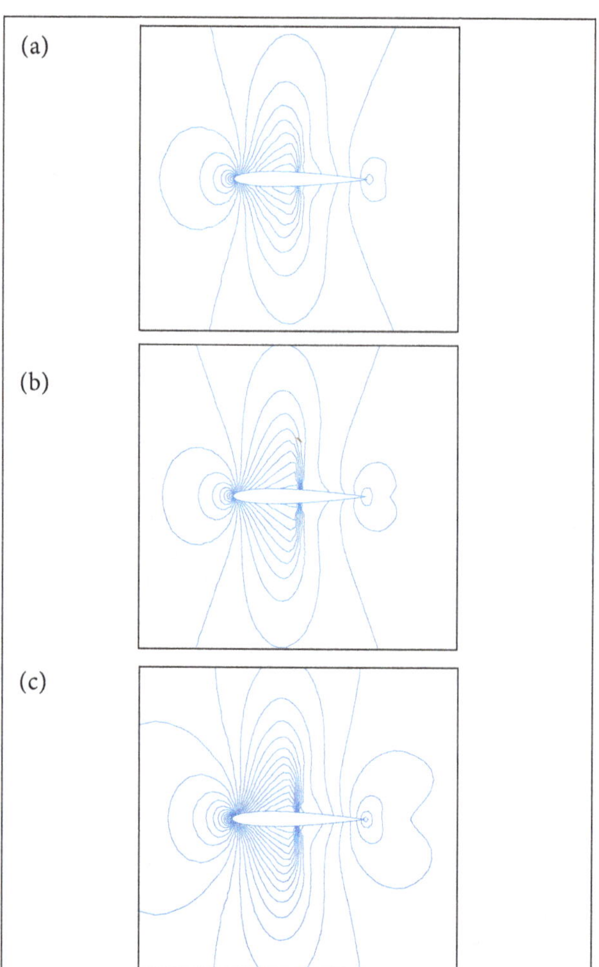

Figure 15. Evenly-spaced density contours on the fine mesh: 30 contours are shown for dimensionless density varying from 0.8 to ranging from 0.8 to 1.6. (a) 1st-order Roe scheme; (b) 2nd-order SVF scheme; (c) 3rd-order SVF scheme

Another relevant simulation is performed to assess the benefits of the curved boundary implementation, namely, the measure of entropy error ϵ_s levels at the airfoil boundary. Since the diffusive flux vectors are 0, there is no physical dissipation mechanism that produces heat in regions of smooth flow, away from shocks. If no external heat is added into the flow, then it is adiabatic. Hence, from the first law of thermodynamics, it follows that entropy, given by

$$s = C_v \ln\left(\frac{p}{\rho^{\gamma}}\right) , \tag{55}$$

is constant throughout the field if no shocks are present. Therefore, the entropy error ϵ_s, defined as

$$\epsilon_s = \frac{p}{p_\infty}\left(\frac{\rho_\infty}{\rho}\right)^{\gamma} - 1 , \tag{56}$$

is a good measure of the accuracy of a numerical solution of the Euler equations.

Figure 16 presents the entropy error generated by the 3rd-order SFV method with linear and curved boundary cells for the coarse mesh, which has only 40 cells to represent the complete airfoil geometry. As expected, the curved boundary approach is able to produce smaller error levels than the linear boundary edges. One can even note, from the figure, that at the $x = 0$ position there is an increase of entropy error due to the presence of the shock wave in this region. This, again, demonstrates that such extension indeed improves the overall accuracy of the high-order SFV method.

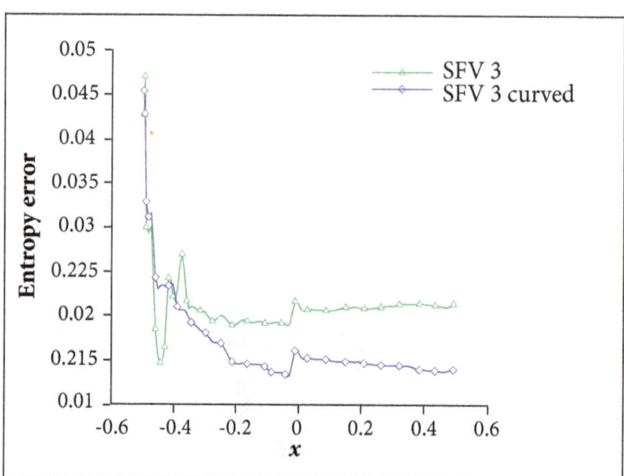

Figure 16. Entropy error for 3rd-order SFV method.

Figure 17 presents the convergence history for the simulations here considered in terms of the L_∞ norm of the continuity equation residue. The convergence history stalls for the SFV method, due to the presence of the limiter. This behavior is typical of solutions which use non-linear limiters (Venkatakrishnan 1995).

Nevertheless, the simulations have reached a steady level, based on the force coefficients. Moreover, Table 5 shows the relative costs of the different methods, normalized by the 2nd-order SFV method. The costs are measured on the fine mesh simulations and provide an overall estimate of the iteration cost associated with high-order solutions. One should observe that the 3rd-order WENO simulation is approximately 4 times more demanding than the 3rd-order SFV scheme. This increase in cost is associated with the dynamic stencil computation characteristic of the WENO solution, whereas the SFV method uses a static computational stencil throughout the simulation.

Table 5. Iteration cost estimates.

Method	Order	Limiter	Time
Roe	1	No	0.09
MUSCL	2	Yes	1.30
SFV	2	Yes	1.00
SFV	3	Yes	3.26
WENO	3	No	12.54

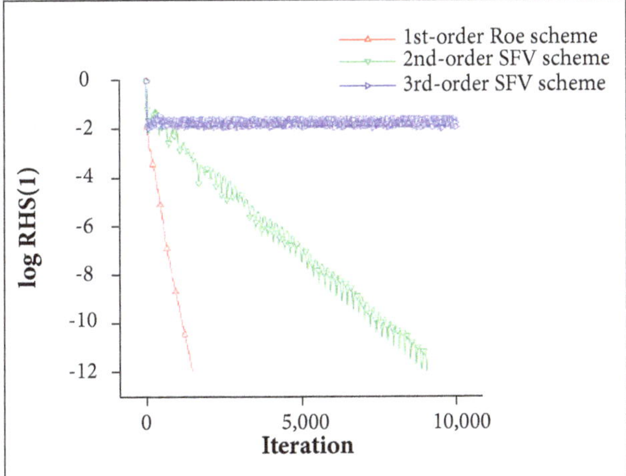

Figure 17. Convergence history for the Roe and SFV schemes.

FORWARD FACING STEP

This test case uses the same geometrical, boundary and flow parameters as the cases studied by Woodward and Colella (1984). The case was first proposed by Emery (1968) for evaluating finite-difference methods, but the results are not comparable to the later studies owing to their low resolution and parametric differences. The forward step test case is designed to resolve complex oblique shock reflections, pertinent to supersonic variable-geometry jet engine intakes. Due to computational constraints faced by Emery (1968), the test case was designed to be numerically simple to set up. The initial conditions are uniform throughout the domain, and the inlet boundary condition is supersonic. Such setup is actually difficult to perform experimentally, due to the unrealistic combination of initial and boundary conditions. However, quantitative results are available for a range of numerical methods (Woodward and Colella 1984), which makes this a good test case for validating the capabilities of the present flow solver, independently of the numerical method used.

The artificial nature of the test setup helps to evaluate the robustness of the spatial discretization algorithm combined with the limiter technique. Due to the strong shock reflection at the lower face of the step during the first few iterations, it is difficult to maintain positivity of pressure and density using various numerical schemes. Furthermore, the edge of the forward step is a singular point of the Prandtl-Meyer expansion fan generated by the flow over the step. The continuity and momentum equations can disobey the second law of thermodynamics through an expansion fan. This introduces numerical difficulty in the form of a nonphysical expansion shock, which at high Mach numbers and small cell sizes can yield negative pressures and densities in the code. The MUSCL reconstruction, for the 2nd-order Roe scheme, for instance, is not able to produce a solution as the simulation diverges after $t = 1.0$ dimensionless time units.

The 2-D configuration is 3 dimensionless length units long and 1 unit wide, with a step of 0.2 units high located at 0.6 units from the configuration inlet. The inflow and outflow boundary conditions are both supersonic, so the solver does not have to account for waves leaving the domain at the entrance boundary, or entering the domain at the exit boundary. Initially, the flow is at $M = 3$ everywhere. As stated in the seminal study of Woodward and Colella (1984), this "admittedly artificial" initial condition makes the problem very easy to set up. Since no physical constants or parameters, such as viscosity, are involved in the Euler equations, space and time units can be eliminated without the need for a dimensionless constant, and the flow is driven by pressure and density ratios. The simulation is run until $t = 4.0$ dimensionless time units, and the resulting shock wave pattern is examined. Two meshes are considered for this simulation, a coarse and a fine one. These meshes are shown in Fig. 18. The coarse mesh has a characteristic length $h = 1/40$, and it has 8,272 cells and 4,134 nodes. The finer mesh has 49,304 cells and 24,650 nodes, as well as a characteristic length $h = 1/100$. Further visualization of the mesh refinement can be seen in Fig. 19, which shows both the coarse and fine meshes for the region of the step. This last figure allows for a better visualization of the level of mesh refinement in both cases.

For the present simulations, the 1st-order Roe method is considered along with the 2nd- and 3rd-order SFV methods. A total variation bounded (TVB) minmod limiter (Shu 1987) is considered in the present study for the 2nd-order SFV scheme in

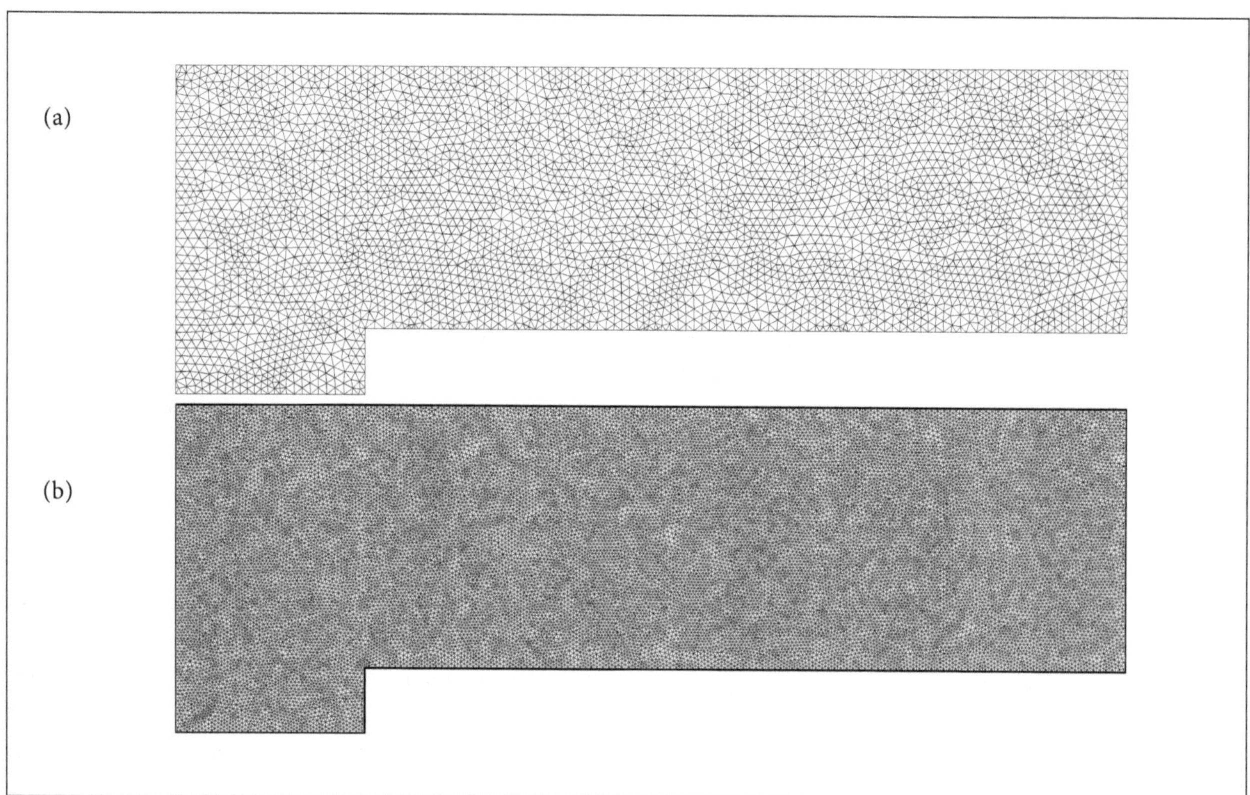

Figure 18. Computational meshes used in the forward facing step configuration simulations. (a) Coarse Mesh; (b) Fine mesh.

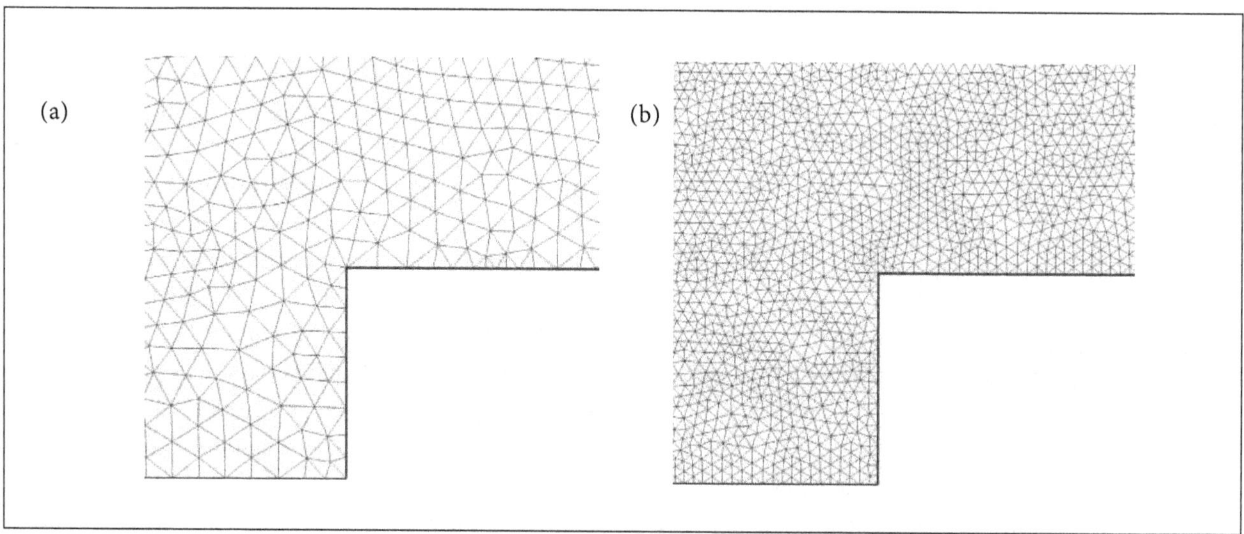

Figure 19. Detail of the domain discretization for the step region. (a) Coarse Mesh; (b) Fine mesh.

order to compare the results with the proposed hierarchical limiter for the 3rd-order SFV method. Note that, for a 2nd-order scheme, the hierarchical limiter would reduce the local reconstruction order to 1st-order accuracy anyway. Density contours for the coarse mesh can be seen in Figs. 20, 21 and 22, respectively, for the cited methods. Thirty evenly-spaced density contour lines are presented, for dimensionless density values ranging from 0.1 to 4.6. One can observe that the SFV schemes present a better resolution of the shear layer compared to the 1st-order method, as well as a slightly improved shock resolution.

In order to investigate the similarities with the 1st-order simulation, the limited cells for the 2nd- and 3rd-order SFV

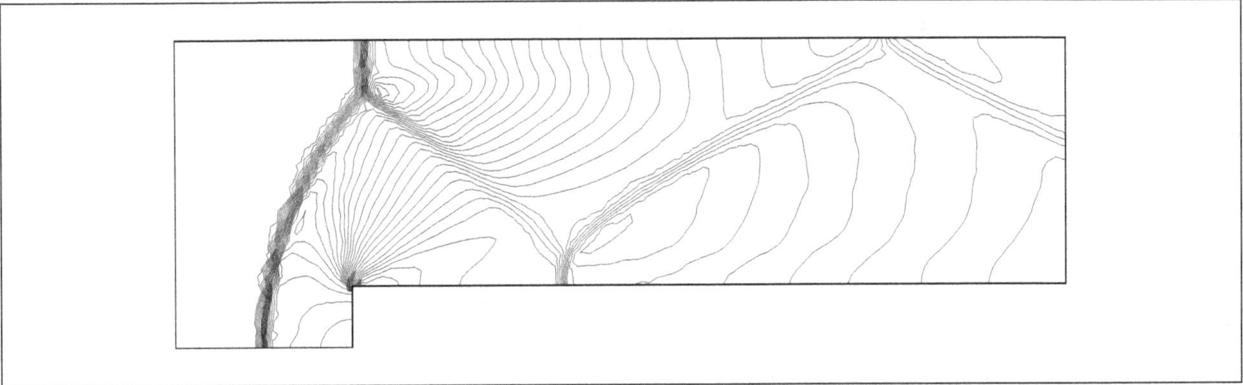

Figure 20. Density contours on the coarse mesh for the 1st-order Roe scheme. Figure shows 30 evenly-spaced dimensionless density contour lines from 0.1 to 4.6.

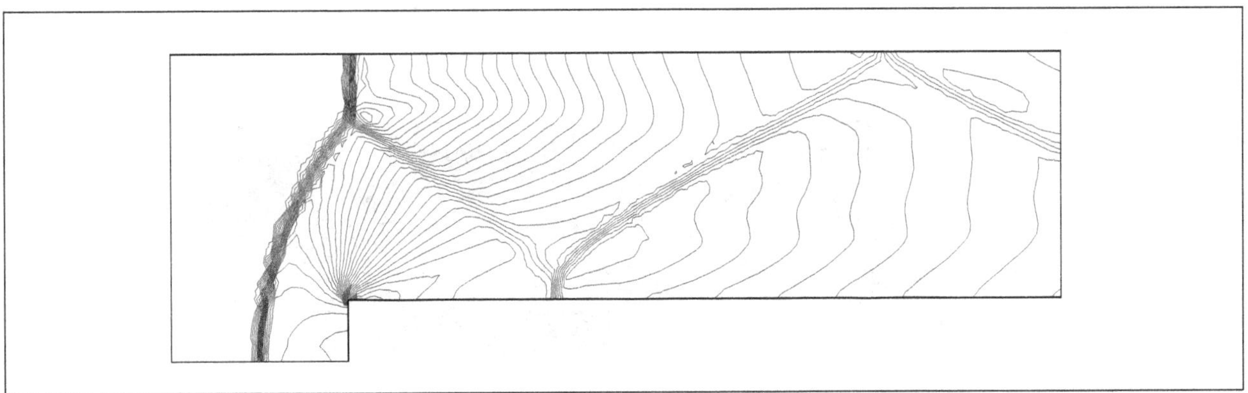

Figure 21. Density contours on the coarse mesh for the 2nd-order SFV scheme with the TVB limiter. Figure shows 30 evenly-spaced dimensionless density contour lines from 0.1 to 4.6.

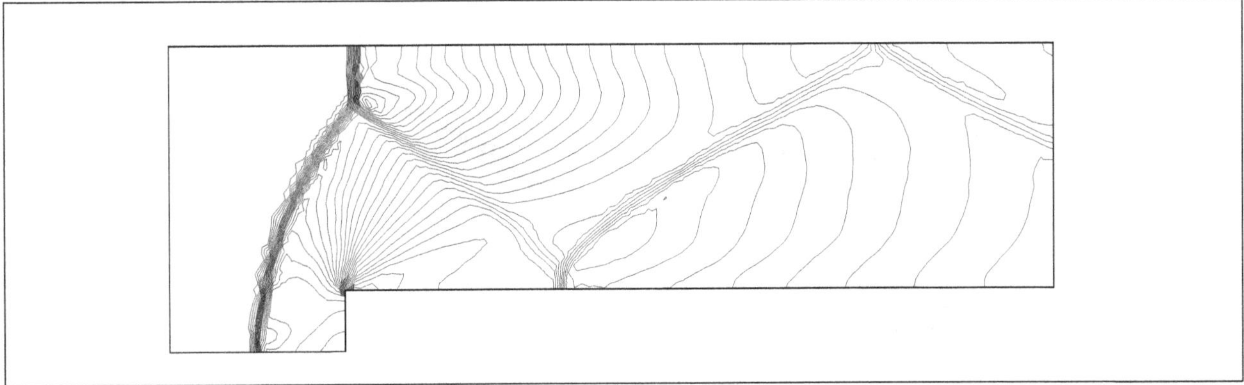

Figure 22. Density contours on the coarse mesh for the 3rd-order SFV scheme with the hierarchical limiter. Figure shows 30 evenly-spaced dimensionless density contour lines from 0.1 to 4.6.

schemes are presented in Figs. 23 and 24. These figures show the cells in which pressure reconstruction was performed at the last time step. For the 2nd-order method, there are too many cells limited by the TVB limiter, drastically reducing the overall solution reconstruction order. For the 3rd-order case, in which the hierarchical limiter is applied, the number of cells marked for limiting is reduced significantly. This confirms the superior behavior of the proposed limiting technique for the SFV method.

The next set of results considers the fine mesh. It is worth mentioning that the actual mesh used in the 3rd-order SFV method simulation has 295,824 control volumes, *i.e.*, 6 times more cells than used in the 1st-order Roe and 3rd-order WENO simulation, because the reconstruction procedure subdivides each original cell, or spectral volume, into 6 new control volumes. This increases the overall simulation costs as reported in Table 6, where the relative computational costs are normalized by those of the 2nd-order SFV scheme. Furthermore, the reader can observe that the 3rd-order WENO simulation is about twice as costly as the 3rd-order SFV scheme. In the present test case, the hierarchical limiter formulation of the SFV method yields a reduction in the cost differences between the present SFV schemes and the WENO solution, in comparison, for instance, with the results observed for the NACA 0012 test case, shown in Table 5. It happens that, due to the unsteady nature of the present problem, the limiter is activated differently at each time step and such behavior causes a penalty in the time step costs of both 2nd- and 3rd-order SFV results.

It is important to observe that, since the 2nd-order SFV method presents a large amount of limited cells due to the TVB limiter formulation, its results are actually representative of a 1st-order scheme. Hence, no results are shown for the fine mesh. Figure 25 shows the dimensionless density contours for the 1st-order Roe and 3rd-order SFV method simulations. Both

Table 6. Cost estimates per time step for the fine mesh.

Method	Order	Limiter	Time
Roe	1	No	0.08
MUSCL	2	Yes	0.25
SFV	2	Yes	1.00
SFV	3	Yes	4.79
WENO	3	No	8.55

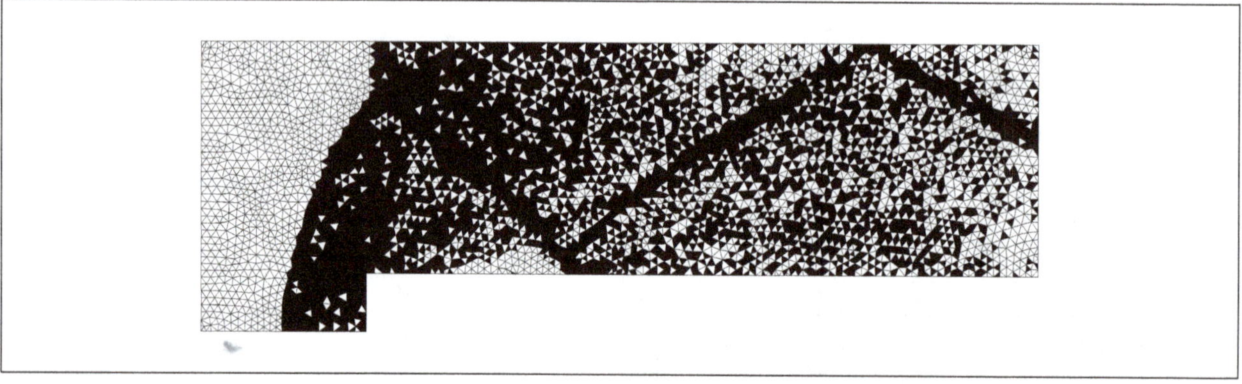

Figure 23. Limited cells for pressure reconstruction at the last time step for the 2nd-order SFV method with the TVB limiter.

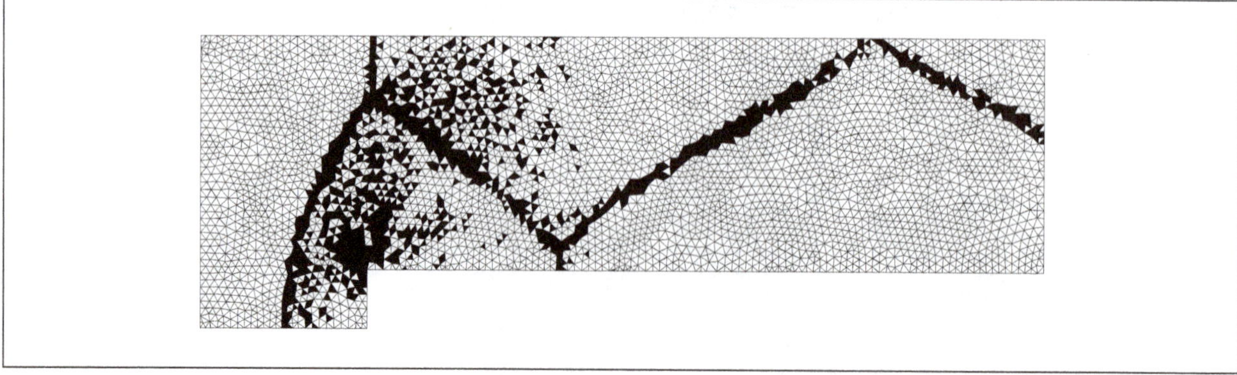

Figure 24. Limited cells for pressure reconstruction at the last time step for the 3rd-order SFV method with the hierarchical limiter.

schemes present a sharp shock resolution, but the SFV method results seem to better capture other flow features such as the shear layer. Again, the limited cells for pressure and internal energy reconstruction at the final time step are shown, respectively, in Figs. 26a and 26b for the 3rd-order SFV method calculations. One can clearly see that the high-order reconstruction is indeed

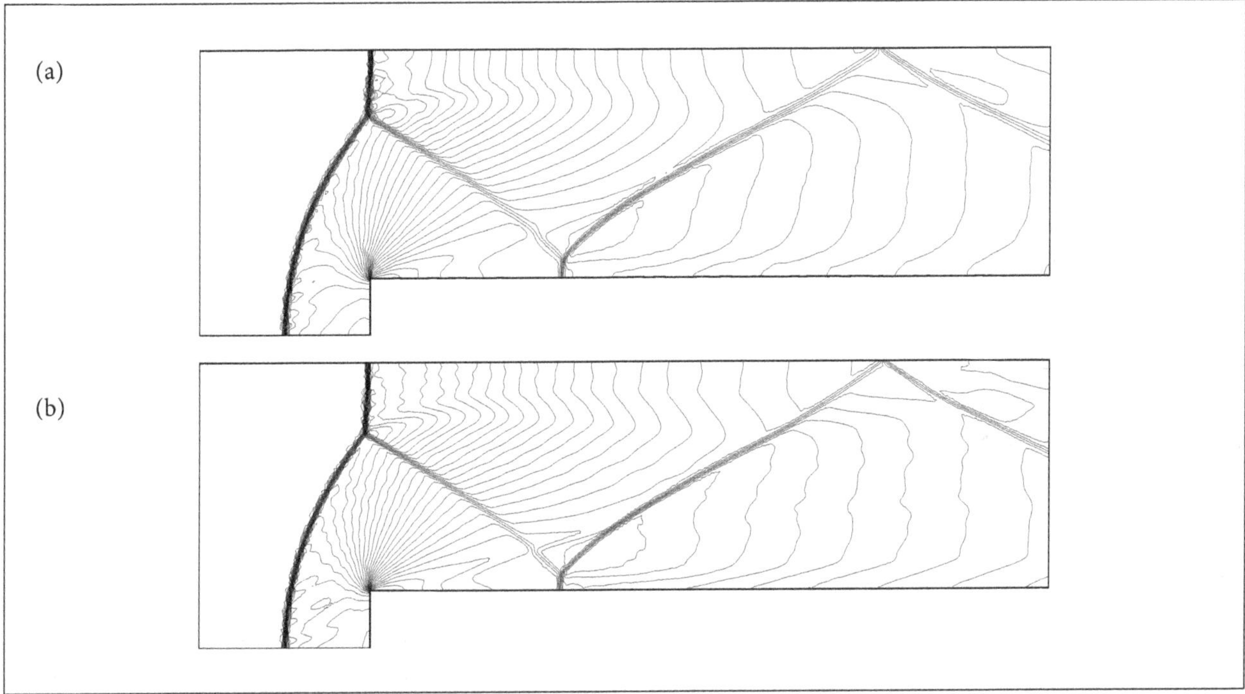

Figure 25. Density contours on the fine mesh for the 1st-order Roe scheme (a) and 3rd-order SFV scheme (b) . Thirty evenly-spaced contour lines for dimensionless density from 0.1 to 4.6.

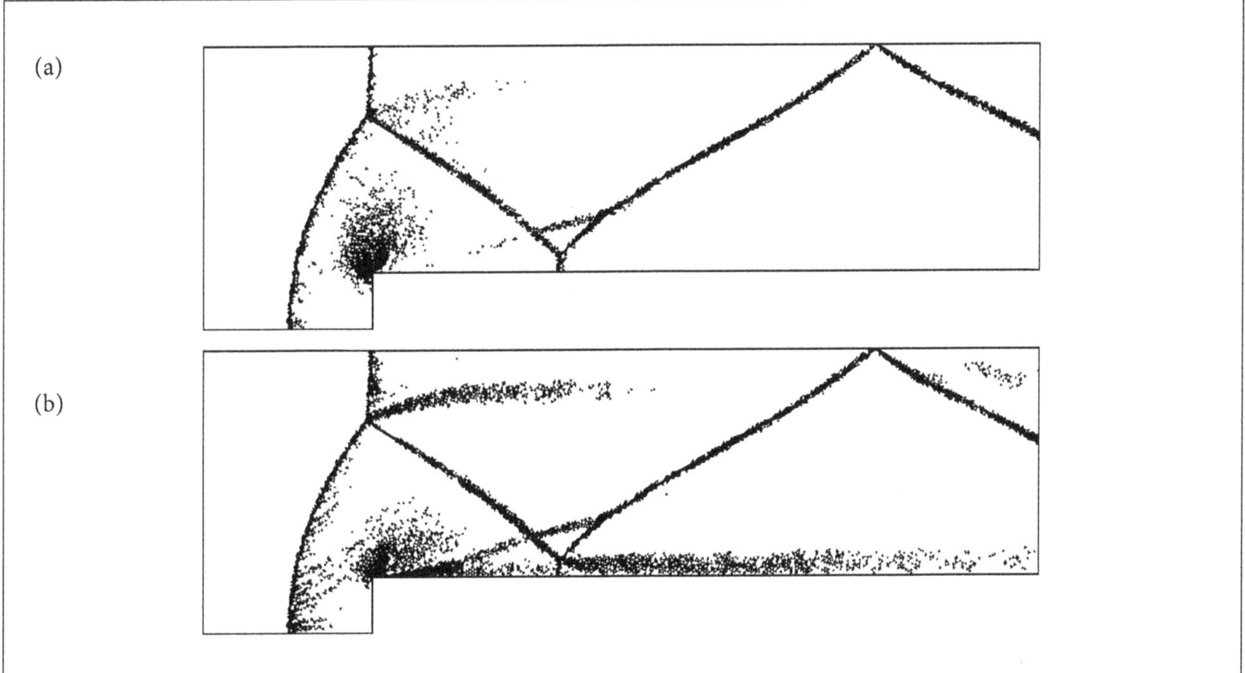

Figure 26 . Limited cells for pressure (a) and internal energy (b) reconstructions at the last time step for the 3rd-order SFV method with the hierarchical limiter on the fine mesh.

happening for the cells away from the discontinuities, since the limiter is only active at the discontinuities themselves.

Finally, the 3rd-order method solution is plotted in Fig. 27 in terms of the density contours at the CV mesh. It is important to understand that this visualization considers data for the CV cells and, therefore, it better represents the actual resolution capabilities of the SFV method. Such definition could be shown in the SV mesh but current visualization tools do not support high-order data visualization. A high-order post-processor would be necessary in order to allow a visualization with this level of resolution at the original SV mesh. Nevertheless, the level of resolution of flow features is improved over the original SV mesh.

DOUBLE MACH REFLECTION

This problem is also a standard test case for high-resolution schemes (Woodward and Colella 1984), and it has been extensively studied by many researchers (Cockburn and Shu 1989; Wang *et al.* 2004). The physical problem is that of a right-moving $M_\infty = 10$ shock wave, perpendicular to the axis of a 30 deg. half-angle wedge, which hits the tip of the wedge at time $t = 0$. Hence, the computational domain for the problem is chosen to be a rectangular region in the intervals $[0, 4] \times [0, 1]$, in the x- and y-directions, respectively. Initially, the right-moving $M_\infty = 10$ shock is positioned at $x = 1/6, y = 0$ and makes a 60 deg. angle with the x-axis. For the bottom boundary, the exact post-shock conditions are imposed, through the Rankine-Hugoniot relations, for the region from $x = 0$ to $x = 1/6$, and a solid wall boundary condition is used for the rest of the lower domain boundary. For the top boundary, the solution is set to describe the exact motion of the $M_\infty = 10$ shock. The left boundary is again set as the exact post-shock conditions, while the right boundary is set as an outflow boundary.

Two meshes are considered for this study, a coarse and a fine grid. The coarse mesh has 3,619 triangular cells, 1,808 nodes and characteristic dimensionless length $h = 1/20$. The fine mesh has 122,941 cells, 61,469 nodes and characteristic length of $h = 1/120$. Both meshes are shown in Fig. 28. It is interesting to mention that, for the SFV method simulations, the total number of CVs in the fine mesh turn out to be 368,823 and 737,646, respectively, for the 2nd- and 3rd-order methods. For the coarse mesh, the 1st-order Roe method is used, along with the 2nd- and 3rd-order SFV method. For the fine mesh, only the 1st-order Roe and 3rd-order SFV method simulations are reported here. For this test case, for instance, the 2nd-order MUSCL-reconstructed Roe scheme, with TVB minmod limiter, was not able to produce a solution. Negative values of pressure and density are found within the computation domain and the simulation diverges. This test case is a difficult one, in the sense that it requires proper boundary specification, on a per-face basis, and it also features a flow with a high level of kinetic energy.

Figure 29 presents numerical density contours for the coarse mesh, using the 1st-order Roe, 2nd- and 3rd-order SFV schemes, Moreover, 30 evenly-spaced contours are shown in each case and the dimensionless density values range from 1.25 to 21.5. For the mesh considered in this study, there is not much difference between the results. It is possible to note, however, that the Mach stem region, to the right end of the images, seems to be better defined for the SFV methods. For such problem, the use of limiters is obviously necessary. Figure 30 presents the limited cells for pressure reconstruction at the final time step for the 3rd-order SFV method. Clearly, only the shock region is indeed marked for limitation, as expected. Although necessary to obtain a numerical solution, the limiter utilization essentially reduces

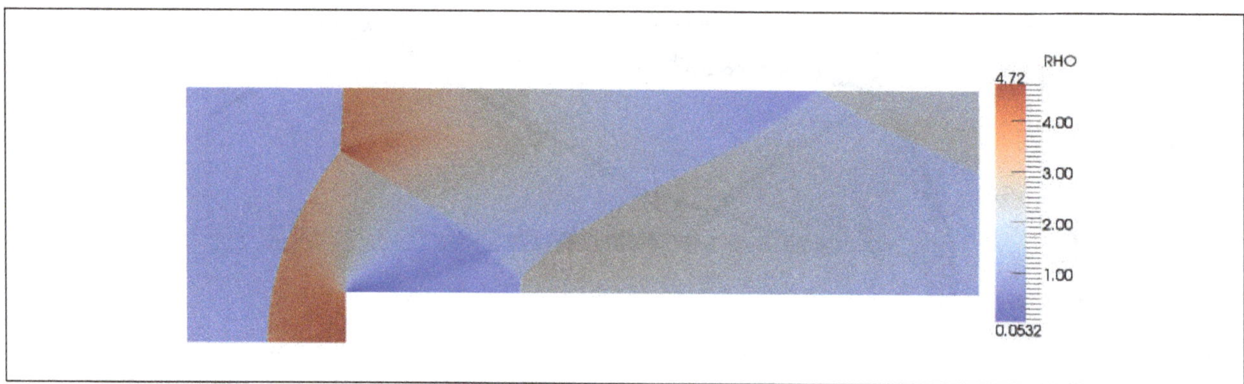

Figure 27. Density contours at the CV mesh level for the 3rd-order SFV method.

the high-order resolution in such regions and it might be responsible for the similarities in the results with the low-order scheme.

Results considering the fine mesh are shown in Figs. 31 which presents dimensionless density contours in the same range used to report the results for the coarse grid. As before, 30 evenly-spaced contours are shown in each case. Clearly, there is much more resolution now, even with the 1st-order method. The shock waves have much improved resolution and the Mach stem corner presents more details than with the coarse mesh simulations. The numerical solution on the fine mesh with the 3rd-order SFV method seems to have spurious oscillations, as one can see in Fig. 31b. The shock waves are clearly better resolved, but the rest of the domain solution seems quite oscillatory and disturbed.

In order to investigate this effect, the limited cells for pressure reconstruction at the last time step are plotted in Fig. 32. One can clearly see that a large number of cells has been selected for pressure limitation. The same behavior is observed for the other variables. This apparent oscillatory behavior has not been previously observed for other test cases and, therefore, the authors suspect that the limiter algorithm has some limitations with regard to simulations with very high Mach numbers and very strong discontinuities.

Further visualization of the results can be seen in Fig. 33, which presents density contours for the solution with the 3rd-order SFV method, but seen at the CV mesh level, that is, with 737,646 cells. As in the previous test case, this visualization considers data for the CV cells across the domain and better represents the actual resolution capabilities of the SFV method. A very sharp main shock wave can be clearly seen in the results, but oscillations in the density distribution are also seen in the post-shock region. Finally, Table 7 shows the relative costs of the different methods, normalized by the 2nd-order SFV method results.

Table 7. Estimates of relative costs per time step.

Method	Order	Limiter	Time
Roe	1	No	0.04
MUSCL	2	Yes	–
SFV	2	Yes	1.00
SFV	3	Yes	3.48
WENO	3	No	5.32

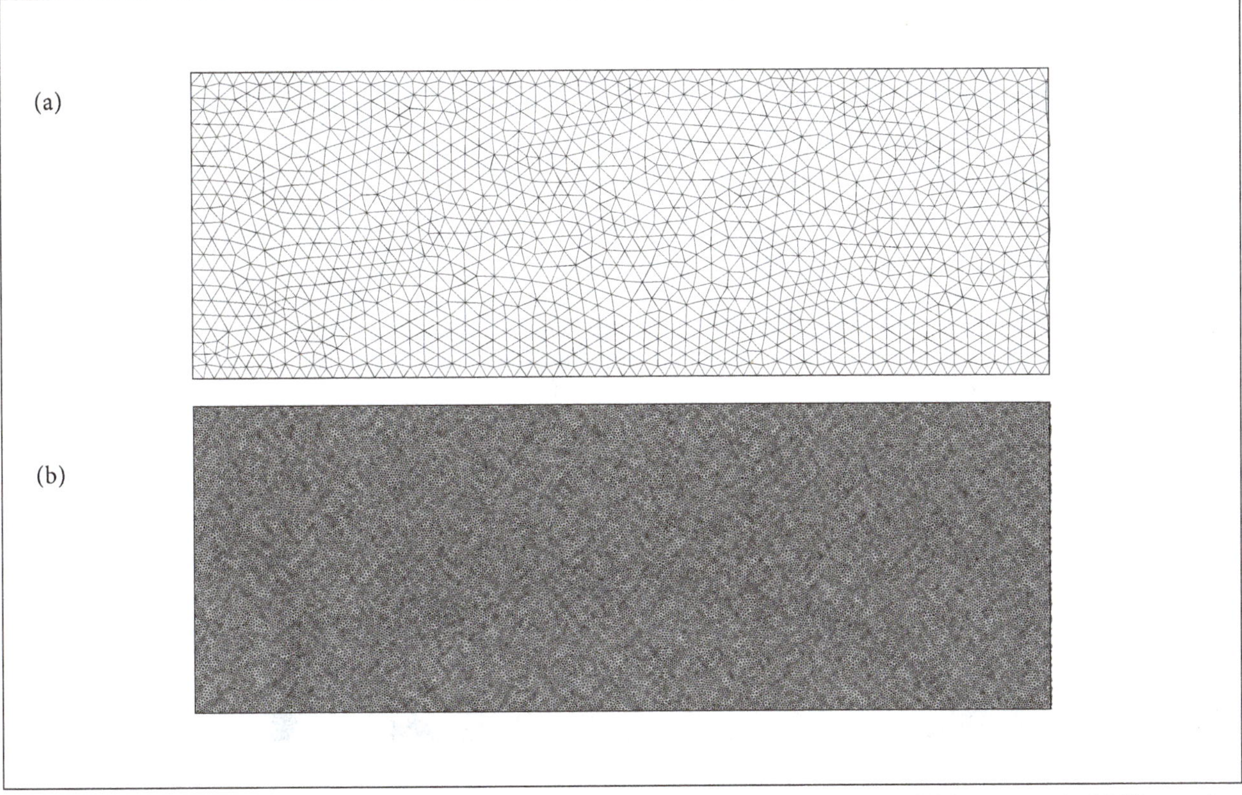

Figure 28. Computational meshes used for the double Mach reflection problem simulations. (a) Coarse mesh; (b) Fine mesh.

Costs are measured on the coarse mesh simulations and provide an overall estimate of the computational requirements associated with such high-order schemes. The 3rd-order WENO solution is about 1.5 times more

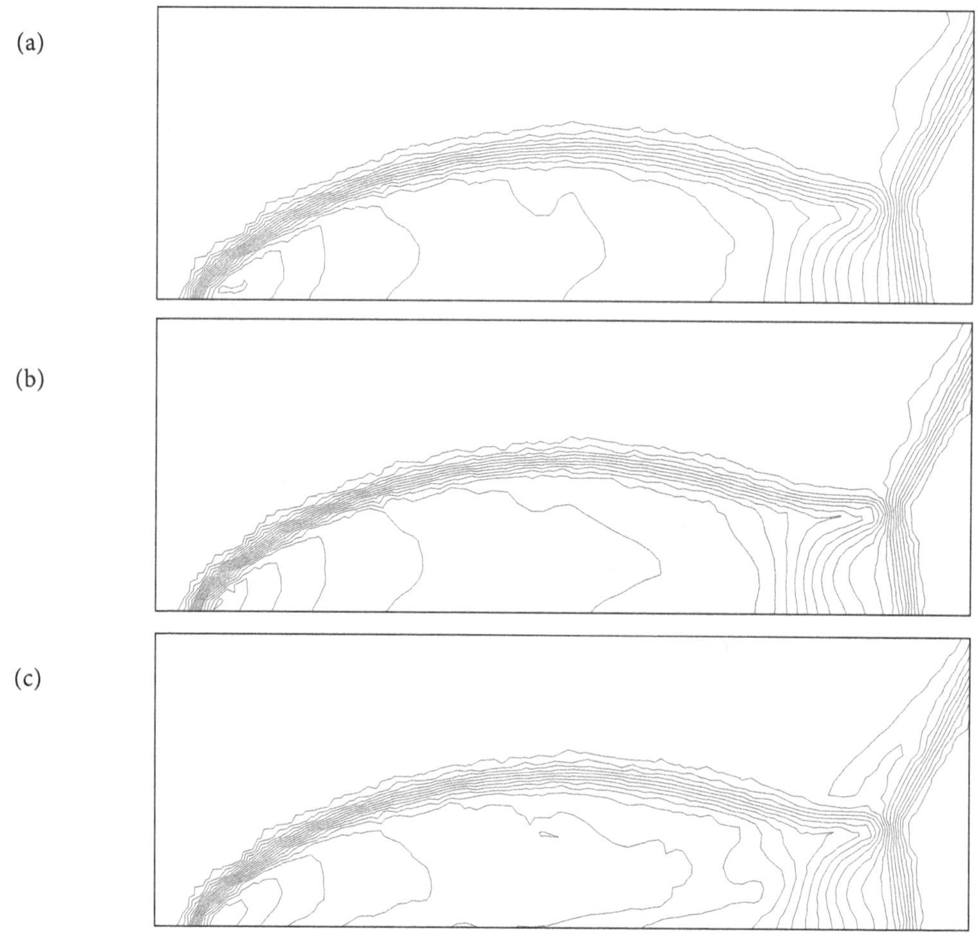

(a)

(b)

(c)

Figure 29. Density contours on the coarse mesh for the 1st-order Roe scheme (a); for 2nd-order SFV scheme(b); and for the 3rd-order SFV scheme (c). Figure shows 30 evenly-spaced dimensionless density contour lines in the range from 1.25 to 21.5.

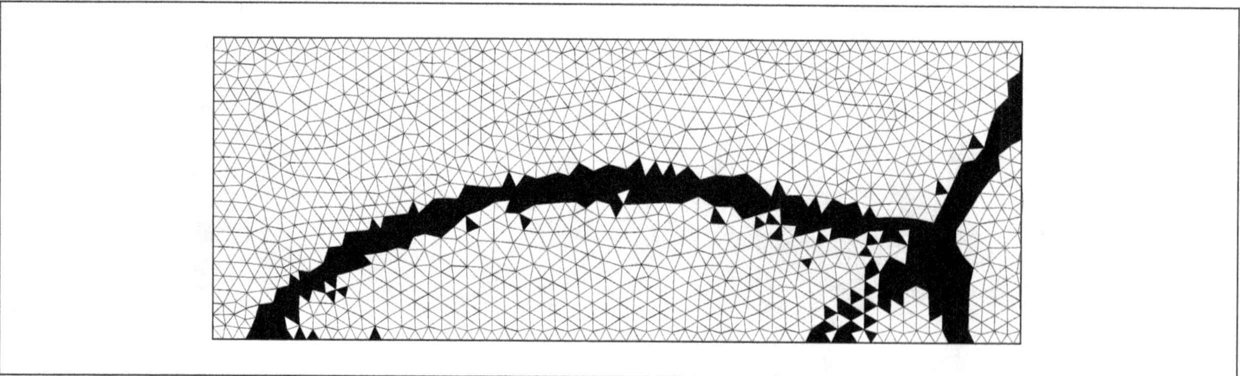

Figure 30. Limited cells for pressure reconstruction at the last time step for the 3rd-order SFV method.

expensive for this problem than the 3rd-order SFV method calculations, considering the same input data. The reader should observe that the hierarchical limiter is active only in the discontinuous region of the solution, as depicted in Fig. 30.

CONCLUDING REMARKS

The application of the high-order SFV method to steady and unsteady inviscid compressible flow simulations is presented. The paper also addresses the use of high-order boundary treatment, which is relevant because it can allow the usage of coarser meshes without giving up in geometric resolution. The present implementation of the limiter technique, which extends, to SFV methods, ideas that have been previously tested on spectral difference schemes, is an important ingredient of the method efficiency. The present limiter reduces the number of limited spectral volumes to a bare minimum, which reduces computational costs and, at the same time, allows for a more uniform high-order solution. Furthermore, a user-input-free limiter implementation contributes to enhance the robustness of the flow solver. Several classical test cases, both steady

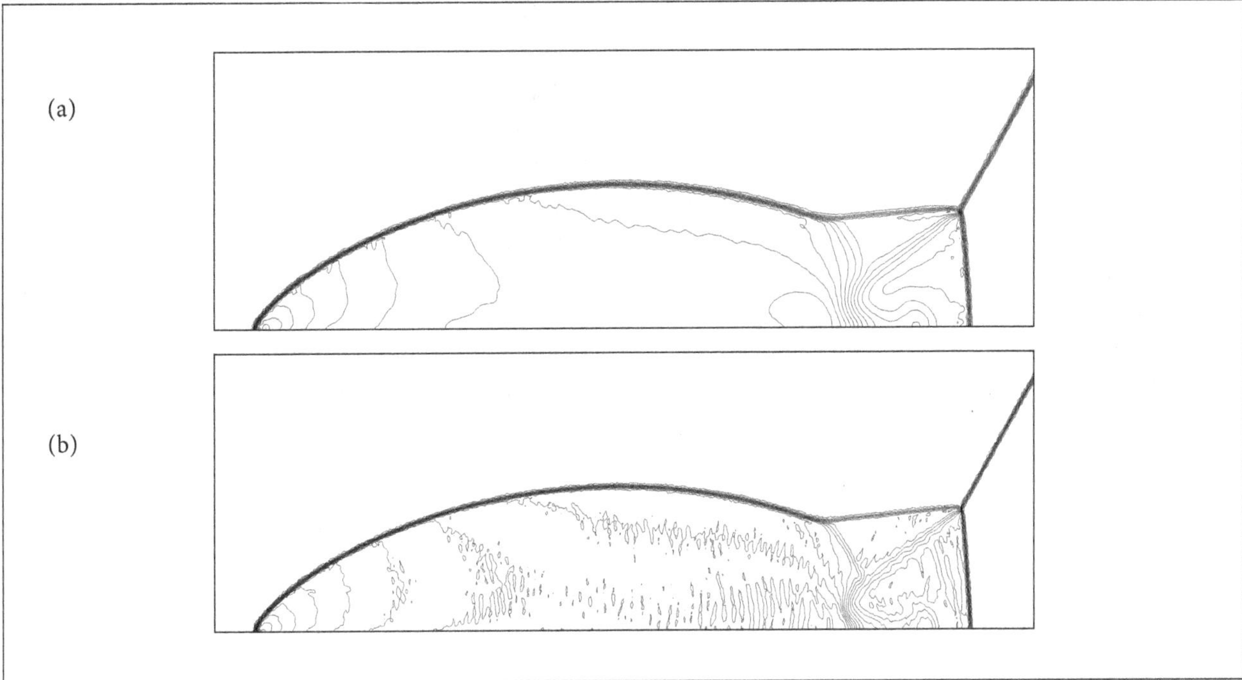

Figure 31. Density contours on the fine mesh for the 1st-order Roe scheme (a) and 3rd-order SFV scheme(b). Figure shows 30 evenly-spaced dimensionless density contour lines in the range from 1.25 to 21.5.

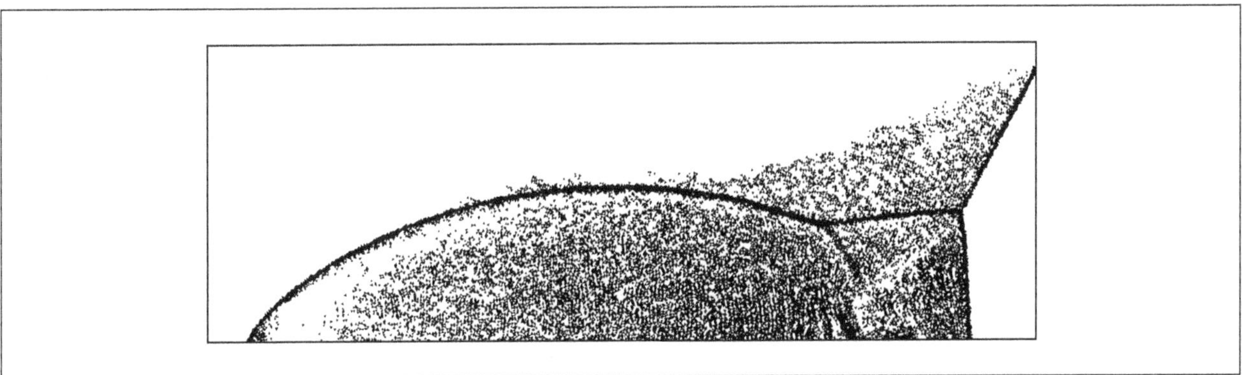

Figure 32. Limited cells for pressure reconstruction at the last time step for the 3rd-order SFV method on the fine mesh.

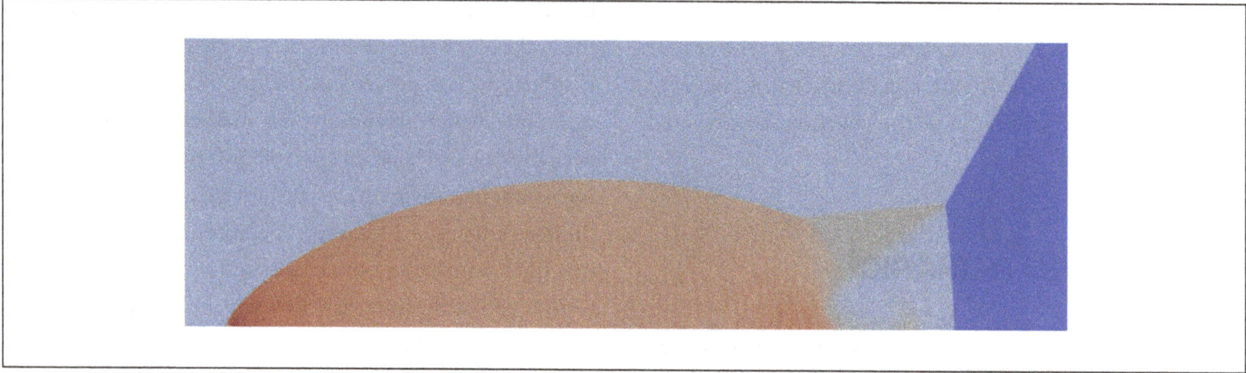

Figure 33. Density contours at the CV mesh level for the 3rd-order SFV method solution in the fine grid.

and unsteady, have been addressed in order to highlight such features.

The main focus of the paper has been on the complex transient problems that stress the numerical method ability to compute flows with strong discontinuities without numerical divergence. The results obtained in the present research for the forward-facing step and the double Mach reflection problems have indicated that the SFV method is able to cope with the strong shock waves and produce good solutions. For both test cases, however, it is clear that the limiter algorithm still has limitations with regard to simulations with very high Mach numbers and very strong discontinuities. It seems that small oscillations are still allowed at the main shock wave, and these are convected downstream by the high-order scheme. Finally, the computational costs of the present high-order implementation are rather modest when compared to the benefits in flow resolution which can be achieved.

ACKNOWLEDGEMENTS

The authors gratefully acknowledge the partial support for this research provided by Conselho Nacional de Desenvolvimento Científico e Tecnológico (CNPq), under the Research Grants No. 309985/2013-7, No. 400844/2014-1 and No. 443839/2014-0. This study is also supported by the Coordenação de Aperfeiçoamento de Pessoal de Nível Superior (CAPES), through a graduate scholarship for the first author. The authors are also indebted to the partial financial support received from Fundação de Amparo à Pesquisa do Estado de São Paulo (FAPESP), under the Research Grant No. 2013/07375-0.

AUTHOR'S CONTRIBUTION

All authors contributed equally for the development of the work reported in the present paper.

REFERENCES

Abgrall R (1994) On essentially non-oscillatory schemes on unstructured meshes: analysis and implementation. J Comput Phys 114(1):45-58. doi: 10.1006/jcph.1994.1148

Barth T, Frederickson P (1990) Higher order solution of the Euler equations on unstructured grids using quadratic reconstruction. Proceedings of the 28th AIAA Aerospace Sciences Meeting; Reno, USA.

Barth T, Jespersen D (1989) The design and application of upwind schemes on unstructured meshes. Proceedings of the 27th Aerospace Sciences Meeting and Exhibit; Reno, USA.

Bigarella EDV, Azevedo JLF (2007) Advanced eddy-viscosity and Reynolds-stress turbulence model simulations of aerospace applications. AIAA J 45(10):2369-2390. doi: 10.2514/1.29332

Bigarella EDV, Azevedo JLF (2012) A study of convective flux schemes for aerospace flows. J Braz Soc Mech Sci & Eng 34(3):314-329. doi: 10.1590/S1678-58782012000300012

Breviglieri C, Azevedo JLF, Basso E (2010a) An unstructured grid implementation of high-order spectral finite volume schemes. J Braz Soc Mech Sci & Eng 32:419-433. doi: 10.1590/S1678-58782010000500001

Breviglieri C, Azevedo JLF, Basso E, Souza MAF (2010b) Implicit high-order spectral finite volume method for inviscid compressible flows. AIAA J 48(10):2365-2376. doi: 10.2514/1.J050395

Breviglieri C, Basso E, Azevedo JLF (2008) High-order unstructured spectral finite volume scheme for aerodynamic applications. Proceedings of the 26th AIAA Applied Aerodynamics Conference; Honolulu, USA.

Cockburn B, Shu CW (1989) TVB Runge-Kutta local projection discontinuous Galerkin finite element method for conservation laws. II: General framework. Math Comp 52:441-435.

Emery A (1968) An evaluation of several differencing methods for inviscid fluid flow problems. J Comput Phys 2(3):306-331. doi: 10.1016/0021-9991(68)90060-0

Friedrich O (1998) Weighted essentially non-oscillatory schemes for the interpolation of mean values on unstructured grids. J Comput Phys 144(1):194-212. doi: 10.1006/jcph.1998.5988

Gruttke WB (1995) The Initial Graphics Exchange Specification (IGES) v. 6.0. Washington: IGES/PDES Organization.

Jiang GS, Shu CW (1996) Efficient implementation of weighted ENO schemes. J Comput Phys 126(1):202-228. doi: 10.1006/jcph.1996.0130

Krivodonova L (2007) Limiters for high-order discontinuous Galerkin methods. J Comput Phys 226(1):879-896. doi: 10.1016/j.jcp.2007.05.011

Liu Y, Vinokur M, Wang ZJ (2006) Spectral (finite) volume method for conservation laws on unstructured grids V: Extension to three-dimensional systems. J Comput Phys 212(2):454-472. doi: 10.1016/j.jcp.2005.06.024

McDevitt J, Okuno AF (1985) Static and dynamic pressure measurements on a NACA 0012 airfoil in the Ames High Reynolds Number Facility. NASA-TP-2485. Moffett Field: NASA Ames Research Center.

Ollivier-Gooch C (1997) Quasi-ENO schemes for unstructured meshes based on unlimited data-dependent least-squares reconstruction. J Comput Phys 133(1):6-17. doi: 10.1006/jcph.1996.5584

Parsani M, van den Abeele K, Lacor C, Turkel E (2010) Implicit LU-SGS algorithm for high-order methods on unstructured grid with *p*-multigrid strategy for solving the steady Navier-Stokes equations. J Comput Phys 229(3):828-850. doi: 10.1016/j.jcp.2009.10.014

Qiu J, Shu CW (2006) A comparison of troubled-cell indicators for Runge-Kutta discontinuous Galerkin methods using weighted essentially nonoscillatory limiters. SIAM J Sci Comput 27(3):995-1013. doi: 10.1137/04061372X

Roe PL (1981) Approximate Riemann solvers, parameter vectors, and difference schemes. J Comput Phys 43(2):357-372. doi: 10.1016/0021-9991(81)90128-5

Shapiro AH (1953) The dynamics and thermodynamics of compressible fluid flow. New York: Wiley.

Shu CW (1987) TVB uniformly high-order schemes for conservation laws. Math Comp 49:105-121.

Sun Y, Wang ZJ, Liu Y (2006) Spectral (finite) volume method for conservation laws on unstructured grids VI: Extension to viscous flow. J Comput Phys 215(1):41-58. doi: 10.1016/j.jcp.2005.10.019

Van Altena M (1999) High-order finite-volume discretisations for solving a modified advection-diffusion problem on unstructured triangular meshes (Master's thesis). Vancouver: University of British Columbia.

Van den Abeele K, Lacor C (2007) An accuracy and stability study of the 2D spectral volume method. J Comput Phys 226(1):1007-1026. doi: 10.1016/j.jcp.2007.05.004

Venkatakrishnan V (1995) Convergence to steady state solutions of the Euler equations on unstructured grids with limiters. J Comput Phys 118(1):120-130. doi: 10.1006/jcph.1995.1084

Wang ZJ (2002) Spectral (finite) volume method for conservation laws on unstructured grids: Basic formulation. J Comput Phys 178(1):210-251. doi: 10.1006/jcph.2002.7041

Wang ZJ, Fidkowski K, Abgrall R, Bassi F, Caraeni D, Cary A, Deconinck H, Hartmann R, Hillewaert K, Huynh HT, Kroll N, May G, Persson P, van Leer B, Visbal M (2013) High-order CFD methods: Current status and perspective. Int J Numer Meth Fluid 72(8):811-845. doi: 10.1002/fld.3767

Wang ZJ, Liu Y (2002) Spectral (finite) volume method for conservation laws on unstructured grids II: Extension to two-dimensional scalar equation. J Comput Phys 179(2):665-697. doi: 10.1006/jcph.2002.7082

Wang ZJ, Liu Y (2004) Spectral (finite) volume method for conservation laws on unstructured grids III: One dimensional systems and partition optimization. J Sci Comput 20(1):137-157. doi: 10.1023/A:1025896119548

Wang ZJ, Liu Y (2006) Extension of the spectral volume method to high-order boundary representation. J Comput Phys 211(1):154-178. doi: 10.1016/j.jcp.2005.05.022

Wang ZJ, Liu Y, Zhang L (2004) Spectral (finite) volume method for conservation laws on unstructured grids IV: Extension to two-dimensional systems. J Comput Phys 194(2):716-741. doi: 10.1016/j.jcp.2003.09.012

Wolf WR, Azevedo JLF (2006) High-order unstructured essentially nonoscillatory and weighted essentially nonoscillatory schemes for aerodynamic flows. AIAA J 44(10):2295-2310. doi: 10.2514/1.19373

Wolf WR, Azevedo JLF (2007) High-order ENO and WENO schemes for unstructured grids. Int J Numer Meth Fluid 55(10):917-943. doi: 10.1002/fld.1469

Woodward P, Colella P (1984) The numerical simulation of two-dimensional fluid flow with strong shocks. J Comput Phys 54(1):115-173. doi: 10.1016/0021-9991(84)90142-6

Xu Z, Liu Y, Shu CW (2009) Hierarchical reconstruction for spectral volume method on unstructured grids. J Comput Phys 228(16):5787-5802. doi: 10.1016/j.jcp.2009.05.001

Yang M, Wang ZJ (2009) A parameter-free generalized moment limiter for high-order methods on unstructured grids. Adv Appl Math Mech 1(4):451-480. doi: 10.4208/aamm.09-m0913

10

Copper Cobalt Magnetic Ceramic Materials Characterization at Terahertz Frequencies

Alan Fernando Ney Boss[1], Antonio Carlos da Cunha Migliano[1,2], Ingrid Wilke[3]

ABSTRACT: This study presents the complex index of refraction and the complex permittivity of a magnetic ceramic material made of copper, cobalt, and iron oxides. The index of refraction and the extinction coefficient of the CuCo-ferrite exhibit an almost frequency independent behavior and were averaged to $n = 3.62 \pm 0.05$ and $k = 0.06 \pm 0.02$, respectively, over the frequency range from 0.2 to 1 THz. The corresponding complex permittivity was $\varepsilon' = 13.12 \pm 0.35$ for the real part and $\varepsilon'' = 0.46 \pm 0.15$ for the imaginary one. The absorption coefficient and the transmittance of the CuCo-ferrite were also determined. The absorption coefficient exhibits a dip at ~0.35 THz, which corresponds to a peak in transmittance at this frequency. The impact of the observations on the potential realization of novel THz electronic devices is discussed.

KEYWORDS: Ferrites, Refractivity, Permittivity, Transmission, Terahertz.

INTRODUCTION

During the last decades, an unexplored gap in the electromagnetic spectrum has been massively studied. This gap is described nowadays as terahertz (THz) frequency range and it is located between microwave and infrared frequencies. Usually, the THz frequency range is defined between 0.1 and 10 THz (Lee 2009), although it is also possible to find definitions between 0.3 and 3 THz (Phillips 2011).

Several methods to achieve these frequencies have been developed and studied. Some sources that generate THz waves are the photomixer (McIntosh *et al.* 1995), the quantum cascade laser (Williams 2007), the microwave frequency multiplier (Li and Yao 2010), the backward wave oscillator (Mineo and Paoloni 2010), the free electron laser (Williams 2002), synchrotron light sources (Roy *et al.* 2006), and so on. Among all these methods, a commonly one used is the THz Time-Domain Spectroscopy (THz-TDS) system, which has the advantage to measure the amplitude and phase of THz electromagnetic radiation in the time domain, allowing a large frequency range to be evaluated in a single run. THz-TDS systems usually utilize photo-conductive antennas or electro-optic crystals excited by femtosecond near-infrared laser pulses as THz radiation sources and detectors, combined with lock-in detection and a time delay stage (Bründermann *et al.* 2012).

THz frequencies may find use in medicine (Siegel 2004), as an alternative to X-rays in certain imaging applications (Hu and Nuss 1995), in security for detecting explosives (Shen *et al.* 2005), in narcotics identification systems (Lu *et al.* 2006), in bio-defense (Woolard *et al.* 2003; Kemp 2011), in chemistry and biology for materials identification (Fischer *et al.* 2005;

1. Departamento de Ciência e Tecnologia Aeroespacial – Instituto Tecnológico de Aeronáutica – Ciências e Tecnologias Espaciais – São José dos Campos/SP – Brazil.
2. Departamento de Ciência e Tecnologia Aeroespacial – Instituto de Estudos Avançados – Divisão de Física Aplicada – São José dos Campos/SP – Brazil.
3. Rensselear Polytechnic Institute – Department of Physics, Applied Physics and Astronomy – Terahertz and Ultrafast Spectroscopy Laboratory – Troy/NY – USA.

Author for correspondence: Alan Fernando Ney Boss | Departamento de Ciência e Tecnologia Aeroespacial – Instituto Tecnológico de Aeronáutica – Ciências e Tecnologias Espaciais | Praça Marechal Eduardo Gomes, 50 - Vila das Acácias | CEP: 12.228-900 – São José dos Campos/SP – Brazil | Email: alan.boss86@gmail.com

Kiwa *et al.* 2007), and other areas such as astronomy (Kulesa 2011; Smirnov *et al.* 2012), space science (Siegel 2007), plasma physics (Tauk *et al.* 2006), and aerospace applications (Petkie *et al.* 2009; Chen 2007).

In aerospace applications, THz radiation can be used to investigate the Radar Cross Section (RCS) of miniature models of airplanes (Iwaszczuk *et al.* 2010; Li *et al.* 2013), which reduces the expenses with real airplanes in anechoic chambers. Also, it is possible to perform nondestructive evaluations of materials used in airplanes, such as foams, paints, and fiberglass composites (Quast *et al.* 2010).

Another type of material that has been studied at THz frequencies and may be applied to aerospace technology are metamaterials, *i.e.*, artificial materials that present negative permittivity or permeability (Capolino 2009). An example of THz metamaterials is presented by Takano *et al.* (2013), where TiO_2 ceramic balls placed in a metallic grid were used to create a material with negative permittivity and permeability.

Although common ceramics have been previously used for THz applications, we could not find major studies at THz frequencies about magnetic ceramic materials, also known as ferrites. The lack of information about the properties of these materials, such as index of refraction and complex permittivity, prevents the design and development of novel THz devices. This has motivated us to characterize ferrites at THz frequencies.

The solid state reaction route is a state-of-the-art method for the fabrication of magnetic ceramics (Brito *et al.* 2009) and achieves the desired composition of a ceramic by weighting powders previously calculated with stoichiometric formula of the ceramic. This process is reproducible if well controlled. Some parameters such as powder granulation, material mixing, and sintering temperature have influence on the material formation and grain size. In this paper, we describe the CuCo-ferrite fabrication process, the THz-TDS system characteristics as well as optical and dielectric properties of a CuCo-ferrite sample measured with a THz-TDS system.

MATERIALS AND METHODS

SAMPLE CHARACTERISTICS

The sample explored in this study is a CuCo ferrite with the stoichiometry $Cu_{0.5}Co_{0.5}Fe_2O_4$. This ferrite was made following the steps described in Fig. 1. Conditions such as weighting the powders and sintering temperature have influence on

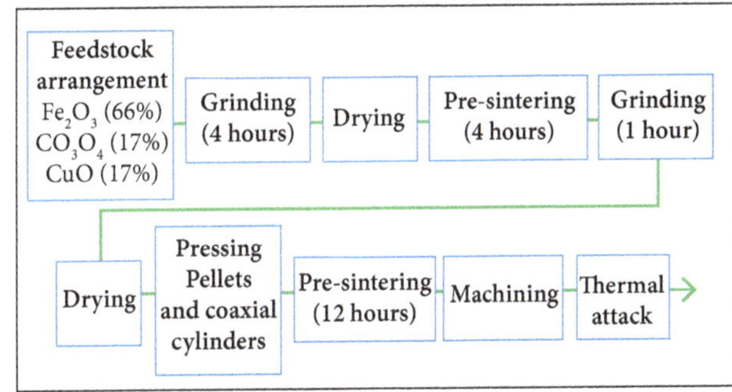

Figure 1. Steps of sample preparation.

the sample characteristics, *i.e.*, a temperature higher than the Curie temperature will melt and fuse the material. Samples were pressed in a cylindrical shape, with a diameter close to 6.4 mm. Sample thickness and flatness were adjusted in the final steps, when the samples were submitted to sandpapering and polishing. After machining the samples, they were submitted to a thermal attack to release internal mechanical stress.

Sample thickness has a direct influence on the THz transmission signal, which requires some precautions during sample preparation. If a sample is too thin it may present multiple reflections and create an etalon effect after a Fourier transformation of the transmitted signal. If the sample is too thick, it may strongly attenuate the transmission signal, which affects the peak to peak time delay analysis.

Sample diameter should also be considered since it will drive the sample holder size and may affect the detection of low frequencies. Also, the sample should be as flat and plane-parallel as possible in order to avoid imprecise measurements of the sample thickness and peak to peak time delay analysis.

Inhomogeneous samples may present regions with different concentrations of materials. This may affect the transmitted signal, which will affect the index of refraction calculation. Scanning Electron Microscope (SEM) and Energy-Dispersive X-ray Spectroscopy (EDS) are useful to visualize the surface and make a semi-quantitative prediction of material distribution.

THZ TIME-DOMAIN SPECTROSCOPY SYSTEM

A THz-TDS transmission system (Perenzoni and Paul 2014) was used in these experiments, but, instead of using parabolic mirrors to guide the THz waves, 50-mm tsurupica lenses were employed. A scheme and a picture of the assembled system is presented in Fig. 2.

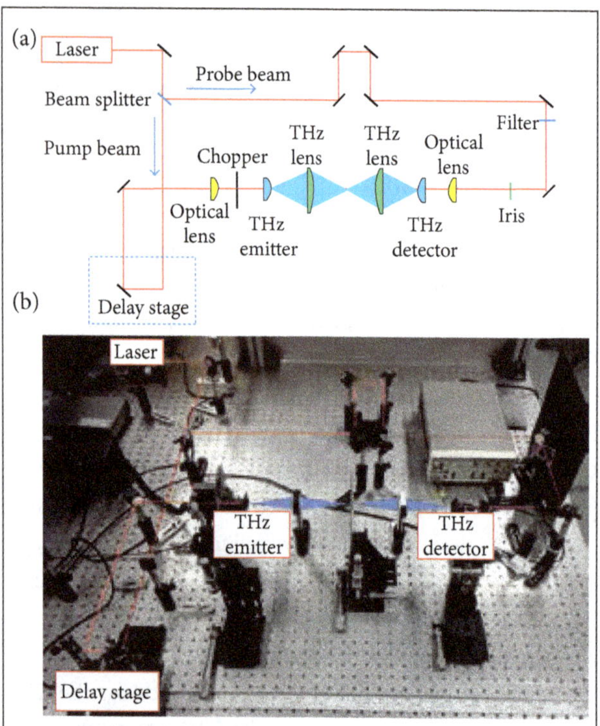

Figure 2. THz-TDS transmission setup using two 50-mm tsurupica THz lenses. (a) Scheme; (b) Picture.

Here, the beam splitter divides the laser beam into pump and probe beams. The pump beam hits the iPCA-21-05-1000-800-h emitter (Batop Electronics 2016b), which has an active antenna area of 1 mm². The probe beam is focused on the 5 µm antenna gap of the PCA-40-05-10-800 detector (Batop Electronics 2016a). In order to attenuate an echo spot caused by the beam splitter an iris was placed right before the optical lens that focuses the probe beam on the detector. This reduces the echo spot and avoids an etalon effect in the analysis. Since the active area of the emitter is bigger than the gap in the detector, the delay stage is placed in the pump beam. This reduces errors caused by laser drift when the delay stage is scanning.

The laser beam was modulated with a mechanical chopper right before the emitter. The modulation frequency is transmitted to the lock-in amplifier, which used this reference frequency to filter the signal from noise through phase-sensitive detection.

Since the sample diameter is about 6.4 mm, the sample holder aperture is about 5 mm. This ensures that there is no THz signal being transmitted directly to the detector. We can estimate the frequencies that will be affected by the sample holder aperture using the Gaussian intensity distribution across the THz radiation beam:

$$f = c \times z/\pi \times \omega^2 \tag{1}$$

where: f is the THz radiation frequency in Hz; c is the speed of light; z is the focal lengths of the THz lens; ω is the THz radiation beam waist.

If we consider that the maximum beam waist will have the size of the aperture, we can estimate that the minimum frequency that will not be affected is approximately ~191 GHz using the 50-mm THz lenses. Frequencies below 191 GHz may be compromised since the focal spot diameter formed by this wavelength ($\lambda = c/f$) will be bigger than the aperture. The maximum frequency is determined by the noise floor of the signal after a Fourier transformation.

The peak to peak time delay is the difference in time of the same peak in the reference signal and in the sample signal. This can be better understood in Fig. 3, where the sample signal has a Δt time difference from the reference signal.

It is possible to perform a rough estimate of the index of refraction of the material under investigation using the time delay Δt considering:

$$\Delta t = (n - 1)d/c \tag{2}$$

where: n is the index of refraction; d is the thickness.

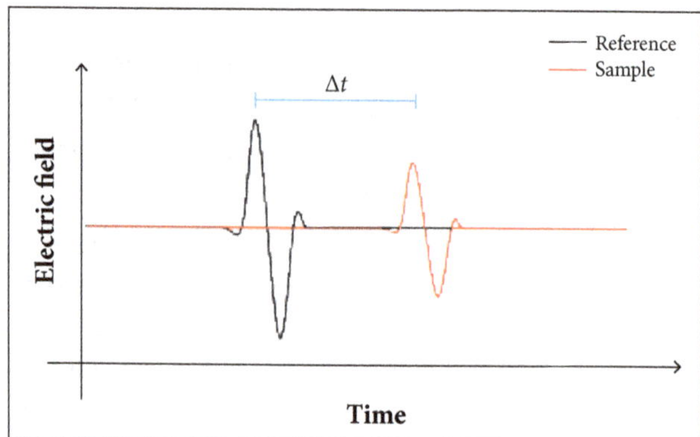

Figure 3. Peak to peak time delay illustration.

RESULTS AND DISCUSSION

Our investigation of the CuCo ferrite starts with an EDS analysis of the sample (Fig. 4). From this semi-quantitative analysis, we confirm that the CuCo ferrite is an inhomogeneous material. Some regions between grains present a concentration of copper and a lack of iron, while

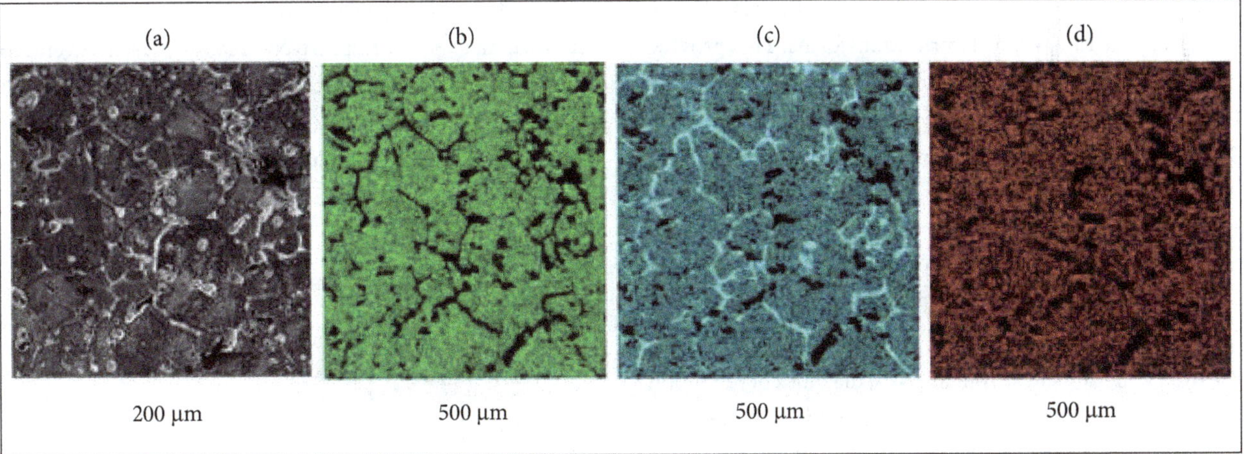

Figure 4. SEM image of the CuCo sample (a), Iron (b), Copper (c), and Cobalt (d) in the same region analyzed with EDS.

cobalt seems equally distributed throughout the sample. Since our THz analysis has a minimum frequency stablished at ~0.2 THz, we can estimate that an area with the size close to the respective wavelength (~1.5 mm) should not be an issue, since the measurement would predict the average composition of the sample. Higher frequencies although may present differences due the material distribution characteristic. In order to ensure a proper characterization, measurements were performed on both sides of the sample and then compared.

THz analysis requires a reference measurement, which is a measurement of the THz radiation pulse passing through the empty sample holder. After the reference measurement, the sample is placed on the sample holder to be first measured and then it is flipped to be measured again. The voltage measured by the lock-in amplifier, which is proportional to the THz electric field (E-field) is sent to a computer with the respective information of the delay stage position. The position is then converted to time and the final data is provided in time (ps) by voltage (V). The reference signal measured with a time resolution of 20 fs as well as the front and back measurements of the sample are visualized in Fig. 5.

We notice a very slight difference between both sample measurements. The time delay of the front and back measurements are 22.173 and 22.162 ps, respectively. Since our sample has 2.54 mm thickness, we estimated a index of refraction of 3.619 for the front side and 3.618 for the back side.

The Fourier transforms of the measured signals are presented in Fig. 6. Here, we estimate the high frequency limit of the measurement for our analysis. The reference signal approaches the noise floor around 1.2 THz, while the sample measurements approach the noise floor at 1 THz. Therefore, we limit our analysis up to 1 THz.

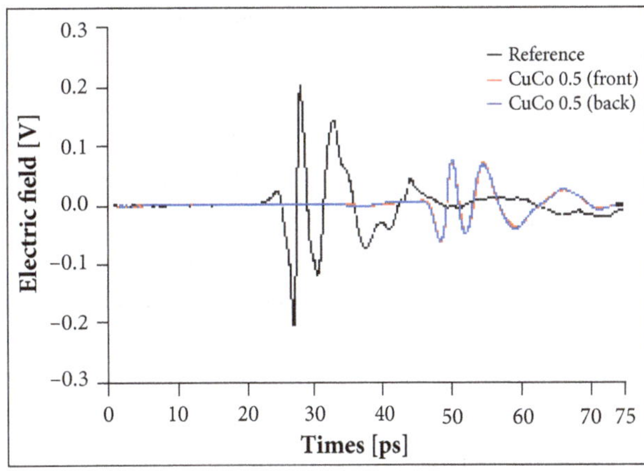

Figure 5. Reference and sample measurements in the time domain.

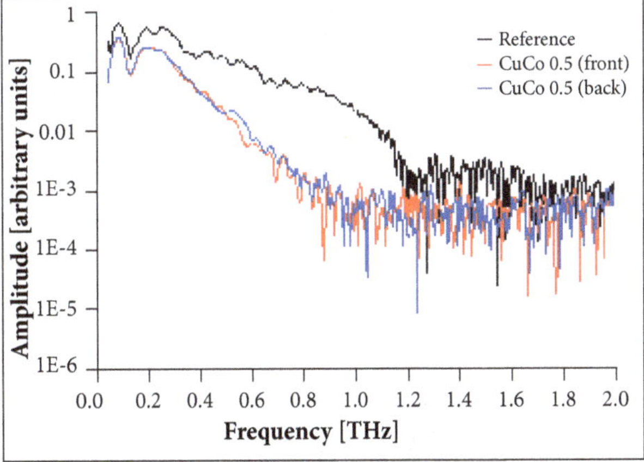

Figure 6. Reference and sample measurements in the frequency domain. From this analysis it was determined that the investigation ranged from 0.2 to 1 THz.

After the Fourier transformation of the signals we calculated the ratio of the Fourier transforms. This provide us the complex transmission coefficient, which is used to calculate the complex index of refraction (Peiponen *et al.* 2013). Figure 7 presents the complex index of refraction for both sample measurements, *i.e.*, the index of refraction and the extinction coefficient for both sides of the CuCo ferrite.

We notice a small difference of the measured index of refraction for the front and back side of the sample, above 0.85 THz. This is attributed to differences in the surface roughness of the front and back side. In our analysis, the index of refraction is considered to be frequency independent within the 0.2 to 1.0 THz frequency band. The extinction coefficient assumes values between $k = 0.05$ to $k = 0.10$.

To validate the assembled system, as well as the THz measurement made with it, we compared the index of refraction in Fig. 7 with the index of refraction calculated from time-domain THz transmission measurements performed with a different THz-TDS system (Fig. 8). This system is a commercial THz-TDS spectrometer that uses two lasers and an optical trigger (Klatt *et al.* 2009) instead of a single laser and a mechanical delay stage. The advantage is a faster measurement with higher resolution. The disadvantage is the high cost of the system, inability to change the system setup and energy per THz-radiation pulse.

Since the index of refraction in Figs. 7 and 8 are close, our further analysis will be only with data provided from the THz-TDS assembled system. This is because the assembled system has a lower resolution over the frequency, which provides a smooth curve over the frequency range.

The complex index of refraction of Fig. 7 is converted to complex permittivity by using the relations real permittivity $\varepsilon' = n^2 - k^2$ and imaginary permittivity $\varepsilon'' = 2nk$ for the real and imaginary parts, respectively. Table 1 presents averaged values over the frequency of the index of refraction, extinction coefficient, real permittivity and imaginary permittivity for both measurements of the CuCo ferrite.

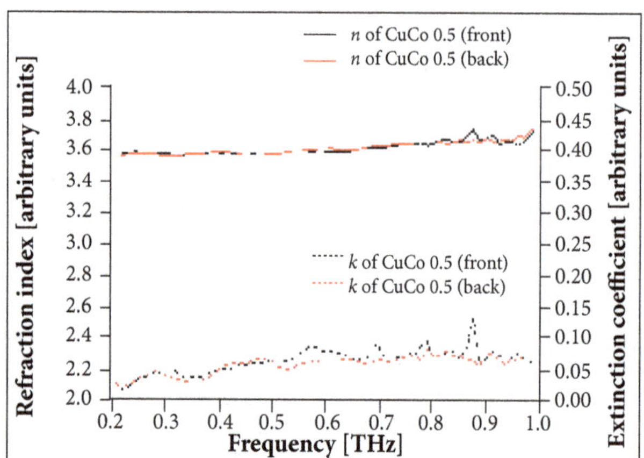

Figure 7. Index of refraction and extinction coefficient of the CuCo sample.

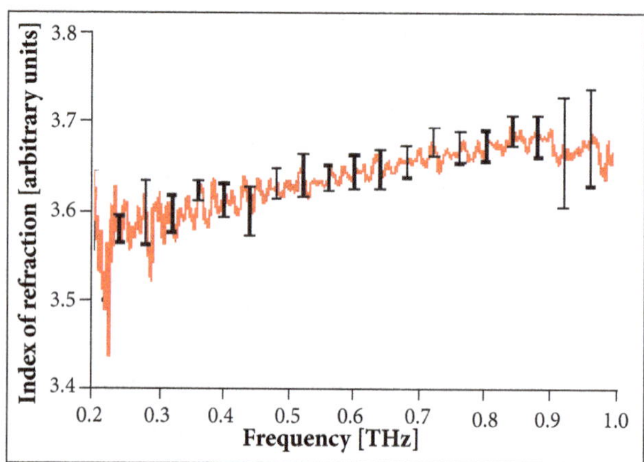

Figure 8. Index of refraction of the CuCo sample measured with THz-TDS spectrometer using optical trigger.

Table 1. Averaged index of refraction, extinction coefficient, real and imaginary permittivity of the CuCo 0.5 ferrite over the frequency from 0.2 to 1 THz.

Parameters	CuCo 0.5 front	CuCo 0.5 back
Index of refraction	3.62 ± 0.05	3.62 ± 0.04
Extinction coefficient	0.06 ± 0.02	0.06 ± 0.02
Real permittivity (ε')	13.12 ± 0.35	13.12 ± 0.31
Imaginary permittivity (ε")	0.46 ± 0.15	0.42 ± 0.11

From the extinction coefficient, the power absorption coefficient is calculated using $\alpha(f) = (4\pi f)k(f)/c$ (Perenzoni and Paul 2014) — Fig. 9. Since the absorption coefficient is calculated considering the frequency dependence of the extinction coefficient, characteristics of the data previously seen in Fig. 7, such as the spike in k for the front measurement above 0.85 THz are observed again in Fig. 9. However, there is a small disturbance at ~0.35 THz on both measurements that was not noticed before.

Using the absorption coefficient and the sample thickness, we can estimate the transmittance using $t(f)$ » $\exp(-\alpha(f)d)$ (Wilke *et al.* 2014). The transmittance of both measurements is plotted in Fig. 10, where we notice that the disturbance at ~0.35 THz becomes more evident. Since this disturbance happens on both measurements, we believe that this may be a characteristic of the sample, which may be explored with others stoichiometries of this ferrite. This may indicate a potential use of this material for THz frequency devices, such as filter.

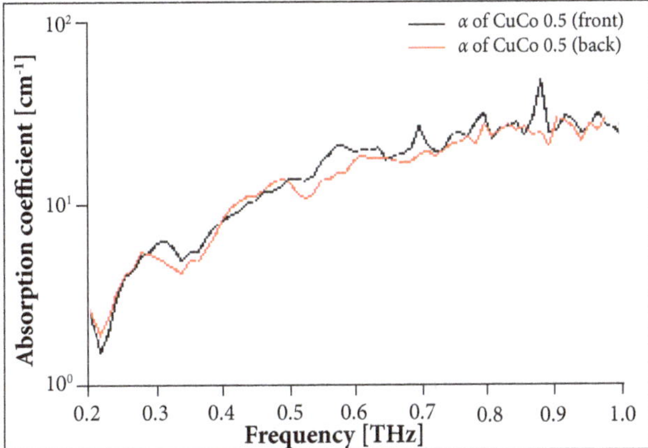

Figure 9. Absorption coefficient in the frequency domain.

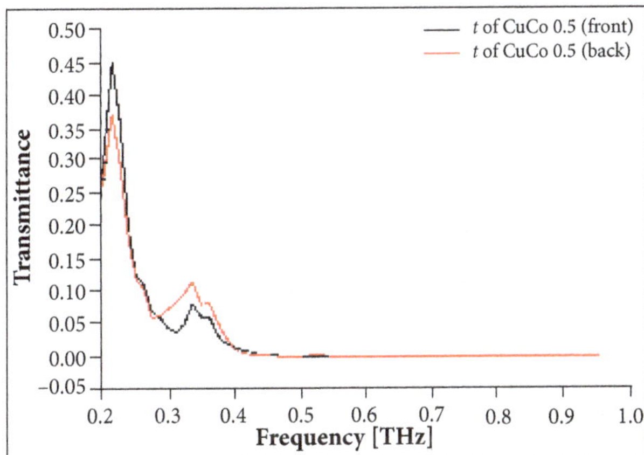

Figure 10. Transmittance in the frequency domain of the sample measurements.

CONCLUSION

In this study, we described the fabrication of copper cobalt ferrite $Cu_{0.5}Co_{0.5}Fe_2O_4$ and the characterization of the dielectric properties of this material in the THz frequency range. The sample was prepared using a state-of-the-art solid state reaction route, and the THz frequency characterization was performed with a THz-TDS transmission system.

Our sample was submitted to EDS analysis to evaluate spatial homogeneity. The EDS measurements revealed sample regions with an excess of copper and lack of iron. The spatial inhomogeneity of the sample composition is not considered to be relevant for the analysis of measurements in the 0.2 to 1 THz frequency band but may become an issue at higher frequencies, where 2 different sites on the sample may present different materials concentrations in the focal spot of the THz radiation beam.

We validate the THz system assembled, as well as the sample's characteristic, by comparing the index of refraction calculated with another THz-TDS system. We also presented the average values for the complex index of refraction and complex permittivity. The index of refraction was around $n = 3.62$ and the extinction coefficient, around $k = 0.06$. The real permittivity was $\varepsilon' = 13.12$ and the imaginary one, below $\varepsilon'' = 0.46$.

Also, the power absorption coefficient and transmittance were calculated and discussed. A small discontinuity in the power absorption coefficient and transmittance close to ~0.35 THz was observed. This may be an indication that other sample stoichiometries may present transmission and absorption characteristics that can be useful for the development of novel THz devices, such as filters.

ACKNOWLEDGEMENTS

Boss AFN thanks the Pró-Estratégia and Demanda Social of the Coordenação de Aperfeiçoamento de Pessoal de Nível Superior (CAPES) for the national scholarships and the Ciências sem Fronteiras of the Conselho Nacional de Desenvolvimento Científico e Tecnológico (CNPq) for the international scholarship. Migliano ACC thanks the Fundação de Amparo à Pesquisa do Estado de São Paulo (FAPESP) for supporting the project 2012//01448-2 and the CT-INFRA 2013 of the Financiamento de Estudos e Projetos (FINEP). Wilke I acknowledges a careful proofreading of the manuscript draft by Charles Khachian from the Rensselaer Polytechnic Institute.

AUTHOR'S CONTRIBUTIONS

Boss AFN and Migliano ACC conceived the idea and discussed about the material preparation; Boss AFN and Wilke I performed the experiments, co-wrote the main text, discussed the results and comments on the manuscript.

REFERENCES

Batop Electronics (2016a) Instruction manual and data sheet PCA-40-05-10-800-x; [accessed 2016 Jul 18]. http://www.batop.com/products/terahertz/photoconductive-antenna/data-sheet/manual_PCA-40-05-10-800.pdf

Batop Electronics (2016b) Instruction manual and data sheet iPCA-21-05-1000-800-h; [accessed 2016 Jul 18]. http://www.batop.com/products/terahertz/photoconductive-antenna/data-sheet/manual_iPCA-21-05-1000-800-h.pdf

Brito VLO, Migliano ACC, Lemos LV, Melo FCL (2009) Ceramic processing route and characterization of a Ni-Zn ferrite for application in a pulsed-current monitor. Progress in Electromagnetics Research 91:303-318. doi: 10.2528/PIER09031603

Bründermann, E, Hübers H-W, Kimmitt MF (2012) Terahertz techniques. Berlin: Springer.

Capolino F (2009) Application of metamaterials. In: Bründermann E, Hübers HW, Kimmitt MF (2012) Terahertz techniques. Berlin: Springer.

Chen CH (2007) Ultrasonic and advanced methods for nondestructive testing and material characterization. Singapore: World Scientific Publishing.

Fischer B, Hoffmann M, Helm H, Modjesch G, Jepsen PU (2005) Chemical recognition in terahertz time-domain spectroscopy and imaging. Semicond Sci Technol 20:S246-S253. doi: 10.1088/0268-1242/20/7/015

Hu BB, Nuss MC (1995) Imaging with terahertz waves. Optic Lett 20(16):1716-1718. doi: 10.1364/OL.20.001716

Iwaszczuk K, Heiselberg H, Jepsen PU (2010) Terahertz radar cross section measurements. Opt Express 18(25):26399-26408. doi: 10.1364/OE.18.026399

Kemp M (2011) Screening mail for powders using terahertz technology. Proc SPIE 8189:8189J. doi: 10.1117/12.898093

Kiwa T, Oka S, Kondo J, Kawayama I, Yamada H, Tonouchi M, Tsukada K (2007) A terahertz chemical microscope to visualize chemical concentrations in microfluidic chips. Jpn J Appl Phys 46(44):L1052-L1054. doi: 10.1143/JJAP.46.L1052

Klatt G, Gebs R, Janke C, Dekorsy T, Bartels A (2009) Rapid-scanning terahertz precision spectrometer with more than 6 THz spectral coverage. Opt Express 17(25):22847-22854. doi: 10.1364/OE.17.022847

Kulesa C (2011) Terahertz Spectroscopy for Astronomy: from comets to Cosmology. IEEE Trans THz Sci Technol 1(1):232-240. doi: 10.1109/TTHZ.2011.2159648

Lee YS (2009) Principles of terahertz science and technology. New York: Springer.

Li HY, Li Q, Xia ZW, Zhao YP, Chen DY, Wang Q (2013) Influence of Gaussian beam on terahertz radar cross section of a conducting sphere. J Infrared Milli Terahz Waves 34:88-96. doi: 10.1007/s10762-012-9950-6

Li W, Yao J (2010) Microwave and terahertz generation based on photonically assisted microwave frequency twelvetupling with large tunability. IEEE Photonics Journal 2(6):954-959. doi: 10.1109/JPHOT.2010.2084993

Lu M, Shen J, Li N, Zhang Y, Zhang C (2006) Detection and identification of illicit drugs using terahertz imaging. J Appl Phys 100:103104. doi: 10.1063/1.2388041

McIntosh KA, Brown ER, Nichols KB, McMahon OB, DiNatale WF, Lyszczarz TM (1995) Terahertz photomixing with diode lasers in low-temperature-grown GaAs. Appl Phys Lett 64:3844-3846. doi: 10.1063/1.115292

Mineo M, Paoloni C (2010) Corrugated rectangular waveguide tunable backward wave oscillator for terahertz applications. IEEE Trans Electron Dev 57(6):1481-1484. doi: 10.1109/TED.2010.2045678

Peiponen KE, Zeitler JA, Kuwata-Gonokami M (2013) Terahertz spectroscopy and imaging. Berlin: Springer-Verlag.

Perenzoni M, Paul DJ (2014) Physics and applications of terahertz radiation. New York: Springer.

Petkie DT, Kemp IV, Benton C, Boyer C, Owens L, Deibel JA, Stoik CD, Bohn MJ (2009) Nondestructive terahertz imaging for aerospace applications. Proc SPIE 7485:74850D. doi: 10.1117/12.830540

Phillips X (2011) Terahertz technology. Delhi: The English Press.

Quast H, Keil A, Loffler T (2010) Investigation of foam and glass fiber structures used in aerospace applications by all-electronic 3D Terahertz imaging. Proceedings of the 35th International Conference on Infrared, Millimeter, and Terahertz Waves; Rome, Italy.

Roy P, Rouzières M, Qi Z, Chubar O (2006) The AILES Infrared Beamline on the third generation Synchrotron Radiation Facility SOLEIL. Infrared Phys Tech 49:139-146. doi: 10.1016/j.infrared.2006.01.015

Shen YC, Lo T, Taday PF, Cole BE, Tribe WR, Kemp MC (2005) Detection and identification of explosives using terahertz pulsed spectroscopic imaging. Appl Phys Lett 86:241116. doi: 10.1063/1.1946192

Siegel PH (2004) Terahertz technology in Biology and Medicine. IEEE MTT-S International Microwave Symposium Digest 52(10):1575-1578. doi: 10.1109/MWSYM.2004.1338880

Siegel PH (2007) THz instruments for space. IEEE Trans Antenn Propag 55(11):2957-2965. doi: 10.1109/TAP.2007.908557

Smirnov AV, Baryshev AM, Pilipenko SV, Myshonkova NV, Bulanov VB, Arkhipov MY, Vinogradov IS, Likhachev SF, Kardashev NS (2012) Space mission Millimetron for terahertz astronomy. Proc SPIE 8442:84424C. doi: 10.1117/12.927184

Takano K, Yakiyama Y, Shibuya K, Izumi K, Miyazaki H, Jimba Y, Miyamaru F, Kitahara H, Hangyo M (2013) Fabrication and performance of TiO_2-ceramic-based metamaterials for terahertz frequency range. IEEE Trans THz Sci Technol 3(6):812-819. doi: 10.1109/TTHZ.2013.2285521

Tauk R, Teppe F, Boubanga S, Coquillat D, Knap W (2006) Plasma wave detection of terahertz radiation by silicon field effects transistors: responsivity and noise equivalent power. Appl Phys Lett 89:253511. doi: 10.1063/1.2410215

Wilke I, Ramanathan V, LaChance J, Tamalonis A, Aldersley M, Joshi PC, Ferris J (2014) Characterization of the terahertz frequency optical constants of montmorillonite. Appl Clay Sci 87:61-65. doi: 10.1016/j.clay.2013.11.006

Williams BS (2007) Terahertz quantum-cascade lasers. Nature Photon 1(517):517-525. doi: 10.1038/nphoton.2007.166

Williams GP (2002) FAR-IR/THz radiation from the Jefferson Laboratory, energy recovered linac, free electron laser. Rev Sci Instrum 73(3):1461-1463. doi: 10.1063/1.1420758

Woolard DL, Brown ER, Samuels AC, Jensen JO, Globus T, Gelmont B, Wolski M (2003) Terahertz-frequency remote-sensing of biological warfare agents. IEEE MTT-S International Microwave Symposium Digest 2:763-766. doi: 10.1109/MWSYM.2003.1212483

Evaluation of Conceptual Midcourse Guidance Laws for Long-Range Exoatmospheric Interceptors

Mohsen Dehghani Mohammad-abadi[1], Seyed Hamid Jalali-Naini[1]

ABSTRACT: This paper presents a comprehensive study on the performance analysis of 8 conceptual guidance laws for exoatmospheric interception of ballistic missiles. The problem is to find the effective thrust direction of interceptor for interception of short-to-super range ballistic missiles. The zero-effort miss and the generalized required velocity concept are utilized for interception of moving targets. By comparison of the 8 conceptual guidance laws, the thrust direction is suggested to be in the direction of generalized velocity-to-be-gained, or constant velocity-to-be-gained direction, rather than to be in the direction along zero-effort miss, or that of linear optimal solution for long-to-super range interception. Even for short coasting ranges, the generalized velocity-to-be-gained may be utilized because of reasonable computational burden for required velocity rather than the numerical computation for zero-effort miss or linear optimal solution with the same miss distance error. In addition, the fuel consumption of the suggested direction has less sensitivity due to estimation error in intercept time. The guidance law based on constant velocity-to-be-gained direction and the optimal solution are suitable for satellites launch vehicles and space missions.

KEYWORDS: Exoatmospheric midcourse guidance, Effective thrust direction, Zero-effort miss, Velocity-to-be-gained, Long-range interceptor, Anti-ballistic guidance.

INTRODUCTION

Exoatmospheric intercept guidance improvements are of high interest in anti-ballistic air defense systems. The main subjects in this area are focused on midcourse and terminal phases of flight for anti-ballistic interceptors. The design considerations for the midcourse guidance are different from the terminal phase one. In the midcourse phase, the on-board trajectory optimization and trajectory shaping are the main issues, whereas the noise contamination and hit probability against very-high speed targets are the key issues for a terminal guidance law (Zarchan 2012).

The literature on exoatmospheric intercept guidance laws can be categorized into intercept guidance laws against moving targets and guidance laws for space missions including ballistic missiles. Since the concepts and guidance algorithms of ballistic missiles are similar to space vehicle guidance laws, we put them in the same category. The early literature on the subject of optimal 2-point guidance for interception of moving targets is based on zero-effort miss (ZEM) in flat-Earth model (Bryson and Ho 1975). In this case, the acceleration command in the optimal energy problems is obtained proportional to ZEM vector. Precisely speaking, the commanded acceleration of optimal energy guidance laws with final constraints in linear systems is obtained in the form of the predicted error vector pre-multiplied by a gain matrix. In the case that the final position vector is only constrained, the solution simplifies to ZEM vector pre-multiplied by a time-varying gain matrix (Rusnak and Meir 1991). In a special case, if the airframe and control systems are assumed to be identical for 3 axes, the matrix gain

1.Tarbiat Modares University – Faculty of Mechanical Engineering – Aerospace Group – Tehran/Tehran – Iran.

Author for correspondence: Seyed Hamid Jalali-Naini | Tarbiat Modares University – Faculty of Mechanical Engineering – Jalal Al Ahmad Street, No 7 | P.O. box14115-111 – Tehran/Tehran – Iran | E-mail: shjalalinaini@modares.ac.ir

simplifies to a scalar, *i.e.* the commanded acceleration becomes proportional to ZEM (Rusnak and Meir 1991). Two classes of explicit guidance laws based on ZEM have been developed with different assumptions for interceptor dynamics and types of target maneuvers (Jalali-Naini 2004).

In spherical-Earth model, in spite of the assumption of a perfect control system, the optimal maneuver is not obtained in the direction of ZEM because of the non-linear nature of the gravitational acceleration. Most literature on exoatmospheric intercept problems utilized the ZEM vector as an effective direction for thrust vectoring of the interceptor (Massoumnia 1995; Feng *et al.* 2009; Li *et al.* 2013). The ZEM can be approximated in an inverse-square gravity field (Li *et al.* 2013; Mohammad-abadi and Jalali-Naini 2016) or numerically computed on-board with a reasonable integration time step as in predictive guidance scheme (Zarchan 2012). As mentioned before, even ZEM is computed exactly in the spherical-Earth model; the acceleration command along the ZEM is not, mathematically, an optimal solution.

On the other hand, guidance laws for space missions are based on the concept of required velocity and velocity-to-be-gained (Battin 1999; Martin 1965, 1966). At a first glance, the concepts of the 2 guidance categories seem to be different. The concepts of required velocity and velocity-to-be-gained can also be utilized or generalized for interception of moving targets (Jalali-Naini and Pourtakdoust 2005; Chen *et al.* 2010). The velocity-to-be-gained vector becomes proportional to ZEM when the gravitational acceleration is assumed to be constant. In a linearized inverse square gravity field, the velocity-to-be-gained vector is obtained in the form of ZEM pre-multiplied by a time-varying gain matrix (Jalali-Naini and Pourtakdoust 2007). Both ZEM and required velocity can be calculated for a linearized gravity field. Several solutions have been presented using a linearized gravity with different assumptions as treated by Newman (1996) and Deihoul and Massoumnia (2003) for interception of ballistic missiles. The ZEM was also obtained for a linear gravity considering control system dynamics and target maneuvers (Jalali-Naini 2008).

Several anti-ballistic guidance schemes were presented based on ZEM, as mentioned earlier. In these guidance schemes, the corrective maneuver is applied in the direction proportional to ZEM, but the guidance gain is modified, manipulated, and/or theoretically or empirically designed. For space missions, Battin (1999) introduced a guidance scheme in order to keep the direction of velocity-to-be-gained constant, and Sokkappa (1966) obtained an approximate optimal guidance assuming Q-matrix to be constant. Circi (2004) compared Battin's formula with the numerical optimal solution for a satellite launch vehicle. For short-range anti-ballistic interception, guidance laws based on ZEM perform well whereas for long-to-super range interception, the maneuvering direction needs to be modified to the direction of velocity-to-be-gained or possibly an optimal one. The question is: what direction should be utilized for what ranges. The present study focuses on quantifying the answer to this question, based on accuracy and some implementation issues. Fortunately, several efficient algorithms are available for on-board computation of required velocity and Q-matrix (Zarchan 2012; Arora *et al.* 2015; Ahn *et al.* 2015).

There is another type of guidance laws for space missions or interception in exoatmosphere, referred to as General Energy Management (GEM) for solid rocket motors without cut-off capability (Zarchan 2012). In this class of guidance schemes, the maneuvering direction is somewhat deviated from a desired direction, ZEM or velocity-to-be-gained, in order to manage the wasting of extra fuel of the rocket so as the space vehicle reaches the required velocity at burnout. Since our study focuses on optimal energy guidance, GEM-type guidance schemes are beyond its scope.

In this article, the performance of the midcourse phase of exoatmospheric interceptors is compared for 8 conceptual guidance schemes. It is assumed that this midcourse phase is followed by a coasting phase. In other words, the interceptor is due to reach near the target position, coasting ballistically, where a small kinetic kill vehicle (KKV) is due to separate in order to intercept its target with minimum effort.

BASIC FORMULATION

The governing equations of motion for a particle P (interceptor or target) with a given acceleration vector, $\mathbf{a}_p(t)$, are given by:

$$\dot{\mathbf{r}}_p = \mathbf{v}_p \tag{1a}$$

$$\dot{\mathbf{v}}_p = \mathbf{a}_p(t) \tag{1b}$$

where: $\mathbf{r}_p$, $\mathbf{v}_p$, and $\mathbf{a}_p$ denote position, velocity, and acceleration vectors at current time t in an inertial reference, respectively; the subscript p also represents the particle P.

The final position (at final time t_f) is obtained by integrating twice with respect to the time as follows:

$$\mathbf{r}_p(t_f) = \mathbf{r}_p(t) + t_{go}\mathbf{v}_p(t) + \int_t^{t_f}\int_t^{\xi}\mathbf{a}_p(\tau)d\tau d\xi \tag{2}$$

where: τ and ξ are dummy indices for time. By converting the preceding double integral to the single one, we have:

$$\mathbf{r}_p(t_f) = \mathbf{r}_p(t) + t_{go}\mathbf{v}_p(t) + \int_t^{t_f}(t_f - \xi)\mathbf{a}_p(\xi)d\xi \tag{3}$$

where: $t_{go} = t_f - t$ is the time-to-go until the final time.

In an exoatmospheric free-flight motion, we have $\mathbf{a}_p = \mathbf{G}_p$, where $\mathbf{G}_p$ is the gravitational acceleration, *i.e.* $\mathbf{G}_p = -\mu\mathbf{r}_p/|\mathbf{r}_p|^3$ for a spherical-Earth model, and μ is the Earth's gravitational constant. Therefore, Eq. 3 may be written in the following form:

$$\mathbf{r}_p(t_f) = \mathbf{r}_p(t) + t_{go}\mathbf{v}_p(t) + \int_t^{t_f}(t_f - \xi)\mathbf{G}_p[\mathbf{r}_p(\xi)]d\xi \tag{4}$$

The preceding equation simplifies for a special case of constant gravity, that is,

$$\mathbf{r}_p(t_f) = \mathbf{r}_p + t_{go}\mathbf{v}_p + \frac{1}{2}\mathbf{G}_p t_{go}^2 \tag{5}$$

The 3-D intercept geometry with respect to an inertial reference ($Oxyz$) is shown in Fig. 1, in which the interceptor I, having velocity $\mathbf{v}_I$, is pursuing its target T, with velocity $\mathbf{v}_T$. The interceptor and target position vectors are denoted by $\mathbf{r}_I$ and $\mathbf{r}_T$, respectively. The relative position $\mathbf{r}$ and velocity $\mathbf{v}$ for the interception problem are defined as:

$$\mathbf{r} = \mathbf{r}_T - \mathbf{r}_I \tag{6a}$$

$$\mathbf{v} = \mathbf{v}_T - \mathbf{v}_I \tag{6b}$$

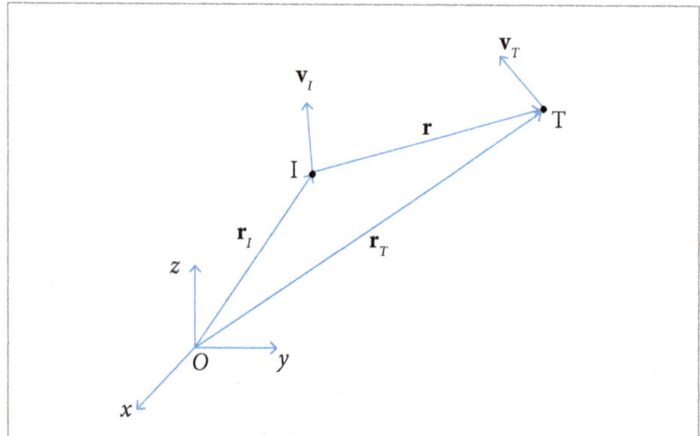

Figure 1. Engagement geometry.

The relative equations of motion are then given by:

$$\dot{\mathbf{r}} = \mathbf{v} \tag{7a}$$

$$\dot{\mathbf{v}} = \mathbf{a}_T - \mathbf{a}_I \tag{7b}$$

where: $\mathbf{a}_T$ and $\mathbf{a}_I$ are the respective target and interceptor accelerations in inertial reference. Using Eqs. 3 and 6a, the final relative position is written as:

$$\begin{aligned}\mathbf{r}(t_f) &= \mathbf{r}_T(t_f) - \mathbf{r}_I(t_f) = \mathbf{r}_T(t) + t_{go}\mathbf{v}_T(t) + \\ &+ \int_t^{t_f}(t_f - \xi)\mathbf{a}_T(\xi)d\xi - \mathbf{r}_I(t) - t_{go}\mathbf{v}_I(t) - \int_t^{t_f}(t_f - \xi)\mathbf{a}_I(\xi)d\xi\end{aligned} \tag{8}$$

Therefore, the relative formulation for miss distance is expressed as follows:

$$\mathbf{r}(t_f) = \mathbf{r}(t) + t_{go}\mathbf{v}(t) + \int_t^{t_f}(t_f - \xi)[\mathbf{a}_T(\xi) - \mathbf{a}_I(\xi)]d\xi \tag{9}$$

In an exoatmospheric free-flight motion, we have $\mathbf{a}_T = \mathbf{G}_T$ and $\mathbf{a}_I = \mathbf{G}_I$, and the substitution yields:

$$\begin{aligned}\mathbf{r}(t_f) &= \mathbf{r}(t) + t_{go}\mathbf{v}(t) + \\ &+ \int_t^{t_f}(t_f - \xi)\{\mathbf{G}[\mathbf{r}_T(\xi)] - \mathbf{G}[\mathbf{r}_I(\xi)]\}d\xi\end{aligned} \tag{10}$$

The solution of the preceding equation is not straightforward. One approach to the problem is the linearization of its non-linear term, *i.e.* gravitational acceleration, that is,

$$\mathbf{G}_I = \mathbf{G}_T + \mathbf{E}(t)\mathbf{r} \tag{11}$$

where:

$$\mathbf{E}(t) = -\left.\frac{\partial\mathbf{G}}{\partial\mathbf{r}_I}\right|_{\text{reference point}}, \quad \mathbf{G} = \mathbf{G}_T - \mathbf{G}_I \tag{12}$$

Therefore, the linearized state-space form for an exoatmospheric interceptor having thrust acceleration, $\mathbf{a}_{th}$ ($\mathbf{a}_I = \mathbf{G}_I + \mathbf{a}_{th}$), is obtained as (Newman 1996):

$$\frac{d}{dt}\begin{bmatrix}\mathbf{r}\\ \mathbf{v}\end{bmatrix} = \begin{bmatrix}\mathbf{0} & \mathbf{I}\\ -\mathbf{E}(t) & \mathbf{0}\end{bmatrix}\begin{bmatrix}\mathbf{r}\\ \mathbf{v}\end{bmatrix} + \begin{bmatrix}\mathbf{0}\\ -\mathbf{a}_{th}\end{bmatrix} \tag{13}$$

where: $\mathbf{I}$ is a 3×3 identity matrix, and

$$\mathbf{E} = \frac{\mu}{R^3}\left[\mathbf{I} - 3\frac{\mathbf{R}\mathbf{R}^T}{R^2}\right], \quad R = |\mathbf{R}| \tag{14}$$

Several assumptions can be made for the parameter **R** such as $\mathbf{R} = \mathbf{r}_I$ (Newman 1996) and $\mathbf{R} = \mathbf{r}_T$, $\mathbf{R} = (\mathbf{r}_I + \mathbf{r}_T)/2$, $\mathbf{R} = \mathbf{r}_T(t_f)$ or $\mathbf{R} = \mathbf{r}_I + \mathbf{r}_T + \mathbf{r}_T(t_f))/3$ as treated by Deihoul (2003). The author claimed that the last relation for **R** gives better results, so it is used in our comparison study.

The solution of the homogenous differential equation ($\ddot{\mathbf{r}} + \mathbf{E}\mathbf{r} = \mathbf{0}$) is obtained in terms of the state-transition matrix, $\mathbf{\Phi}(t,t_0)$, as treated by Newman (1996) when **E** is assumed to be a constant matrix, that is,

$$\mathbf{r}(t) = \mathbf{\Phi}_{11}(t,t_0)\mathbf{r}(t_0) + \mathbf{\Phi}_{12}(t,t_0)\mathbf{v}(t_0) \tag{15}$$

where: $\mathbf{\Phi}_{ij}$ is a 3 × 3 submatrix ($i, j = 1, 2$) in the following partitioned matrix form:

$$\mathbf{\Phi}(t,t_0) = \begin{bmatrix} \mathbf{\Phi}_{11} & \mathbf{\Phi}_{12} \\ \mathbf{\Phi}_{21} & \mathbf{\Phi}_{22} \end{bmatrix} \tag{16}$$

and

$$\mathbf{\Phi}_{11} = \mathbf{\Phi}_{22} = \mathbf{I} - \frac{1}{2!}\mathbf{E}(t - t_0)^2 + \frac{1}{4!}\mathbf{E}^2(t - t_0)^4 - \dots \tag{17}$$

$$\mathbf{\Phi}_{12} = \mathbf{I}(t - t_0) - \frac{1}{3!}\mathbf{E}(t - t_0)^3 + \frac{1}{5!}\mathbf{E}^2(t - t_0)^5 - \dots \tag{18}$$

$$\mathbf{\Phi}_{21} = -\mathbf{E}(t - t_0) + \frac{1}{3!}\mathbf{E}^2(t - t_0)^3 - \frac{1}{5!}\mathbf{E}^3(t - t_0)^5 + \dots \tag{19}$$

Therefore, the linearized relation of the miss distance vector in terms of the current states is given by:

$$\mathbf{r}(t_f) = \mathbf{\Phi}_{11}(t_f, t)\mathbf{r} + \mathbf{\Phi}_{12}(t_f, t)\mathbf{v} \tag{20}$$

ZERO-EFFORT MISS

The ZEM at the current time, $\mathbf{ZEM}(t)$, is the distance that the interceptor would miss its target position if the interceptor made no corrective maneuver ($\mathbf{a}_{th}$) after the time t (Zarchan 2012), that is,

$$\mathbf{ZEM}(t) = \mathbf{r}_I^*(t_f) - \mathbf{r}_I(t_f)\,|_{\mathbf{a}_{th}(\xi)=\mathbf{0} \text{ for } \xi \geq t} \tag{21}$$

where: $\mathbf{r}_I^*(t_f)$ is the desired final position of the interceptor.

For an exoatmospheric interceptor ($\mathbf{a}_I = \mathbf{G}_I + \mathbf{a}_{th}$), we have:

$$\mathbf{r}_I(t_f)\,|_{\mathbf{a}_{th}(\xi)=\mathbf{0} \text{ for } \xi \geq t} = \mathbf{r}_I + t_{go}\mathbf{v}_I + + \int_t^{t_f}(t_f - \xi)\mathbf{G}[\mathbf{r}_I(\xi)\,|_{\mathbf{a}_{th}=\mathbf{0}}]d\xi \tag{22}$$

The substitution of Eq. 22 into Eq. 21 yields:

$$\mathbf{ZEM}(t) = \mathbf{r}_I^*(t_f) - \mathbf{r}_I - t_{go}\mathbf{v}_I - \int_t^{t_f}(t_f - \xi) \mathbf{G}[\mathbf{r}_I(\xi)\,|_{\mathbf{a}_{th}=\mathbf{0}}]d\xi \tag{23}$$

For a free-falling target, *i.e.* $\mathbf{a}_T = \mathbf{G}_T$, using Eq. 10, the ZEM relation can also be expressed in relative coordinates as follows:

$$\mathbf{ZEM}(t) = \mathbf{r} + t_{go}\mathbf{v} + \int_t^{t_f}(t_f - \xi) \{\mathbf{G}[\mathbf{r}_T(\xi)] - \mathbf{G}[\mathbf{r}_I(\xi)\,|_{\mathbf{a}_{th}=\mathbf{0}}]\}d\xi \tag{24}$$

where: $\mathbf{r}_I^*(t_f) = \mathbf{r}_T(t_f)$.

The preceding relation simplifies for a special case of constant gravity or for the case that the interceptor is assumed to be near its target as treated by Massoumnia (1995), that is,

$$\mathbf{ZEM}(t) = \mathbf{r} + t_{go}\mathbf{v} \tag{25}$$

Using Eq. 20, the linearized ZEM relation is given by:

$$\mathbf{ZEM}_{\text{Lin}}(t) = \mathbf{\Phi}_{11}(t_f - t)\mathbf{r} + \mathbf{\Phi}_{12}(t_f - t)\mathbf{v} \tag{26}$$

Two different definitions of ZEM are utilized in the guidance theory. In basic definition, the ZEM is defined as a miss distance vector without further control effort. The intercept time is not imposed to the intercept problem and it is the time of the nearest distance between an interceptor and its target without further control effort. The second definition is based on a specified final time and comes from linear optimal guidance laws with the assumption of a fixed final time. The ZEM vector for the basic definition is, here, denoted by $\mathbf{ZEM}_{\min}$ whereas it is denoted by **ZEM** for the second definition. The final time, the time of the nearest distance denoted by $t^*_{f\text{ZEM}}$, for the basic definition is obtained by $\partial\,|\,\mathbf{ZEM}\,|\,/\,\partial t_f = 0$. For example, in a special case of constant gravity, from Eq. 25, we have:

$$\frac{\partial\,|\,\mathbf{ZEM}\,|}{\partial t_{go}} = 0 \Rightarrow t^*_{go_{\text{ZEM}}} = -\frac{\mathbf{r} \cdot \mathbf{v}}{|\,\mathbf{v}\,|^2} \tag{27}$$

where: $t^*_{go\text{ZEM}} = t^*_{f\text{ZEM}} - t$.

It is worth noting that the component of ZEM perpendicular to the interceptor-target line-of-sight (LOS), $\mathbf{ZEM}_{\text{PLOS}}$, may be replaced for **ZEM** in a guidance formulation as treated by Zarchan (2012).

GENERALIZED REQUIRED VELOCITY

The required velocity, $\mathbf{v}_R$, for Lambert's problem is defined as an instantaneous velocity, required to satisfy the final position constraint in a specified final time (Battin 1999). This concept is well-known in space missions and surface-to-surface applications. The implementation of guidance laws based on the required velocity may be implicit or explicit.

The required velocity concept may be generalized for an intercept problem against a moving target in the endoatmosphere considering interceptor dynamics. The interceptor desired velocity $\mathbf{v}^*$ is the velocity that makes ZEM equal to zero. This desired velocity is referred to as generalized required velocity (Jalali-Naini and Pourtakdoust 2005). The interceptor dynamics is, here, assumed to be perfect, the interceptor moves in the exoatmosphere, and a moving target is considered. For brevity, we use the term "required velocity" instead of "generalized required velocity".

For example, the required velocity for the case of constant gravity is obtained from Eq. 23 as:

$$\mathbf{v}_R = \frac{\mathbf{r}_T(t_f) - \mathbf{r}}{t_{go}} - \frac{1}{2}\mathbf{G}t_{go} \tag{28}$$

Therefore, the relation between the required velocity and ZEM for the case of constant gravity is simply obtained as:

$$\mathbf{ZEM} = \mathbf{v}_g t_{go} \tag{29}$$

where: $\mathbf{v}_g$ is referred to as the velocity-to-be-gained ($\mathbf{v}_g = \mathbf{v}_R - \mathbf{v}_I$).

For a spherical-Earth model, the required velocity causes Eq. 23 equal to zero, that is,

$$\begin{aligned}\mathbf{0} = \mathbf{r}_T(t_f) - \mathbf{r}_I - t_{go}\mathbf{v}_R - \\ - \int_t^{t_f} (t_f - \xi)\mathbf{G}\{\mathbf{r}_I[\mathbf{r}_I(t), \mathbf{v}_R(t), \xi]\}d\xi\end{aligned} \tag{30}$$

The preceding relation may be solved approximately for the guidance problem, which is beyond the scope of the present study.

CONCEPTUAL GUIDANCE LAWS

It is necessary to determine the effective direction of interceptor thrust vector for short-to-super range exoatmospheric intercept problem. The thrust direction is determined by a conceptual guidance law (GL). After calculation of the effective thrust direction, a steering law is needed to convert the errors into commended body rates. The interceptor is assumed to be non-throttleable with thrust cutoff capability. The conceptual guidance laws are, here, categorized in 5 main classes, namely, guidance laws based on ZEM, guidance laws based on linearized ZEM, guidance laws based on generalized velocity-to-be-gained, guidance laws based on constant direction for velocity-to-be-gained, and optimal solution.

GUIDANCE LAWS BASED ON ZERO-EFFORT MISS

In this case, the thrust acceleration is assumed to be applied perfectly in the direction of ZEM. Three guidance schemes may be considered regarding to 2 definitions of ZEM and also the component of ZEM perpendicular to LOS as follows:

$$\mathbf{a}_{th} = |\mathbf{a}_{th}|\,\mathbf{e}_{\mathrm{ZEM}} \quad \text{for } |\mathbf{ZEM}| > \varepsilon \tag{31}$$

$$\mathbf{a}_{th} = |\mathbf{a}_{th}|\,\mathbf{e}_{\mathrm{ZEM}_{\min}} \quad \text{for } |\mathbf{ZEM}_{\min}| > \varepsilon \tag{32}$$

$$\mathbf{a}_{th} = |\mathbf{a}_{th}|\,\mathbf{e}_{\mathrm{ZEM}_{\mathrm{PLOS}}} \quad \text{for } |\mathbf{ZEM}_{\mathrm{PLOS}}| > \varepsilon \tag{33}$$

where: $\mathbf{e}_{\mathrm{ZEM}}$, $\mathbf{e}_{\mathrm{ZEM}_{\min}}$ and $\mathbf{e}_{\mathrm{ZEM}_{\mathrm{PLOS}}}$ are the unit vectors along **ZEM**, $\mathbf{ZEM}_{\min}$, and $\mathbf{ZEM}_{\mathrm{PLOS}}$, respectively; ε is an allowable miss distance, determined from practical considerations.

In each guidance law, the powered phase of flight is terminated when its corresponding $|\mathbf{ZEM}|$ becomes equal or less than ε and then it is followed by a coasting phase until intercept. The component of **ZEM** normal to LOS is calculated by:

$$\mathbf{ZEM}_{\mathrm{PLOS}} = \mathbf{ZEM} - (\mathbf{ZEM}\cdot\mathbf{e}_r)\mathbf{e}_r \tag{34}$$

where: $\mathbf{e}_r = \mathbf{r}/r$ is the unit vector along LOS ($r = |\mathbf{r}|$).

To calculate the time-to-go until intercept in $\mathbf{ZEM}_{\mathrm{PLOS}}$ relation, the component of ZEM along LOS is imposed to be zero. For a special case of a free-falling target in a flat-Earth model, the relation $t_{go} = -r/\dot{r}$ zeros out the LOS component of ZEM.

GUIDANCE LAWS BASED ON LINEARIZED ZERO-EFFORT MISS

Here, conceptual guidance laws are given using linearized relations for the problem. The first is similar to Eq. 31, but with linearized relation for ZEM, that is,

$$\mathbf{a}_{th} = |\mathbf{a}_{th}|\,\mathbf{e}_{\mathrm{ZEM}_{\mathrm{Lin}}} \quad \text{for } |\mathbf{ZEM}_{\mathrm{Lin}}| > \varepsilon \tag{35}$$

where: $\mathbf{e}_{\mathrm{ZEM}_{\mathrm{Lin}}}$ is the unit vector along $\mathbf{ZEM}_{\mathrm{Lin}}$.

The second conceptual guidance is based on the linear optimal guidance law (OGL) obtained by Deihoul and Massoumnia (2003). Their linearized OGL may be expressed in the following form:

$$\mathbf{U}_{\mathrm{Lin}} = \mathbf{M}(t_{bgo}, t_{go}, r_T, r_I)\mathbf{ZEM}_{\mathrm{Lin}} \tag{36}$$

where: $\mathbf{U}_{\mathrm{Lin}}$ is the optimal thrust vector for a throttleable rocket motor; $\mathbf{M}$ is a 3 × 3 matrix that causes the thrust vector to deviate from $\mathbf{ZEM}_{\mathrm{Lin}}$ direction and change its magnitude as well; t_{bgo} is the time-to-go until burnout.

The relation of matrix $\mathbf{M}$ is available in Deihoul and Massoumnia (2003). The thrust direction in the second conceptual guidance is applied in the direction of $\mathbf{U}_{\mathrm{Lin}}$ for non-throttleable rocket motors as follows:

$$\mathbf{a}_{th} = |\mathbf{a}_{th}|\, \mathbf{e}_{\mathbf{U}_{\mathrm{Lin}}} \quad \text{for } |\mathbf{ZEM}_{\mathrm{Lin}}| > \varepsilon \tag{37}$$

To compare the performances of the guidance schemes precisely, a third relation based on the exact calculation of ZEM in a spherical-Earth model is written as follows:

$$\mathbf{a}_{th} = |\mathbf{a}_{th}|\, \mathbf{e}_{\mathrm{U}} \quad \text{for } |\mathbf{ZEM}| > \varepsilon \tag{38}$$

where: $\mathbf{e}_{\mathrm{U}}$ is the unit vector along $\mathbf{U}$ calculated by:

$$\mathbf{U} = \mathbf{M}(t_{bgo}, t_{go}, r_T, r_I)\mathbf{ZEM} \tag{39}$$

where $\mathbf{ZEM}$ is computed numerically.

GUIDANCE LAW BASED ON VELOCITY-TO-BE-GAINED

In space missions a class of guidance laws is based on required velocity with the desired thrust acceleration along the velocity-to-be-gained. The required velocity concept and velocity-to-be-gained can be generalized for interception of a moving target, as mentioned before. The conceptual guidance law is then given by:

$$\mathbf{a}_{th} = |\mathbf{a}_{th}|\, \mathbf{e}_{v_g} \quad \text{for } |\mathbf{v}_g| > \varepsilon_v \tag{40}$$

where: $\mathbf{e}_{v_g}$ is the unit vector along the velocity-to-be-gained; $\boldsymbol{\varepsilon}_v$ is an allowable velocity-to-be-gained error.

It is worth noting that the preceding conceptual guidance law is equivalent to Eq. 31 for constant gravity assumption as is evident from Eq. 29.

GUIDANCE BASED ON CONSTANT DIRECTION OF VELOCITY-TO-BE-GAINED

An effective direction of thrust acceleration in space missions is obtained by satisfying the relation $\dot{\mathbf{v}}_g \times \mathbf{v}_g = \mathbf{0}$ as follows (Battin 1999):

$$\mathbf{a}_{th} = \mathbf{b} + (q - \mathbf{b} \cdot \mathbf{e}_{v_g})\mathbf{e}_{v_g} \tag{41}$$

where:

$$q = \sqrt{|\mathbf{a}_{th}|^2 - |\mathbf{b}|^2 + (\mathbf{b} \cdot \mathbf{e}_{v_g})^2} \tag{42}$$

$$\mathbf{b} = -\mathbf{Q}\mathbf{v}_g, \; \mathbf{Q} = \frac{\partial \mathbf{v}_R}{\partial \mathbf{r}_I} \tag{43}$$

This conceptual guidance law causes the direction of velocity-to-be-gained to be fixed in inertial space.

RESULTS AND DISCUSSION

To compare the performance of conceptual guidance laws, a nonlinear flight simulation is utilized. The interceptor and its target are taken as particles in vertical planar motion with perfect dynamics, *i.e.* the thrust acceleration is assumed to be exactly in the desired direction of computed thrust direction, without any error or delay.

First, guidance laws (Eqs. 31 – 33) with different ZEM definitions, *i.e.* **ZEM**, $\mathbf{ZEM}_{\mathrm{min}}$, and also the normal component of **ZEM**, $\mathbf{ZEM}_{\mathrm{PLOS}}$, are compared in a flat-Earth model with constant gravity. The interceptor is located at origin (0,0) with $a_{th} = 50$ m/s^2 ($a_{th} = |\mathbf{a}_{th}|$). The fuel consumption, $\Delta V = |\mathbf{a}_{th}|\, t_{co}$, of the mentioned conceptual guidance laws are shown in Figs. 2 – 4, where t_{co} is the thrust cut-off time, applied when $|\mathbf{ZEM}| \leq \varepsilon$. In these figures, the solid lines indicate the fuel consumption when the thrust vector is applied along **ZEM** direction. The circle and square signs indicate the values of fuel consumption when the thrust direction is applied along $\mathbf{ZEM}_{\mathrm{min}}$ and $\mathbf{ZEM}_{\mathrm{PLOS}}$, respectively, for their corresponding resulted final times. In Fig. 2, the fuel consumption is depicted versus predetermined final time for 2 cases of an initial 0 velocity, $\mathrm{v}_I(0) = 0$, and a vertical velocity of $\mathrm{v}_{I_z}(0) = 1$ km/s for a stationary target at $\mathbf{r}_T = [500 \quad 0]^{\mathrm{T}}$km. As seen in Fig. 2, the fuel consumption depends on the value of the final time when the thrust vector is applied along ZEM.

Also, the minimum fuel consumption is not occurred necessarily when $\mathbf{ZEM}_{min}$ is utilized; however, it may occur for a special case. It should be noted that an appropriate final time (or time-to-go until intercept) is estimated in practice.

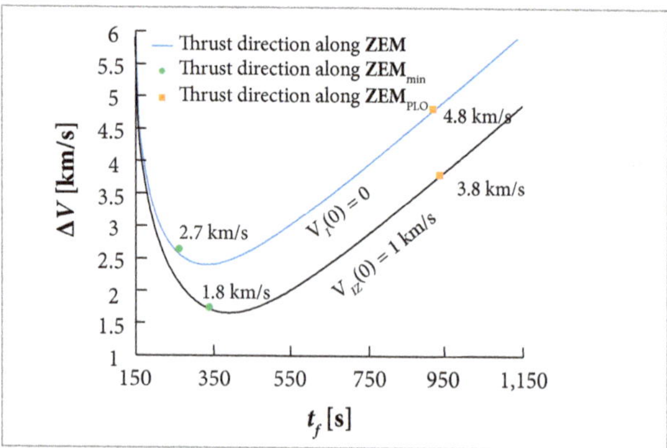

Figure 2. Fuel consumption versus final time for interception of a stationary target in flat-Earth model.

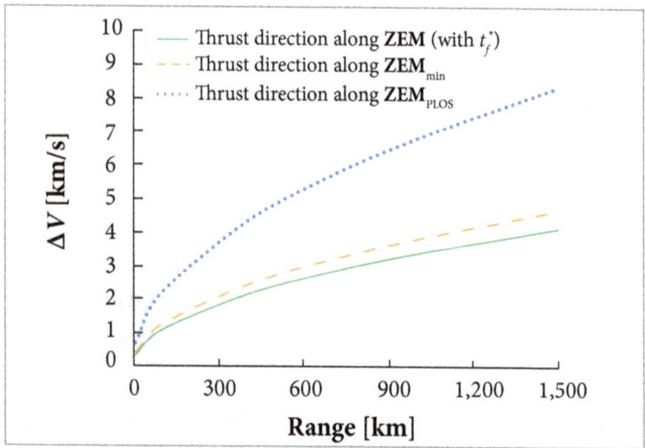

Figure 3. Fuel consumption versus range for interception of a stationary target in flat-Earth model.

In order to investigate more precisely, the fuel consumption is drawn *versus* range in Fig. 3 for the 3 mentioned guidance schemes when the interceptor launches from rest. Other initial values and parameters are similar to Fig. 2. As shown in Figs. 2 and 3, the fuel consumption is highly increased using guidance law (Eq. 33) based on $\mathbf{ZEM}_{PLOS}$. In Fig. 3, the minimum fuel consumption for the case of thrust direction along **ZEM** is computed by setting the optimum value, t_f^*, for intercept time. In other words, t_f^* is the intercept time for minimum ΔV when the thrust acceleration is imposed along **ZEM**. The value of ε is, here, chosen as 1 m due to numerical errors.

In the next step, the performances of the conceptual guidance laws are compared in Figs. 4a and 4b for a free-falling target at an initial altitude of 100 km, having a minimum required velocity to hit the origin in a flat-Earth model. Initial values and parameters are similar to Fig. 2, except target position. The target range is taken 100 and 500 km for Figs. 4a and 4b, respectively. A vertical solid line has been drawn for each of Figs. 4a and 4b, showing the maximum possible final time (t_{fmax}), *i.e.* the time of hitting origin (0,0) by the free-falling target. The region at the right-hand side of this vertical line is not acceptable, because the interception of a free-falling target occurs behind the origin in negative altitudes. This is an important difference between the two cases of stationary and free-falling targets.

According to our analysis for a flat-Earth model, the capture criteria for the conceptual guidance law based on $\mathbf{ZEM}_{PLOS}$ are highly restricted comparing to the conceptual guidance law based on **ZEM**, at least, for non-throttleable rockets.

We are now to study the performance of conceptual guidance laws (Eqs. 31, 32, and 40) for a spherical-Earth model in Figs. 5 – 14 against stationary targets. The interceptor fires

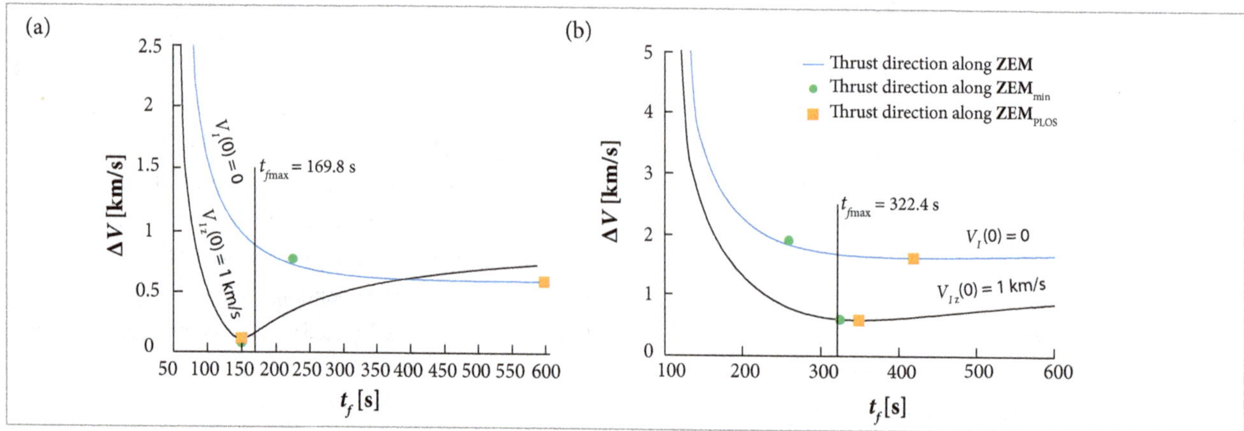

Figure 4. Fuel consumption versus final time for interception of a free-falling target in flat-Earth model. (a) initial range = 100 km; (b) initial range = 500 km.

from position $\mathbf{r}_{I_0} = [6400 \quad 0]^T$ km with a velocity of 1 km/s along near vertical horizon (89°) and $a_{th} = 50$ m/s². The initial distance of target from the earth center is taken $r_T = 6{,}500$ km.

First, the fuel consumptions of these guidance laws are plotted in Figs. 5a; 5b and 5c for a target at range angles of 10; 40 and 70°, respectively. As seen in the figures, if the thrust direction is applied along the direction of $\mathbf{v}_g$, the fuel consumption is considerably reduced for long-range applications. To investigate the trajectory of the interceptor based on ZEM direction, four scenarios with different final times are selected. These points, namely, S_1; S_2; S_3 and S_4 are assigned in Fig. 5a for different specified final times when the range angle is 10°.

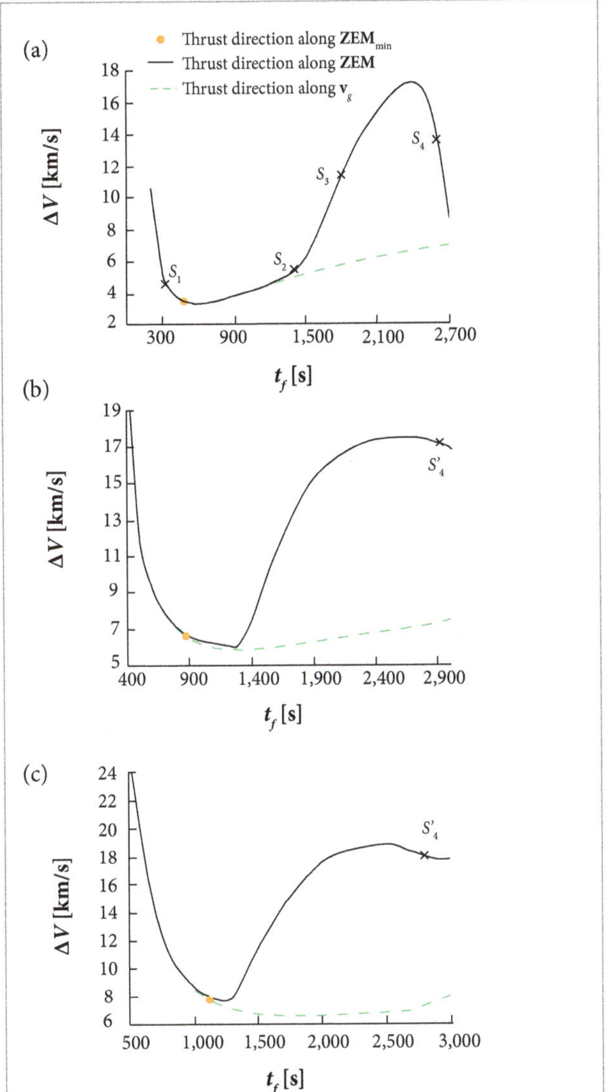

Figure 5. Fuel consumption versus final time for interception of a stationary target in spherical-Earth model. (a) range angle = 10°; (b) range angle = 40°; (c) range angle = 70°.

The typical interceptor trajectories of the mentioned scenarios are illustrated in Figs. 6a, 6b, 6c, and 6d for the points S_1, S_2, S_3, and S_4, respectively, and compared to interceptor trajectories based on $\mathbf{v}_g$ direction with the same final times. The interceptor trajectories based on ZEM and $\mathbf{v}_g$ are viewed by solid and dashed lines, respectively. The thrust cut-off time, t_{co}, is also observed for each trajectory in the figures. In the case of ZEM-based trajectory, increasing total flight time causes an extra revolution of trajectory around the earth center, as shown in Fig. 6d. The maximum limit of total flight time to avoid this phenomena is assigned with S_4 in Fig. 5a and with S_4' in Figs. 5b and 5c with different range angles. The typical interceptor trajectories with final times larger than that of S_4' in Figs. 5b and 5c, are similar to Fig. 6d for ZEM-based trajectories. Moreover, Fig. 7 shows that the fuel consumption of $\mathbf{v}_g$-based guidance law is less sensitive to the estimation of total flight time. In this figure, the total flight time is considered about the total flight time for

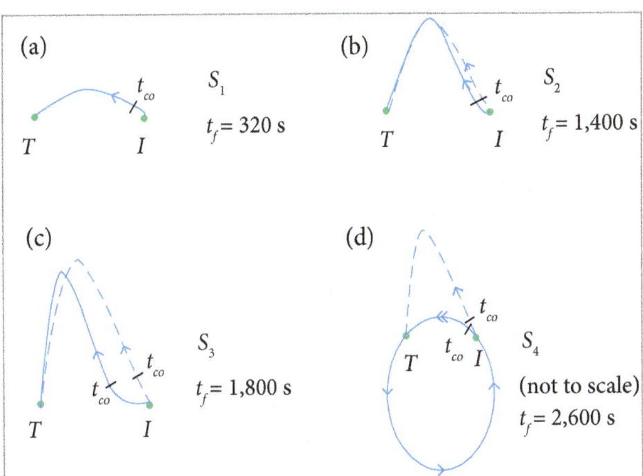

Figure 6. Typical interceptor trajectories for initial range angle = 10° and different values of total flight times (solid line: thrust vector along ZEM direction; dashed line: thrust vector along $\mathbf{v}_g$ direction).

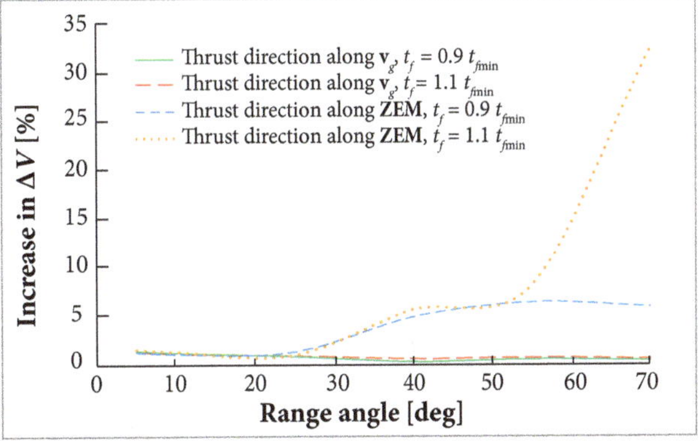

Figure 7. Increase in ΔV for guidance laws based on $\mathbf{v}_g$ and ZEM.

minimum ΔV for each guidance scheme, and it is denoted by $t_{f_{min}}$, which is obtained for each range.

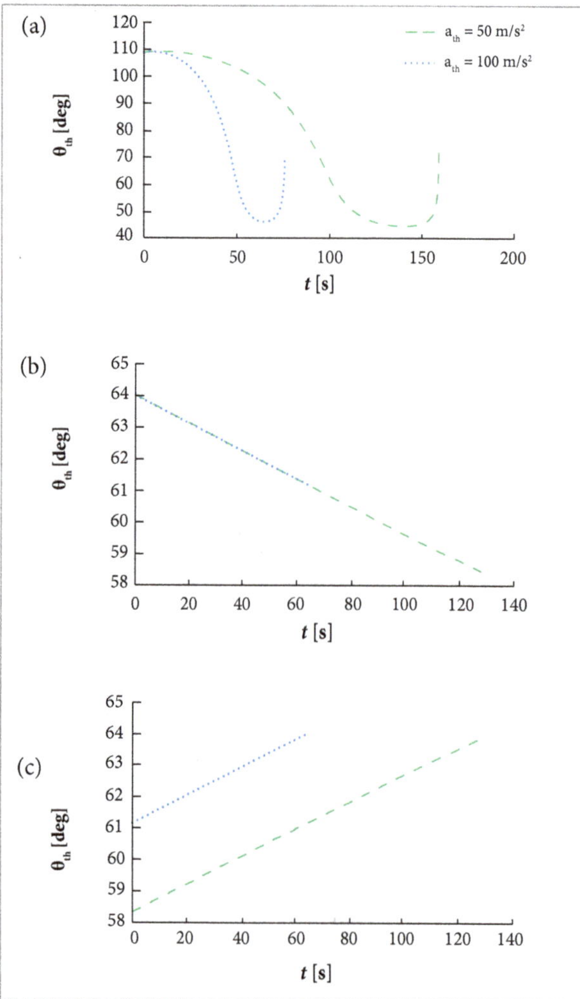

Figure 8. Thrust angle profiles with respect to the equatorial plane for interception of a stationary target for minimum energy orbit with initial range angle = 10°. (a) thrust vector along ZEM direction; (b) thrust vector along $\mathbf{v}_g$ direction; (c) constant $\mathbf{v}_g$ direction.

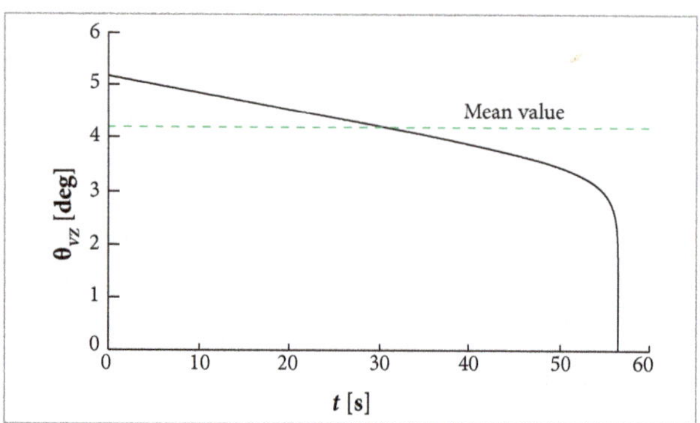

Figure 9. History of the angle between $\mathbf{v}_g$ and ZEM for thrust vector along ZEM direction (minimum energy orbit with a range angle of 10°).

The total rotation of the thrust direction of conceptual guidance laws (Eqs. 31, 32, and 40) can be seen in Figs. 8a, 8b and 8c for different values of thrust acceleration of 50 and 100 m/s² when the range angle is 70°. The final times are chosen

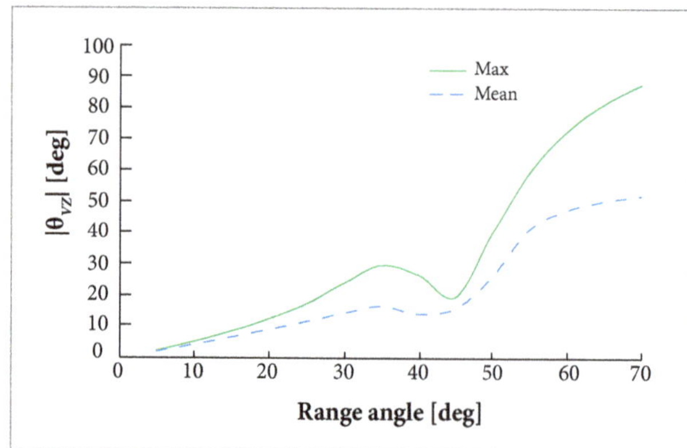

Figure 10. Max and mean of the absolute value of the angle between $\mathbf{v}_g$ and ZEM versus range angle for thrust vector along ZEM direction (minimum energy orbit).

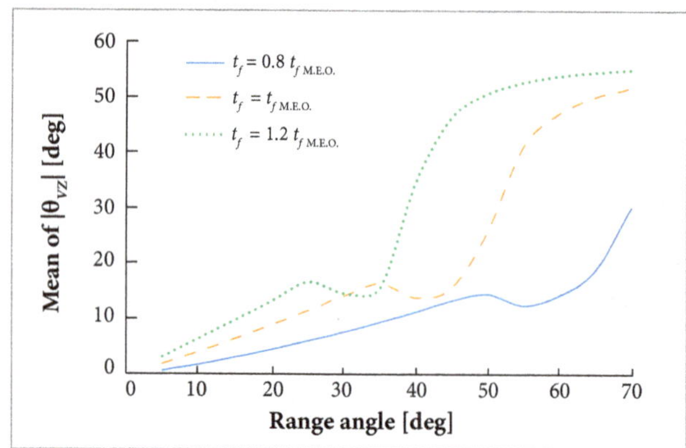

Figure 11. Mean of the absolute value of the angle between $\mathbf{v}_g$ and ZEM versus range angle for thrust vector along ZEM direction with different total final times.

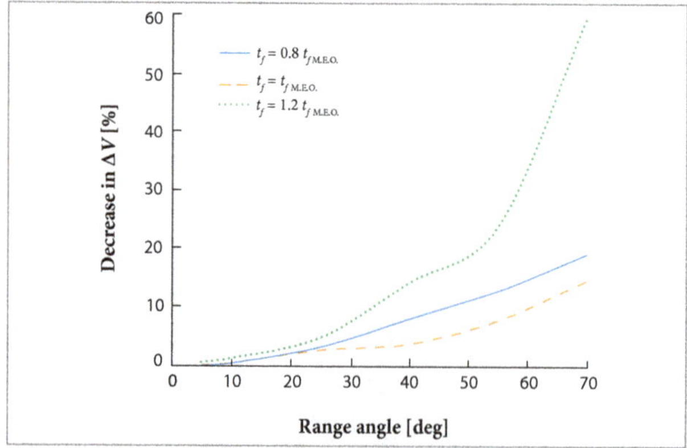

Figure 12. Decrease in ΔV for $\mathbf{v}_g$- based guidance with respect to ZEM-based guidance for minimum and non-minimum energy orbit.

for minimum energy orbit. Other parameters and initial values are similar to Fig. 5. In these figures, θ_{th} is the angle of the thrust acceleration with respect to the equatorial plane. As can be seen in these figures, the guidance law based on **ZEM** has a larger total rotation of the thrust vector than those of the two guidance laws based on $\mathbf{v}_g$. Also, it is observed that the rate of change of θ_{th} is nearly constant for the two guidance laws based on $\mathbf{v}_g$. This property phenomena can be used for implementation of the guidance laws based on $\mathbf{v}_g$.

An important question comes from the implementation point of view: what is the typical value of the angle between **ZEM** and $\mathbf{v}_g$? If this value is large enough to overwhelm the control system tracking error and estimation error of required velocity in the presence of target tracking error, the performance study of the guidance law can go ahead for this purpose.

First, the angle between $\mathbf{v}_g$ and **ZEM**, denoted by θ_{VZ}, is depicted in Fig. 9 *versus* time for an interceptor when its thrust acceleration is applied along **ZEM** for a range angle of 10°. The maximum and the mean of $|\theta_{VZ}|$, $\int_0^{t_{co}} |\theta_{VZ}|\, dt/t_{co}$, are observed by solid and dashed lines, respectively. The maximum and mean values of $|\theta_{VZ}|$ are shown in Fig. 10 *versus* range angle. The final times in Figs. 9 and 10 are chosen for minimum energy orbit, *i.e.* minimizing the required velocity. To investigate more precisely, the mean value of $|\theta_{VZ}|$ is plotted in Fig. 11 *versus* range angle for a deviation of ±20% with respect to the final time of minimum energy orbit. As expected, the values of θ_{VZ} are large enough to overcome noisy measurements and control system tracking error for medium-to-super range applications. The effect of this deviation on ΔV can be viewed in Fig. 12 where the comparison is made with respect to the fuel consumption of the ZEM-based guidance law.

The effect of target altitude is investigated in Fig. 13. First, the fuel consumption is plotted *versus* target radial position for a range angle of 10° for the minimum energy orbit. As can also be seen in Fig. 5a, there is little difference between the fuel consumptions of $\mathbf{v}_g$- and **ZEM**-based guidance laws. This difference is increased by increasing the range angle. For example, the difference is shown in Fig. 13b for the range angle of 40°. Moreover, the maximum value of the angle between $\mathbf{v}_g$ and **ZEM** versus target radial position can be viewed in Figs. 14a and 14b for minimum and non-minimum energy orbits,

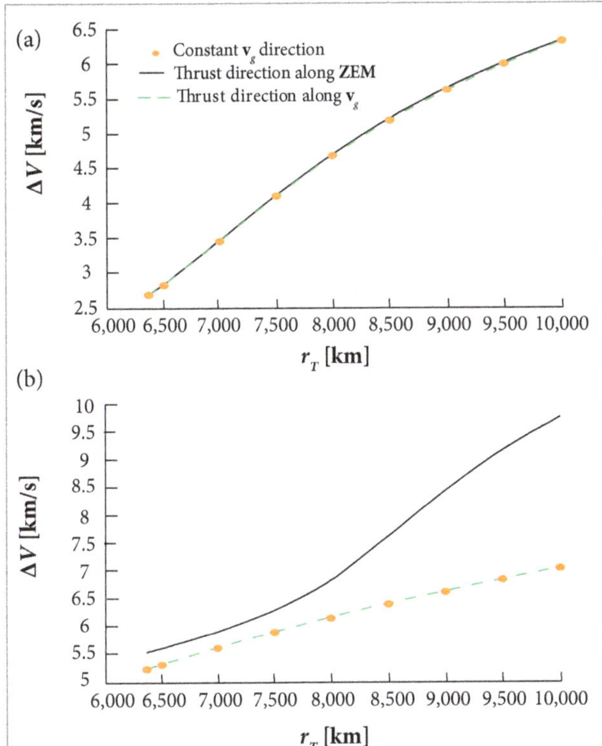

Figure 13. Fuel consumption versus target radial position for interception of a stationary target in spherical-Earth model. (a) range angle = 10°; (b) range angle = 40°.

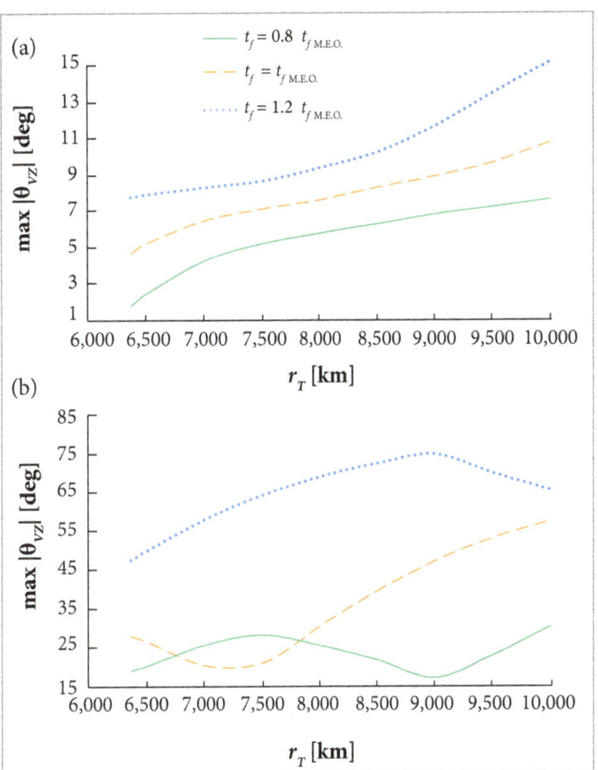

Figure 14. Max of the absolute value of the angle between $\mathbf{v}_g$ and ZEM versus target radial position for thrust vector along ZEM direction. (a) range angle = 10°; (b) range angle = 40°.

e.g. $t_f = (1\pm 0.2)t_{fM.E.O.}$, respectively, where M.E.O. is the mean energy orbit.

After a preliminary study of conceptual guidance laws for flat-Earth model and also stationary targets in spherical-Earth model, we focus on free-falling targets in the case of spherical-Earth model. The interceptor initial position, velocity, and acceleration due to thrust is taken similar to the studied case of spherical-Earth model with stationary targets.

First, the guidance law based on linearized ZEM, Eq. 26, is compared to the guidance based on the direction computed by linear optimal guidance law (Eq. 36) as obtained by Deihoul and Massuomnia (2003). The miss distance and fuel consumption of the two guidance schemes are plotted in Figs. 15 and 16 *versus* final time for a target at a range angle of 40°, having the required velocity of minimum energy to hit the initial position of the interceptor. The initial distance of target from the Earth center is taken $r_T = 6{,}500$ km. These 2 guidance laws produce nearly the same results; however, the little differences in the results cannot appear properly in Figs. 15 and 16 because of the scale of these figures. These results are obtained by setting an optimized value of $t_b = 60$ s in Eq. 37 for the gain matrix **M**. This analysis turns out the gain matrix **M** do not give a significant improvement on the performance of the ZEM-based guidance schemes for an initial range angle less than 40°.

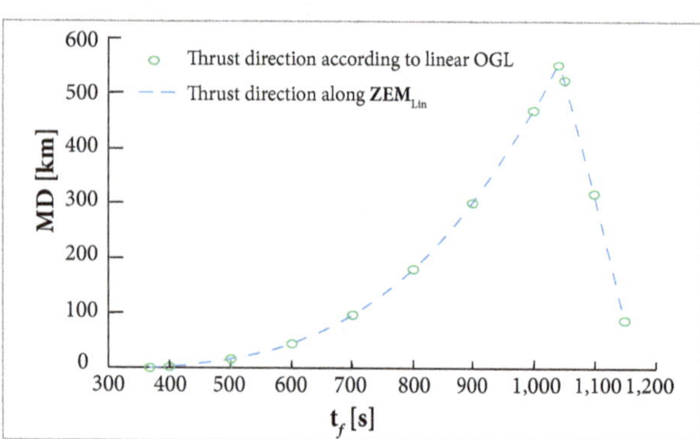

Figure 15. Miss distance (MD) versus final time for interception of a free-falling target in spherical-Earth model using linearized ZEM relation (initial range angle = 40°).

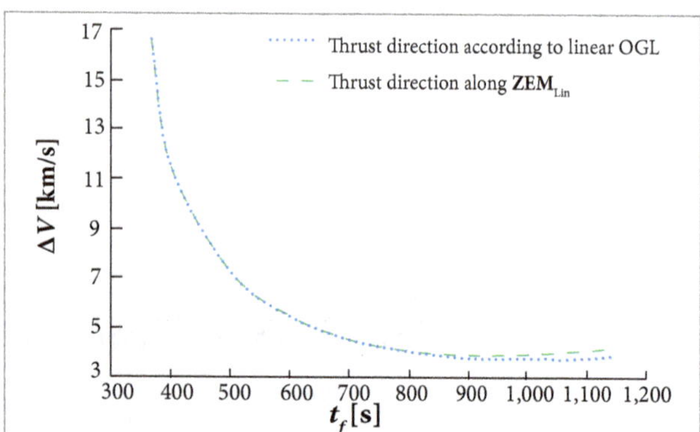

Figure 16. Fuel consumption versus final time for interception of a free-falling target in spherical-Earth model (initial range angle = 40°).

To study more precisely, the initial distance of the target from the earth center increases to 7,500 km. First, the value of t_b is chosen 100 s by the performance analysis based on Fig. 17 for an initial range angle of 90° with 3 different values of total final times. In the next step, the fuel consumptions of four conceptual guidance laws, *i.e.* guidance laws based on $\mathbf{ZEM}_{\text{min}}$, based on **ZEM**, based on the direction of linear optimal problem, and based on velocity-to-be-gained are compared together. For this purpose, the fuel consumption is depicted *versus* final time in Figs. 18a; 18b and 18c for 3 different initial range angles of 60°; 90° and 120°, respectively. The value of t_b is optimized for each range angle. Other parameters and initial values are taken as before. For a typical range comparison, the interceptor travels 29.7° (3,301 km) for an initial range angle of 90° (10,002 km) as plotted in Fig. 19 for a total flight time of 1,600 s. The thrust direction along the velocity-to-be-gained produces less fuel consumption among the other mentioned guidance laws as shown in the Fig. 18. For short range applications, the angle between ZEM and velocity-to-be-gained is negligible as shown in Fig. 20. The angle between the 2 directions increases by increasing the initial range angle. The effect of matrix gain of **M** is also increased by increasing the range angle, but for these ranges the direction of the velocity-to-be-gained produces a considerable decrease on fuel consumption. Besides, the estimation error of time-to-go for guidance laws based on velocity-to-be-gained has less sensitivity than that of ZEM-based guidance schemes.

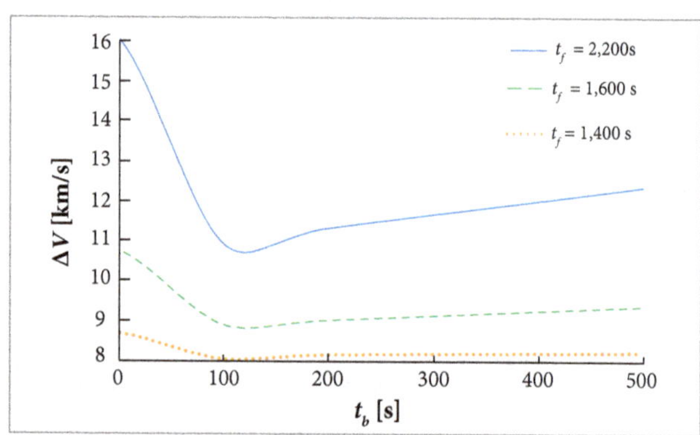

Figure 17. Fuel consumption versus burnout time for interception of a free-falling target in spherical-Earth model, when thrust vector is along linearized ZEM direction (initial range angle = 90°).

The angle of thrust direction with respect to the equatorial plane is plotted in Fig. 21 *versus* time for 3 guidance laws. As can be seen, there is a little difference between the velocity-to-be-gained direction and the direction of guidance laws based on constant $\mathbf{v}_g$ direction. The maximum difference for an initial range angle of 70° is about 5.26°. In addition, the rate of change of thrust direction for guidance laws based on velocity-to-be-gained are very small comparing to that of the guidance laws based on ZEM. To investigate more precisely the difference between the thrust angles under the 2 conceptual guidance laws based on velocity-to-be-gained, θ_{BV}, is shown in Fig. 22 in three forms, *i.e.* max of $|\theta_{BV}|$, mean of $|\theta_{BV}|$, and mean of θ_{BV} for a minimum energy orbit and a stationary target. Initial values for interceptor and its target are the same as Figs. 5 – 7. As seen in Fig. 22, the mean value of θ_{BV} is 1.11° for an initial

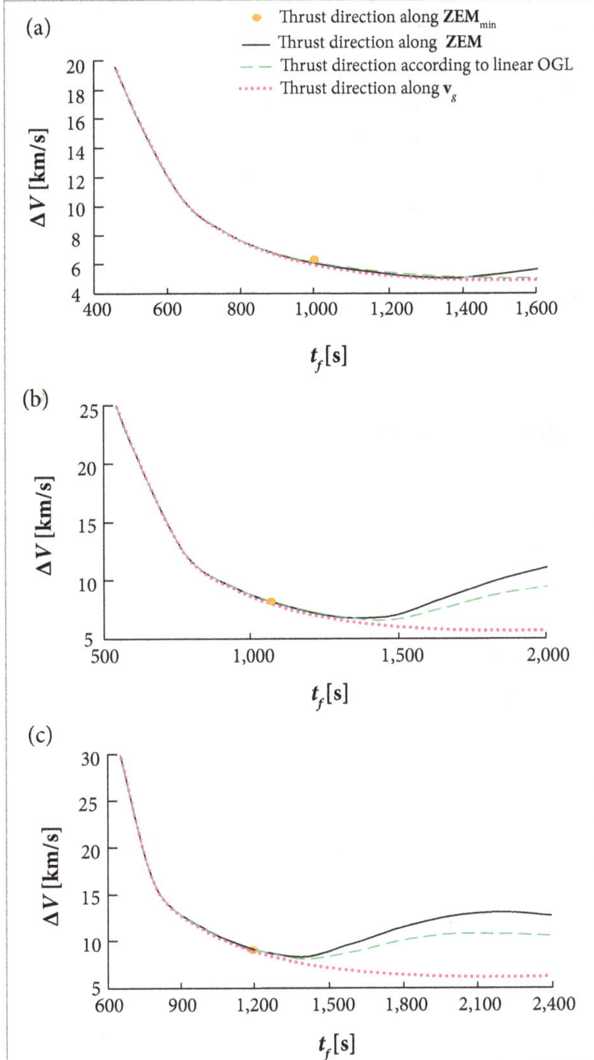

Figure 18. Compassion of fuel consumption of four conceptual guidance laws for interception of a free-falling target in spherical-Earth model. (a) initial range angle = 60°; (b) initial range angle = 90°; (c) initial range angle = 120°.

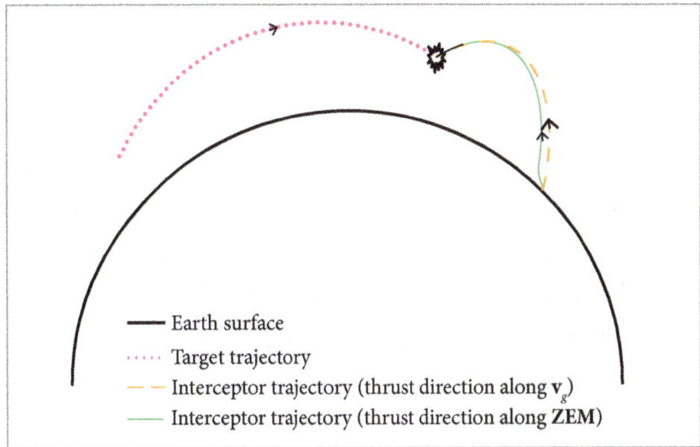

Figure 19. Interceptor and its target trajectories for initial range angle of 90° and total flight time of 1,600 s.

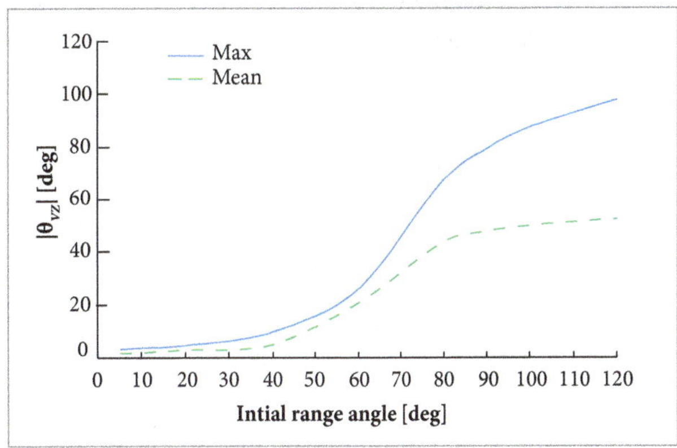

Figure 20. Maximum and mean of the absolute value of the angle between $\mathbf{v}_g$ and ZEM versus initial range angle for interception of a free-falling target (thrust vector along ZEM direction, minimum energy orbit).

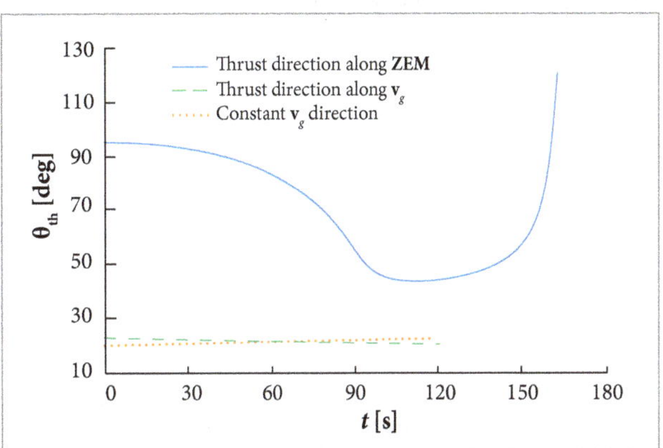

Figure 21. Comparison of thrust angle with respect to the equatorial plane for 3 conceptual guidance laws (initial range angle = 70°; a_{th} = 50 m/s²).

range angle of 70°. Their fuel consumptions are compared in Fig. 23 versus initial range angle for minimum and non-minimum energy orbit. For example, the guidance based on the constant $\mathbf{v}_g$ direction causes a decrease of 0.32% in fuel consumption for minimum energy orbit with respect to the guidance law based on $\mathbf{v}_g$ direction for a range angle of 180°. As seen in Fig. 22, below the initial range angle of 10.3°, max $|\theta_{BV}|$ is less than 1°. However, improving the performance of a guidance law is always of interest by a modified formulation without any additional hardware or extra cost. Fortunately, several iterative and approximate methods are available in literature to calculate the required velocity and Q-matrix for Lambert's problem.

For space missions, Sokkappa (1966) developed a near optimal guidance in closed-loop for throttleable spacecraft assuming constant Q-matrix; however, the Q-matrix was updated for onboard computation. Our simulation results show that the maximum difference between the thrust angles of Sokkappa's solution and guidance based on constant $\mathbf{v}_g$ direction is less than 0.5° when the initial range angle is less than 180° (the initial values and parameters are similar to Fig. 22). In addition, Sokkappa compared his optimal solution with the guidance law based on constant $\mathbf{v}_g$ direction for a case of injection from an earth orbit of 100 nautical miles to pass through an inertial point of 180,000 nautical miles. The guidance law based on constant $\mathbf{v}_g$ direction produces ΔV = 3,328.6 m/s, whereas Sokkappa's solution has a 34.2 m/s decrease in ΔV. Also, the performance of guidance law based on constant $\mathbf{v}_g$ direction and numerical optimal solution was compared by Circi (2004) for a satellite launch vehicle. The optimum solution has a decrease of 39 kg for a payload mass of 1,734 kg under the guidance law based on constant $\mathbf{v}_g$ direction for a perigee of 150 km.

It is worth noting that the guidance laws based on velocity-to-be-gained are applicable provided that a required velocity can be defined. For example, a required velocity cannot be defined for a fixed-final-time problem when final position and velocity vectors are both constrained. The impact angle of the conceptual guidance schemes is not considered in this investigation. It may be accomplished using appropriate choice of final time.

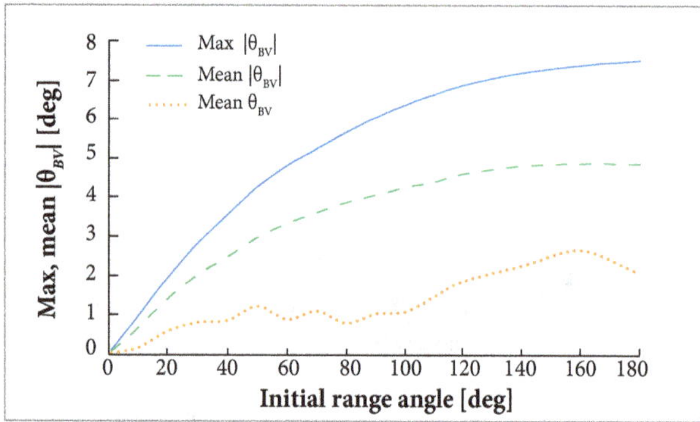

Figure 22. Maximum and mean of the absolute value, and mean value of the angle between the thrust direction along $\mathbf{v}_g$ and constant $\mathbf{v}_g$ direction versus range angle for interception of a stationary target for minimum energy orbit.

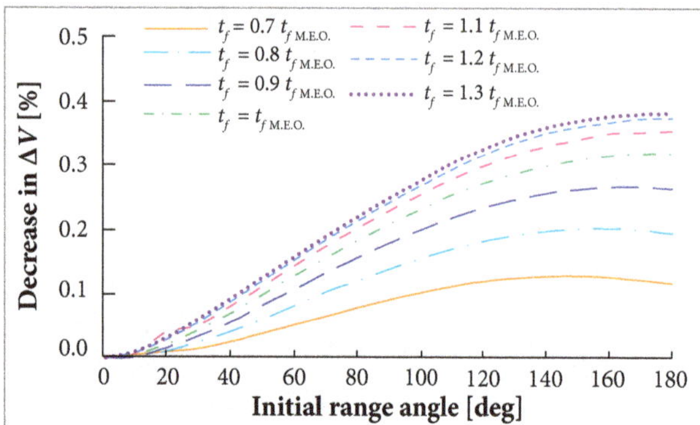

Figure 23. The percentage decrease in ΔV for the guidance law based on constant $\mathbf{v}_g$ direction with respect to $\mathbf{v}_g$ direction for minimum and non-minimum energy orbit.

CONCLUSIONS

This study suggests the effective thrust direction of an exoatmospheric interceptor for interception of short-to-super range moving targets with final position constraint. This has accomplished using a comprehensive study on conceptual guidance laws with stationary, moving, and free-falling targets. The first class of guidance law is based on ZEM. Three guidance schemes are considered in the first class, regarding to 2 definitions of ZEM and the component of ZEM perpendicular to line-of-sight. The capture criteria of the guidance scheme based on the perpendicular component of ZEM are highly restricted for non-throttleable rockets. The second class of conceptual guidance laws are based on linearized formulation, *i.e.* linearized ZEM and linear optimal control theory for throttleable rockets when its computed thrust direction is applied to non-throttleable rockets. The third class of guidance laws is based on the generalized required velocity and generalized velocity-to-be-gained. Two guidance schemes are considered in the third class. In the first scheme, the thrust acceleration is applied along the velocity-to-be-gained vector, whereas the second scheme tries to avoid the rotation of the velocity-to-be-gained in an inertial space.

The suggested direction of thrust acceleration is along the generalized velocity-to-be-gained, defined based on the generalized required velocity for interception of moving targets. For short-range application, the same results have been achieved when the thrust acceleration applied in the direction of ZEM or in the direction of the generalized velocity-to-be-gained. Increasing the range angle, the difference in performance is appeared. For long-range interception, the suggested thrust direction requires less amount of fuel rather than conceptual guidance laws based on ZEM or linearized formulations. The guidance performance is not improved as expected using a ZEM vector multiplied by a gain matrix to deviate the thrust direction from ZEM vector, as obtained in linearized optimal solutions.

If the intercept time is chosen a bit larger than the minimum energy orbit due to tactical consideration, *e.g.* salvo firing, adjustment of impact angle, etc., the suggested direction will have a significant fuel savings rather than guidance laws based on ZEM. Moreover, the performance of the suggested conceptual guidance law has less sensitive to the estimation error of final time. The guidance scheme based on constant velocity-to-be-gained direction may improve negligibly the fuel performance of the interceptor in the presence of noise for suborbital interception. Finally, the optimal solution does not give a better performance when the target position and velocity are contaminated by noise for suborbital interception such as a ballistic target; however, it improves the performance for a satellite launch vehicle and possibly for interception of orbital targets.

AUTHOR'S CONTRIBUTION

Mohamed-abadi MD performed the numerical solutions and prepared the figures. The idea, assumptions, classifications, and framework belong to Jalali-Naini SH, who wrote the manuscript. Both authors discussed the results and commented on the manuscript.

REFERENCES

Ahn J, Bang J, Lee SI (2015) Acceleration of zero-revolution Lambert's algorithms using table-based initialization. J Guid Control Dynam 38(2):335-342. doi: 10.2514/1.G000764

Arora N, Russell RP, Strange NJ, Ottesen D (2015) Partial derivatives of the solution to the Lambert boundary value problem. J Guid Control Dynam 38(9):1563-1572. doi: 10.2514/1.G001030

Battin RH (1999) An introduction to the mathematics and methods of astrodynamics. Revised edition. Reston: American Institute of Aeronautics and Astronautics.

Bryson AE, Ho YC (1975) Applied optimal control. New York: Hemisphere.

Chen FL, Xiao Y, Chen W (2010) Guidance based on velocity-to-be-gained surface for super-range exoatmospheric intercept. Acta Aeronautica et Astronautica Sinica 31(2):342-349. In Chinese.

Circi C (2004) Hybrid methods and Q-guidance for rocket performance optimization. Proc IME G J Aero Eng 218(5):353-359. doi: 10.1243/0954410042467040

Deihoul AR (2003) Anti ballistic optimal midcourse guidance law (PhD thesis). Tehran: Sharif University of Technology. In Persian.

Deihoul AR, Massoumnia MA (2003) A near optimal midcourse guidance law based on spherical gravity. Scientia Iranica 10(4):436-442.

Feng C, Yelun X, Wanchun C (2009) Guidance based on zero effort miss for super-range exoatmospheric intercept. Acta Aeronautica et Astronautica Sinica 30(9):1583-1589. In Chinese.

Jalali-Naini SH (2004) Modern explicit guidance law for high-order dynamics. J Guid Control Dynam 27(5):918-922. doi: 10.2514/1.5902

Jalali-Naini SH (2008) Generalization of zero-effort miss equations in atmospheric guidance laws with application to midcourse flight (PhD thesis). Tehran: Sharif University of Technology. In Persian.

Jalali-Naini SH, Pourtakdoust SH (2005) Modern midcourse guidance laws in the endoatmosphere. Proceedings of the AIAA Guidance, Navigation and Control Conference and Exhibit; San Francisco, USA.

Jalali-Naini SH, Pourtakdoust SH (2007) A unified approach to intercept guidance laws. Proceedings of the 6th Iranian Aerospace Society Conference; Tehran, Iran.

Li LG, Jing WX, Gao CS (2013) Design of midcourse trajectory for tactical ballistic missile intercept on the basis of zero effort miss. App Mech Mater 397-400:536-545. doi: 10.4028/www.scientific.net/AMM.397-400.536

Martin FH (1965) Closed-loop near-optimum steering for a class of space missions (PhD thesis). Cambridge: Massachusetts Institute of Technology.

Martin FH (1966) Closed-loop near-optimum steering for a class of space missions. AIAA J 4(11):1920-1927. doi: 10.2514/3.3819

Massoumnia MA (1995) Optimal midcourse guidance law for fixed-interval propulsive maneuvers. J Guid Control Dynam 18(3):465-470. doi: 10.2514/3.21410

Mohammad-abadi MD, Jalali-Naini SH (2016) Approximate solution of zero-effort-miss under gravitational acceleration inversely proportional to the cubic distance. Modares Mech Eng 16(4):135-144. In Persian.

Newman B (1996) Strategic intercept midcourse guidance using modified zero effort miss steering. J Guid Control Dynam 19(1):107-112. doi: 10.2514/3.21586

Rusnak I, Meir L (1991) Modern guidance law for high-order autopilot. J Guid Control Dynam 14(5):1056-1058. doi: 10.2514/3.20749

Sokkappa BG (1966) On optimal steering to achieve required velocity. Proceedings of the 16th International Astronautical Congress; Athens, Greece.

Zarchan P (2012) Tactical and strategic missile guidance. 6th edition. Reston: American Institute of Aeronautics and Astronautics.

12

Reinforced Transparencies for Aerospace Application – Case Description

Melis De Bruyn Neto[1], Rita de Cássia Mendonça Sales[1,2], Koshun Iha[1], José Atílio Fritz Fidel Rocco[1]

ABSTRACT: This paper describes the polycarbonate acrylic laminated development that can be applied in aeronautics and aerospace transparencies. The case studied is a laminated double-curved transparency (bubble form) used in an observation side window of a military aircraft. Side windows need strength and specific characteristics, similar to windshields, allowing the perfect visualization and image capture. Laminated transparencies composed by different materials have better qualities than the monolithic ones. This kind of transparency can offer high mechanical and chemical resistance, high transparency, no fragmentation and easy maintenance or recovery. A significant amount of information about materials and processes was jointed in order to build the reinforced transparency and validate this study. The final results were analyzed based on two points of view: mechanic resistance and, especially, optical quality.

KEYWORDS: Transparency, Laminated polycarbonate, Acrylic, Window.

INTRODUCTION

Whenever it is necessary to see or capture images through a protective barrier, it arises the need to define it and build it as effectively and efficiently as possible. The problem that comes from this observation is to measure all the variables that can influence the requirements and, mostly important, to define the materials that can be used, as well as the methods and processes that will lead to the manufacture of a quality product. The materials chosen, suitable for the construction of reinforced transparencies for aerospace industry, as windshields, must to be taken into account, as well as all loads they are subjected to, including structural terms, mechanical and thermal loads (Fam and Rizkalla 2006). In addition, superior optical characteristics are also essential (Fixler 1977).

The modern engineering and architecture "design" requires glazing materials that offer high levels of safety and high-performance mechanical properties. These properties include: ballistic resistance; wind loads; explosion; and physical attacks resistance. In some applications, noise level reduction, solar radiation resistance, and thermal barrier behavior are also desirable. The laminated glazing is the union of multiple layers of various materials such as glass, polymeric films, resins, and flexible sheets of transparent polymers (polycarbonate and/or acrylic), usually applied to obtain complex geometric shapes (Fixler 1977; Smith *et al.* 1996).

It is necessary a prior knowledge of the physicochemical properties of the engineering materials to make the right choice for each application. To the right choice, it is necessary to know the load or the effort to which the transparency must resist and the kind of effort that will be neutralized by the transparent barrier. However, it should be clear that there will always be an effort

1. Departamento de Ciência e Tecnologia Aeroespacial – Instituto Tecnológico de Aeronáutica – Divisão de Ciências Fundamentais – São José dos Campos/SP – Brazil.
2. Centro Estadual de Educação Tecnológica Paula Souza – Faculdade de Tecnologia de São José dos Campos - Prof. Jessen Vidal – São José dos Campos/SP – Brazil.

Author for correspondence: Rita de Cássia Mendonça Sales | Centro Estadual de Educação Tecnológica Paula Souza – Faculdade de Tecnologia de São José dos Campos - Prof. Jessen Vidal | Avenida Cesare Mansueto Giulio Lattes, 1.350 – Eugênio de Melo | CEP: 12.247-014 – São José dos Campos/SP – Brazil | Email: rita.sales@fatec.sp.gov.br

that can overcome the barrier, because there is no definitive mechanical barrier. Thus, the barrier will be "strengthened" until reaching the desired resistance and including a safety factor.

Besides the mechanical strength, transparent barriers must have optical quality, being able to see or capture images by means of high light transmission level, with a minimum deviation or absorption, until the images reach the observer's eye or the image capture device.

Reinforced transparencies are not only intended to prevent a mechanical failure, but to perform with maximum efficiency and preferably not releasing fragments that reach whom or what will be protected by the transparent barrier. The objective of this study was to describe the information needed to design and to construct a reinforced transparency for aerospace application made with polymeric materials, which is part of the "observation window" of a military patrol aircraft.

MATERIALS AND METHODS

MATERIALS CHOICE

The materials chosen for the observation window were based on MIL-PRF-5425E (1998) and MIL-P-46144C (1986). The chosen structural materials, in sheet form, were: acrylic or polymethylmethacrylate - PMMA (ASTM D4802-02) (Modified "cast" acrylic ACRYLITE® 249, CYRO Ind, USA); polycarbonate - PC (Fox and Christopher 1962; LeGrant and Bendler 2000) (LEXAN® 9034, GE-Plastics, USA); and, for bonding, in film form, polyurethane - PU (KRYSTALFLEX®PE399, Huntsman, USA), which is the material that allows a good adhesion between acrylic and polycarbonate.

The following sequence was used to prepare the specimens: two outer layers with 3.18 mm thickness of acrylic (PMMA); a central PC layer with 6.35 mm thickness; and between acrylic and PC layers, two aliphatic PU layers with 1.27 mm thickness were inserted. In other words, the final composition was: PMMA + PU + PC + PU + PMMA (Fig. 1).

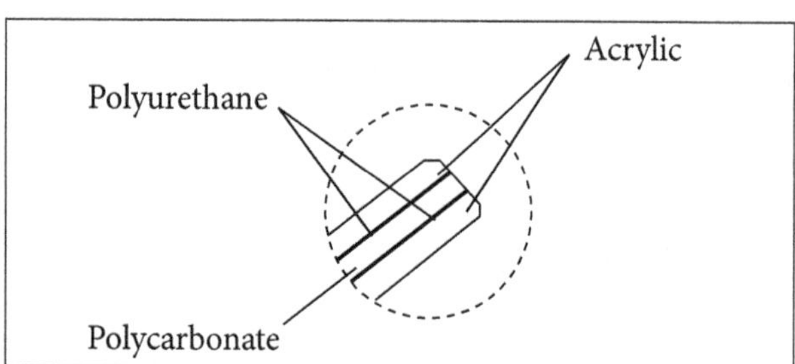

Figure 1. Materials sequence in specimens section.

TEMPERATURE RESISTANCE TEST

Two samples with nominal size (20 × 10 cm) were prepared and laminated in the same way of the shearing test samples. The samples were submitted to 100 °C for 50 min, using a circulating air oven (Kamp).

SPECIMENS PREPARATION

All the raw material plates were cut into dimensions much larger than the finished parts, using a circular saw. After this process, the plates were dehumidified at 100 °C, during 6 h in an air circulating oven.

The dehumidified materials were laminated in the sequence previously defined. A vacuum bag was prepared around the laminated (all the bagging materials are from Airtech®, USA; vacuum film and sealant tape). The set, formed by the laminate and the vacuum bag, was placed in an autoclave (FERLEX®, Brazil) to be bonded. To ensure the bonding material, the set formed by the laminate and the vacuum bag was heat-treated to 120 °C for 2 h at a 2 °C/min heating rate and pressured at 0.7 MPa inside the autoclave. The temperature and pressure were maintained during 3 h; then, it was cooled until the ambient temperature was reached. At the end, the pressure was released and the cycle, finished.

THERMOFORMING PROCESS

After autoclave process, the set was placed on a tool designed for the observation window manufacture. The tooling (Fig. 2) was heated to 180 °C at a 6 °C/min heating rate. After the laminate reached the melting point, it was blown at a pressure of 0.14 MPa to achieve its definitive new shape. The specimen was then cooled at 2 °C/min.

The part trimming was performed in a conventional manner using bandsaw and sander. The polishing was done using fine finishing hand tools and then through sandpapers and finer abrasives, until it reached the desired polishing. The specimen, properly polished, was placed against a reticule grid to check possible optical distortions.

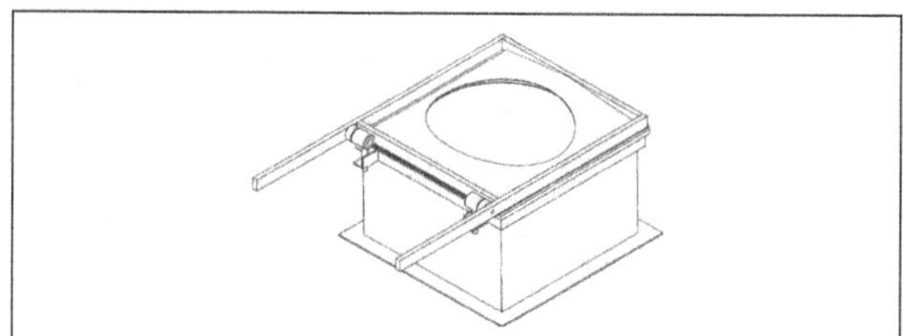

Figure 2. Thermoforming tool.

SHEARING TESTS IN PLANE

Six samples with nominal size (4.0 × 5.0 cm; section area: 20 cm^2) were prepared using the following materials: an acrylic sheet, with 3.18 mm thickness, adhered to a PC sheet, with 6.35 mm thickness. For the sheets adhesion, it was used an aliphatic PU crystal adhesive film with 1.27 mm thickness.

The samples were laminated in the same way of the observation window. They were tested after 48 h of stabilization. The shearing test was performed in plane (Fig. 3) using MIL-P-25690B (1995) as general references. It was used an electromechanical testing machine for tension and compression Tinus Olsen brand, model H100KS, with maximum capacity of 100 kN. The test speeds were 3 and 5 mm/min.

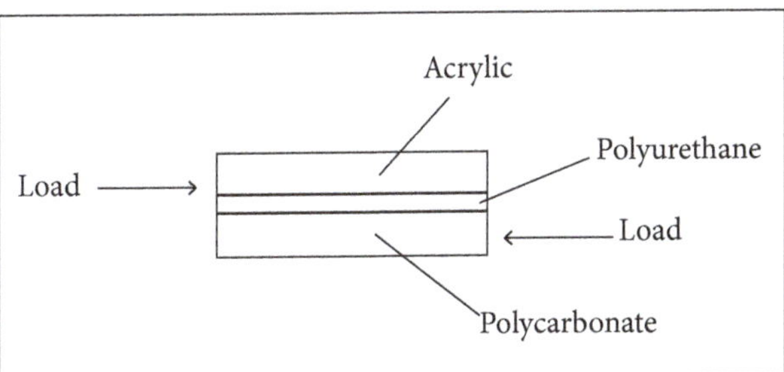

Figure 3. Schematic diagram of load applied in samples during the shearing tests in plane.

Table 1. Acrylic and polycarbonate mechanical properties.

Material	Modified "Cast" Acrylic	Polycarbonate
Standard	MIL-PRF-5425E 1998	MIL-P-46144C 1986
Commercial type	ACRYLITE® 249	LEXAN® 9034
Provider	CYRO Ind, USA	GE-Plastics, USA
Poisson's ratio	0.35	0.37
Tensile strength	80.3 MPa	65.5 MPa
Maximum elongation	4.4 %	110.0 %
Tensile Modulus	2.76 GPa	2.39 GPa
Compressive strength	124.1 MPa	86.18 MPa
Flexural strength	113.76 MPa	93.08 MPa
Shear strength (ultimate)	62.05 MPa	68.95 MPa
Izod impact (notched)	21.35	640.5 - 854

RESULTS AND DISCUSSION

MATERIALS CHOICE

Nowadays, most of civil and military aircraft windows are made with acrylic (PMMA) due to its transparency, availability, strength and ease maintenance, fulfilling all the aviation requirements and standards (Blass 1985; LaPluma and Bridenbaugh 1988). This material also meets the requirements for external surfaces of reinforced transparency. However, in case of violent impact, it can shatter, causing sudden aircraft depressurization. To satisfy the high impact strength requirement, the best polymer is PC, already used in military fighter aircraft canopies (Wiser 1971). Both materials have high light transmission levels and they are thermoformable. To connect these two materials, the aliphatic PU films (Plepis 1991; Vilar 1991) can be used, providing the adhesion between layers. Table 1 shows the properties of the chosen materials for the manufacture of aircraft observation window.

After materials selection, it was carried out a temperature resistance test to verify the laminate behavior at 100 °C. There was no appearance of undesirable bubbles, yellowing, delamination or any other conduct that could compromise the functions of the laminate or its general appearance. It was also verified that the temperature would not compromise the good light transmission without distortion or deformation.

SHAPE DETERMINATION

The final shape of the transparency is better defined by the obtaining process than by a geometry defined through engineering software (Fig. 4). The method used to obtain the transparency form was the thermoforming by blowing compressed air (Blass 1985), since it allows the material to expand freely until the desired format is obtained.

The expansion process by free blowing determines the shape and cooperates with the optical quality, avoiding deformities, and it does not cause undesirable distortions in contact with the tool — they are visible only at the edges of transparency. The result by thermoforming is an elliptical paraboloid shape (Fig. 5), caused by laminated thinning (more in the center than in the edges), because of area increasing and thickness loss.

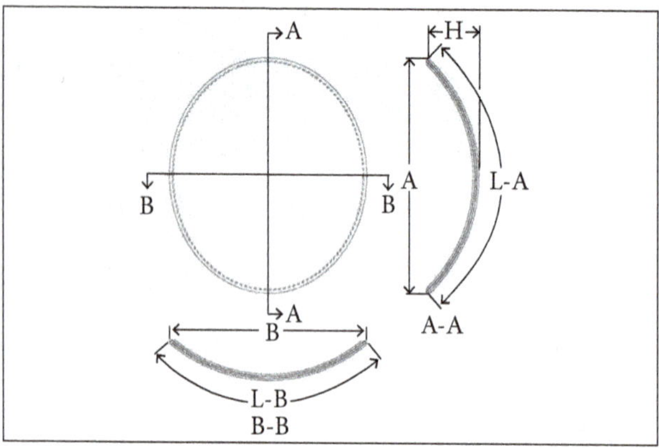

Figure 4. Orthogonal views and transparency section. A-A: Height; B-B: Width; L-A: Vertical vision angle; L-B: Horizontal vision angle; H: Depth.

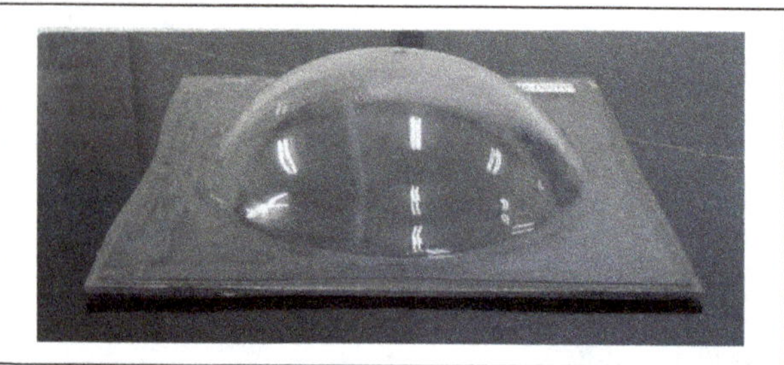

Figure 5. Laminated after thermoforming made with an appropriate tool.

OPTICAL DISTORTIONS

Samples submitted for verification did not have relevant distortions, detected by the human eye, which could compromise the good appearance of the grid images captured through the transparence (Fig. 6). However, it was detected the presence of a slight optical power due to thickness variation at the center of the part, resulting from thinning, occurred during the thermoforming process, which creates a kind of diverging lenses. The contact with the thermoforming tool produces contact marks that are manifested in the form of distortion of the edge part; however, these distortions do not affect the results, when transparency is in use, since it does not become apparent because this region is covered by the transparency frame.

The optical distortions verification (Fixler 1977) is an important parameter in reinforced transparencies (Pardini and Peres 1996). The reinforced transparency, besides being considered a barrier to prevent an effort to continue, should allow observing or capturing images through it with the best possible quality, not adding deviations image distortions, retain the light wavelength or filter visible spectrum.

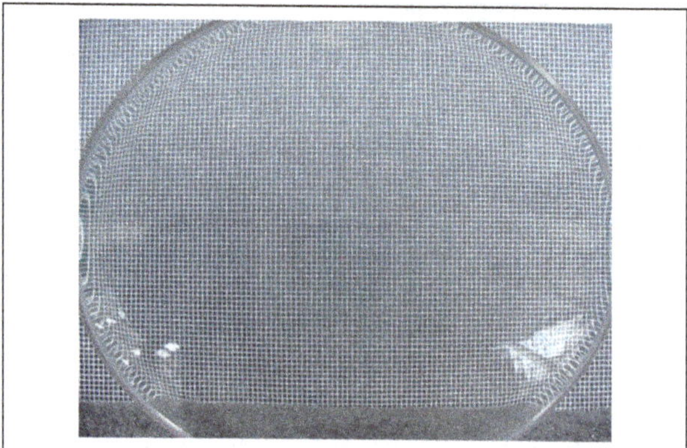

Figure 6. Transparency view without distortions in the central area.

All the images captured during the transparencies flight test had an excellent clarity and no visible distortions, which confirmed the prior results obtained with the grid.

SHEARING RESULTS IN PLANE

Aircraft side windows have a significant temperature differential between inner and outer layer, as much higher as the aircraft flying altitude. The internal temperature is usually around 20 °C, and the outer temperature can be less than −54 °C, so the dimensional variation between layers produces high big shear forces that must be compensated by the polyurethane layer elasticity. In this case, the PU adhesion between acrylic and/or polycarbonate must be verified. The shear test indicates the degree of adhesion acquired by the laminated.

Each trial lasted until a little over 2 min, since the total displacement was around 6 mm. Tables 2 and 3 contain the results obtained in the shear test.

The test speed of 3 mm/min was appropriate, according to the behavior observed, which does not present a premature rupture and produces a clear "load × deflection" curve. The standard deviation in general demonstrates a short range of values, or more reliability in the results obtained. The test speed of 5 mm/min does not produce results with small variation, since the values obtained were two or three times higher than the values obtained with test speed of 3 mm/min.

The maximum shear stress average demonstrates a satisfactory adhesion degree between layers in both speed tests and was greater than the expected.

A more detailed analysis of the displacement data between the sheets of acrylic and polycarbonate may provide an order of magnitude for the lengthening ability of the polyurethane adhesive, but the PU manufacturer (Huntsman, USA) indicates

Table 2. Results obtained in the shear test at the speed of 3 mm/min.

Sample	Maximum load (kgf)	Breaking load (kgf)	Displacement (mm)	Shear stress (kgf/cm²)
S1	1,010.2	982.7	6.2	49.1
S3	1,034.3	967.7	6.4	48.4
S5	972.8	896.7	6.3	44.8
Average	1,005.8	949.0	6.3	47.5
Standard deviation	31.0	45.9	0.1	2.3

Table 3. Results obtained in the shear test at the speed of 5 mm/min.

Sample	Maximum load (kgf)	Breaking load (kgf)	Displacement (mm)	Shear stress (kgf/cm²)
S2	1,059.2	1,027.5	6.2	51.4
S4	881.4	793.3	5.8	39.7
S6	951.1	913.7	5.7	45.7
Average	963.9	911.5	5.9	45.6
Standard deviation	89.6	117.1	0.3	5.9

500% of elongation at the KRYSTALFLEX® PE399 datasheet (Huntsman 2015).

All samples tested were broken or showed detachment from plastic layers. The S2 had supported the major load, and S4, the minor one. Figure 7 shows a typical curve of "load × deflection" generated by S2 test.

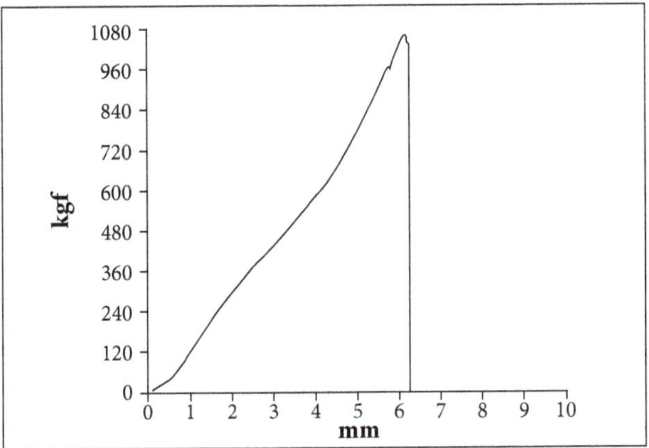

Figure 7. Curve "load × deflection" of the S2.

CONCLUSIONS

The objective was achieved by implementing the proposed road map, which was to organize some of the information needed to design and to construct a reinforced transparency for an aerospace application, which is part of the "observation window" of a military patrol aircraft, as well as to validate such information through the construction of the transparency and analysis of the results.

The combination of the techniques used and the layers union in an appropriate and sequential manner was essential for the construction of transparency, which was successfully completed. Any failure would not allow the following step.

The optical results allow to state that the material choices were appropriate, especially on the composition and position of them. The acrylic usage on the internal and external faces of transparency had ensured the necessary optical quality. The fabrication process chosen, by the free expanding thermoforming, guaranteed a regular format of the visible area because there were no contacts with any tool.

All samples presented a satisfactory adhesion degree between layers during the shearing tests in plane. The parameters of speed test were greater than the expected, but this did not affect the results; the samples showed mechanical resistance parameters required by the aeronautical industry. Therefore, it was possible to demonstrate the technological knowledge necessary for the development and manufacture of reinforced aerospace transparencies for this type of application.

REFERENCES

Blass A (1985) Processamento de polímeros. Florianópolis: UFSC.

Fam A, Rizkalla S (2006) Structural performance of laminated and unlaminated tempered glass under monotonic transverse loading. Construct Build Mater 20(9):761-768. doi: 10.1016/j.conbuildmat.2005.01.051

Fixler SZ (1977) Thermsstructural and material considerations in the design of the F-14 Aircraft Transparencies. J Aircraft, 14(3):257-264.

Fox DW, Christopher WF (1962) Polycarbonates. New York: Reinhold Publishing Corporation.

Huntsman (2015) KRYSTALFLEX® PE399 data sheet on line. [accessed 2015 Oct 15]. http://www.huntsman.com/polyurethanes

LaPluma PT, Bridenbaugh JC (1988) Specifications and measurement procedures for aircraft transparencies. Wright-Patterson AFB, OH: Armstrong Aerospace Medical Research Laboratory.

LeGrant DG, Bendler JT (2000) Handbook of polycarbonate science and technology. New York: Marcel Dekker.

Pardini LC, Peres RJC (1996) Tecnologia de fabricação de pré-impregnados para compósitos estruturais utilizados na indústria aeronáutica. Polímeros 6(2):32-42.

Plepis AMD (1991) Caracterização térmica e viscoelástica de resinas poliuretanas derivadas do óleo de mamona (Master's thesis). São Paulo: Universidade de São Paulo.

Smith FC, Moloney LD, Matthews FL, Hodges J (1996) Fabrication of woven carbon fibre/polycarbonate repair patches. Compos Appl Sci Manuf 27(11):1089-1095. doi: 10.1016/1359-835X(96)00070-X

Vilar WD (1991) Química e tecnologia dos poliuretanos. 2nd ed. Rio de Janeiro: Vilar Consultoria Técnica Ltda.

Wiser GL (1971) Transparency applications of polycarbonates. Aircraft Eng 43(8):18-20. doi: 10.1108/eb034797

13

A Comparative Study of Four Feedback Linearization Schemes for Motion Planning of Fixed-Wing Unmanned Aerial Vehicles

Hossein Bonyan Khamseh[1]

ABSTRACT: In this paper, different feedback linearization schemes are studied to address the motion planning problem of fixed-wing unmanned aerial vehicles. For a unmanned aerial vehicle model with second-order dynamics, several schemes are studied to make the vehicle (i) fly over and (ii) make a loitering around the objective position. For each scheme, comparisons are made to illustrate the advantages and disadvantages. Lyapunov stability analysis is used to prove the stability of the proposed schemes, and simulation results for some case studies are included to show their feasibility.

KEYWORDS: Feedback linearization, Unmanned aerial vehicle, Motion planning.

INTRODUCTION

In recent years, unmanned aerial vehicles (UAVs) have gained increasing attention for various missions such as remote sensing of agricultural products (Costa *et al.* 2012), forest fire monitoring (Casbeer *et al.* 2006), search and rescue (Almurib *et al.* 2011), transmission line inspection (Li *et al.* 2013) and border monitoring (Beard *et al.* 2006). To this date, various approaches have been employed to address the motion planning of UAVs to reach, fly over or loiter around an objective position. As an example, in Frew *et al.* (2008) and Lawrence *et al.* (2008), vector fields with a stable limit cycle centered on the target position were constructed. In the mentioned studies, the authors employed a Lyapunov vector field guidance (LVFG) law to bring the UAV to an observation "orbit" around the target. Also, in Gonçalves *et al.* (2011), a vector field approach was used to bring several non-holonomic UAVs to a static curve embedded in the 3-D space. In Gonçalves *et al.* (2010), vector fields were determined such that a robot converged to a time-varying curve in n-dimensions and circulated it. In Hsieh *et al.* (2008), decentralized controllers were proposed to bring a number of robotic agents to generate desired simple planar curves, while avoiding inter-agent collision. In Hsieh *et al.* (2007), the controllers were modified such that the robots converged to a star-shaped pattern and, once on the objective curve, circulated it. In Bonyan Khamseh *et al.* (2014), based on the concept of flight corridor, a decentralized coordination strategy was proposed to bring a team of fixed-wing UAVs to a circular orbit, while avoiding inter-UAV collision. In Hafez *et al.* (2013), model predictive control was used to create a

1.Universidade Federal de Minas Gerais – Escola de Engenharia Elétrica – Departamento de Engenharia Elétrica – Belo Horizonte/MG – Brazil.

Author for correspondence: Hossein Bonyan Khamseh | Universidade Federal de Minas Gerais - Escola de Engenharia Elétrica - Departamento de Engenharia Elétrica Av. Pres. Antônio Carlos, 6627 – CEP: 31 270-901 – Pampulha - Belo Horizonte/MG – Brazil | Email: h.bonyan@gmail.com

$$\dot{\boldsymbol{E}} = \dot{\boldsymbol{\Gamma}} + \ddot{\boldsymbol{\Gamma}} = \dot{\boldsymbol{\Gamma}} + \boldsymbol{A} + B\boldsymbol{u} \tag{7}$$

Similar to Kanchanavally *et al.* (2006), if we define $\boldsymbol{u} = B^{-1}(-\boldsymbol{A} - \dot{\boldsymbol{\Gamma}} + \boldsymbol{v})$ and , it is easy to see that:

$$\dot{\boldsymbol{E}} = K\boldsymbol{E} \tag{8}$$

where: $K \in R^{3\times3}$ and eigenvalues of K have negative real parts.

In order to study the error dynamics given by Eq. 8, one may consider the following Lyapunov function:

$$V = \frac{1}{2}\boldsymbol{E}^T\boldsymbol{E} \tag{9}$$

The time derivative of the above positive definite V is given by:

$$\dot{V} = \dot{\boldsymbol{E}}^T\boldsymbol{E} = \boldsymbol{E}^T\boldsymbol{K}^T\boldsymbol{E} < 0 \tag{10}$$

Therefore the error converges to zero. Regarding the internal dynamics with a vector of relative degree of [2 2 2], one needs to propose two more transformations to complete the diffeomorphism. The internal dynamics is given by Eq. 11:

$$\begin{cases} \eta_7 = x_3 \rightarrow \dot{\eta}_7 = x_4 \\ \eta_8 = x_4 \rightarrow \dot{\eta}_8 = \frac{1}{I}u_1 \end{cases} \tag{11}$$

For a UAV with a forward-looking camera, the only possible motion where the error goes to zero is when the UAV flies on a straight line over the objective position. Therefore, as $t \rightarrow \infty$, $x_4 \rightarrow 0$. With $x_4 \rightarrow 0$, it is easy to conclude that x_3 will be bounded and therefore it is not going to cause undesirable effects. Simulation results verifying the feasibility of this approach will be given in "Simulations" section.

UAV WITH SIDE-LOOKING CAMERA, SCHEME #2

In this subsection, a UAV with variable-angle side-looking camera is considered. The equations of motion of this configuration are identical to those given in Eq. 1 and therefore are not repeated here. For a side-looking camera, the output is given by:

$$\boldsymbol{\Gamma} = \begin{bmatrix} x_1 + L\sin x_3 \\ x_2 - L\cos x_3 \\ x_5 \end{bmatrix} \tag{12}$$

where the first two elements of Γ represent x and y positions of the footprint of the camera (see Fig. 2).

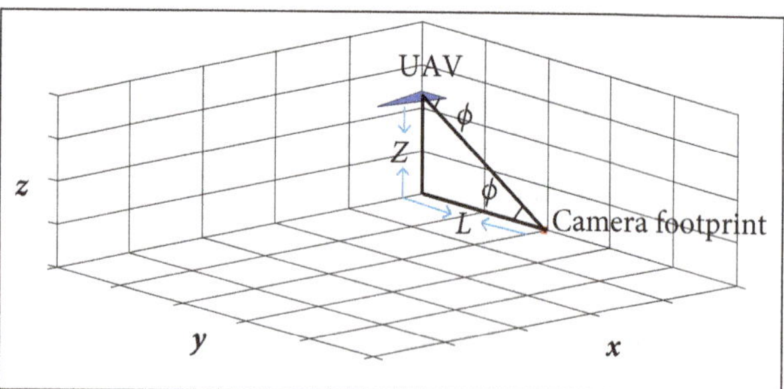

Figure 2. The UAV and its camera footprint in scheme #2.

For the given output, one can verify that $[r_1\ r_2\ r_3] = [2\ 2\ 2]$. Therefore, an equation identical to that given in Eq. 3 is obtained, in which:

$$A_1 = -vx_4\sin x_3 - 2x_6x_8\frac{1}{\sin^2 x_7}\sin x_3 + \\ + 2x_5(x_8)^2\frac{\cos x_7}{\sin^3 x_7}\sin x_3 + 2x_6\cot x_7\, x_4\cos x_3 - \\ - 2x_5x_8\frac{1}{\sin^2 x_7}x_4\cos x_3 - L(x_4)^2\sin x_3$$

$$A_2 = vx_4\cos x_3 + 2x_6x_8\frac{1}{\sin^2 x_7}\cos x_3 - \\ - 2x_5(x_8)^2\frac{\cos x_7}{\sin^3 x_7}\cos x_3 + 2x_6\cot x_7\, x_4\sin x_3 - \\ - 2x_5x_8\frac{1}{\sin^2 x_7}x_4\sin x_3 + L(x_4)^2\cos x_3$$

$$A_3 = 0$$

$$B = \begin{bmatrix} \frac{L}{I}\cos x_3 & \frac{1}{m}\cot x_7\sin x_3 & -\frac{1}{J}\frac{x_5}{\sin^2 x_7}\sin x_3 \\ \frac{L}{I}\sin x_3 & -\frac{1}{m}\cot x_7\cos x_3 & \frac{1}{J}\frac{x_5}{\sin^2 x_7}\cos x_3 \\ 0 & \frac{1}{m} & 0 \end{bmatrix}$$

One can verify that $\det(B) = -(x_5)^2\cos x_7/m{\cdot}I{\cdot}J\sin^3 x_7$. This determinant can be zero if $x_5 = 0$ or $\cos x_7 = 0$. With the reasoning given in the previous subsection (see Eqs. 4 – 6), one can conclude that $x_5 \neq 0$, i.e. the altitude cannot be zero. For a UAV with constant forward velocity and a side-looking camera, the only possible motion where the error goes to zero is when the UAV loiters around the objective position, with a fixed loitering radius. Therefore, with a non-zero loitering radius, it is easy to conclude that $\cos x_7 \neq 0$. Also, $\det(B) \rightarrow \infty$ if $x_7 \rightarrow k\pi$. Yet, $x_7 \rightarrow k\pi$

means that L → ∞, which is not common in practical scenarios. Therefore, it is concluded that $\det(B) = -(x_5)^2 \cos x_7/m{\cdot}I{\cdot}J \sin^3 x_7 \neq 0$ for practical applications, and thus the system given by Eq. 1 with output given by Eq. 12 is input-output linearizable.

For this scheme, the Lyapunov stability analysis is identical to that given by Eqs. 7 – 10. Therefore, it can be concluded that the error dynamics asymptotically converges to zero. Regarding the internal dynamics with a vector of relative degree of [2 2 2], one needs to propose two more transformations to complete the diffeomorphism. The internal dynamics is given by and in Eq. 13:

$$\begin{cases} \eta_7 = x_3 \rightarrow \dot{\eta}_7 = x_4 \\ \eta_8 = x_4 \rightarrow \dot{\eta}_8 = \frac{1}{I} u_1 \end{cases} \tag{13}$$

For a UAV loitering around a given objective position with constant (finite) forward velocity, x_4 will be bounded and cannot go to infinity. Also, in a loitering motion, x_3, i.e. the heading angle, can be shown by $2k\pi + \theta'$ where θ' is a finite value and therefore the internal dynamics will not cause undesirable effects in our approach. Simulation results regarding this scheme will be given in "Simulations" section.

An important drawback of this method is that one cannot explicitly control the final loitering radius of the UAV. Therefore, in the next scheme, we try to explicitly define the loitering radius as one of the system outputs.

UAV WITH SIDE-LOOKING CAMERA, SCHEME #3

In this section, we modify the equations of motion given in Eq. 1 in a manner that a new useful scheme is obtained. In the control affine form, the new dynamic equations of motion are given by Eq. 14:

$$\begin{bmatrix} \dot{x}_1 \\ \dot{x}_2 \\ \dot{x}_3 \\ \dot{x}_4 \\ \dot{x}_5 \\ \dot{x}_6 \\ \dot{x}_7 \\ \dot{x}_8 \end{bmatrix} = \begin{bmatrix} \dot{r}_x \\ \dot{r}_y \\ \dot{\theta} \\ \dot{\omega} \\ \dot{r}_z \\ \dot{v}_z \\ \dot{L} \\ \ddot{L} \end{bmatrix} = \begin{bmatrix} v\cos x_3 \\ v\sin x_3 \\ x_4 \\ 0 \\ x_6 \\ 0 \\ x_8 \\ 0 \end{bmatrix} + \begin{bmatrix} 0 & 0 & 0 \\ 0 & 0 & 0 \\ 0 & 0 & 0 \\ \frac{1}{I} & 0 & 0 \\ 0 & 0 & 0 \\ 0 & \frac{1}{m} & 0 \\ 0 & 0 & 0 \\ 0 & 0 & 1 \end{bmatrix} \begin{bmatrix} u_1 \\ u_2 \\ u_3 \end{bmatrix} = \boldsymbol{f} + \boldsymbol{g}\boldsymbol{u} \tag{14}$$

Here, the main difference is that L and $\dot{L}$ are explicitly considered to be state variables. One may define the system output as:

$$\boldsymbol{\Gamma} = \begin{bmatrix} x_1^2 + x_2^2 \\ x_5 \\ x_7 \end{bmatrix} \tag{15}$$

where the first element of Γ is the square of the distance of the UAV from the origin of the coordinate system, i.e. the stationary objective position.

It can be readily seen that $[r_1\ r_2\ r_3] = [3\ 2\ 2]$. Therefore, in a compact form, one can write:

$$\begin{bmatrix} \dddot{\Gamma}_1 \\ \ddot{\Gamma}_2 \\ \ddot{\Gamma}_3 \end{bmatrix} = \boldsymbol{A} + B\boldsymbol{u} \tag{16}$$

where:

$$\boldsymbol{A} = \begin{bmatrix} -2vx_4^2(x_1\cos x_3 + x_2\sin x_3) \\ 0 \\ 0 \end{bmatrix},$$

$$B = \begin{bmatrix} \frac{-2v}{I}(x_1\sin x_3 - x_2\cos x_3) & 0 & 0 \\ 0 & \frac{1}{m} & 0 \\ 0 & 0 & 1 \end{bmatrix}$$

One can verify that $\det(B) = -2v_c/m{\cdot}I\,(x_1 \sin x_3 - x_2 \cos x_3)$. An important disadvantage is that this determinant can be zero if $x_1 \sin x_3 - x_2 \cos x_3 = 0$, i.e. when the UAV is either flying radially inward or radially outward. Yet, if the heading of the UAV does not fall within this region, the UAV can converge to a loitering motion around the origin of the coordinate system. On the other hand, the advantage of this scheme is that, depending on the value of R_1, i.e. the first element of the reference signal, one can come up with scenarios in which the UAV converges to a loitering radius either smaller or greater than the initial one. Also, a second advantage is that one can explicitly control L, as R_3. Therefore, for $R_3 < 2\sqrt{R_1}$, $R_3 = 22\sqrt{R_1}$ and $R_3 > 2\sqrt{R_1}$, one can define scenarios in which the UAV loiters around the origin while the camera footprint sweeps a circle with the radius smaller than, equal to or greater than $\sqrt{R_1}$. This is schematically shown in Figs. 3a to 3c.

Regarding the stability of the system, if we define the error as:

$$\begin{bmatrix} E_1 \\ E_2 \\ E_3 \end{bmatrix} = \left(\begin{bmatrix} \Gamma_1 \\ \Gamma_2 \\ \Gamma_3 \end{bmatrix} - \begin{bmatrix} R_1 \\ R_2 \\ R_3 \end{bmatrix} \right) + \left(\begin{bmatrix} \ddot{\Gamma}_1 \\ \dot{\Gamma}_2 \\ \dot{\Gamma}_3 \end{bmatrix} - \begin{bmatrix} \ddot{R}_1 \\ \dot{R}_2 \\ \dot{R}_3 \end{bmatrix} \right) \tag{17}$$

for a stationary reference signal, one will have $\dot{\boldsymbol{R}} = \ddot{\boldsymbol{R}} = 0$. Differentiating Eq. 17, one has:

$$\begin{bmatrix} \dot{E}_1 \\ \dot{E}_2 \\ \dot{E}_3 \end{bmatrix} = \begin{bmatrix} \dot{\Gamma}_1 \\ \dot{\Gamma}_2 \\ \dot{\Gamma}_3 \end{bmatrix} + \begin{bmatrix} \dddot{\Gamma}_1 \\ \ddot{\Gamma}_2 \\ \ddot{\Gamma}_3 \end{bmatrix} \tag{18}$$

Now, if we define $\boldsymbol{u} = B^{-1}(-\boldsymbol{A} - \dot{\Gamma} + \upsilon)$ and $\upsilon = K\boldsymbol{E}$, it is easy to see that:

$$\dot{\boldsymbol{E}} = K\boldsymbol{E} \tag{19}$$

where: K is a matrix with eigenvalues which have negative real parts.

In order to study the error dynamics given by Eq. 19, one may consider the following Lyapunov function:

$$V = \frac{1}{2}\boldsymbol{E}^T\boldsymbol{E} \tag{20}$$

The time derivative of the above positive definite V is given by:

$$\dot{V} = \dot{\boldsymbol{E}}^T\boldsymbol{E} = \boldsymbol{E}^T\boldsymbol{K}^T\boldsymbol{E} < 0 \tag{21}$$

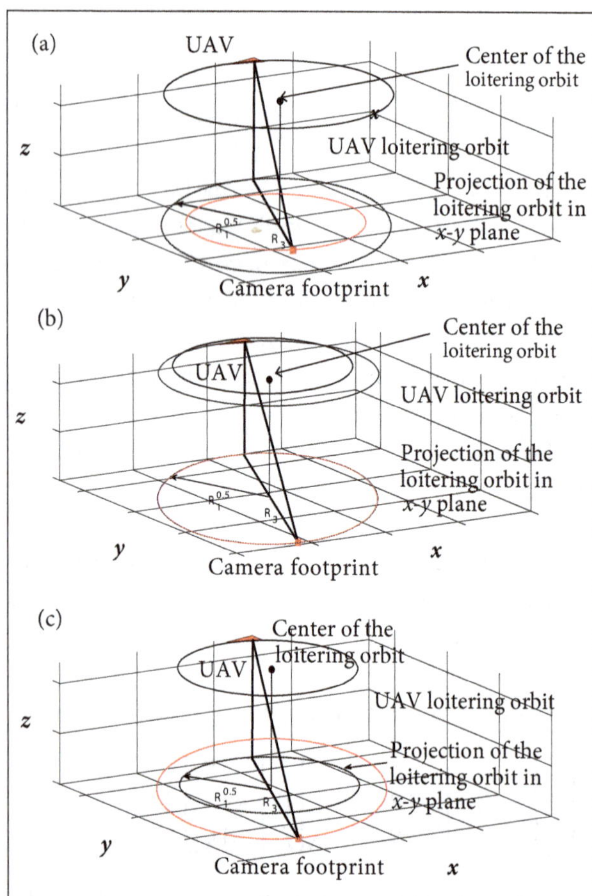

Figure 3. Three different scenarios in scheme #3.

Therefore the error dynamics is asymptotically stable. Regarding the internal dynamics with a vector of relative degree of [3 2 2], one needs to propose one more transformation to complete the diffeomorphism. The internal dynamics is given by $\dot{\eta}_8$ in Eq. 22:

$$\eta_8 = x_4 \rightarrow \dot{\eta}_8 = \frac{1}{I}u_1 \tag{22}$$

For a UAV loitering around a given objective position with constant (finite) forward velocity, x_4 will be bounded and cannot go to infinity. Therefore, the internal dynamics will not cause undesirable effects in our approach. Simulation results regarding this scheme will be given in "Simulations" section.

UAV WITH SIDE-LOOKING CAMERA AND ONE VIRTUAL FORWARD-LOOKING CAMERA, SCHEME #4

In this subsection, we modify the previous scheme in the sense that the UAV can loiter around the origin with a desirable radius while avoiding the singularity problem of scheme #3. Convergence to a loitering radius (i) smaller or (ii) greater than the initial radius is studied separately.

Convergence to a Loitering Radius Smaller than the Initial One

In this scenario, it is initially assumed that the UAV is equipped with a virtual forward-looking camera. From geometry, one can find two tangent lines (and their corresponding tangency points) between the initial position of the UAV and the circle with the reference radius. In the first phase, the UAV can choose one of the tangency points as its virtual objective position and fly over it, according to scheme #1, discussed earlier. Assuming that the UAV flies over the tangent line, its heading will be perpendicular to the radius of the objective circle as it reaches the virtual objective position. As the UAV reaches the tangency point, it switches to scheme #3. The advantage here is that, in the second phase, it is ensured that the heading of the UAV is far from inward-outward direction and therefore scheme #3 can bring the UAV to loiter around the objective position, with desirable radius. This is schematically shown in Fig. 4.

The details of feedback linearization, control laws and stability analysis of scheme #1 and scheme #3 were discussed in the previous subsections and are not repeated here. Simulation results of this scheme will be given in "Simulations" section.

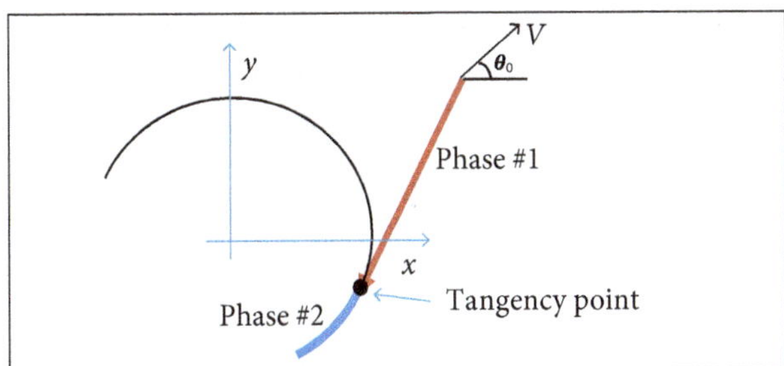

Figure 4. Scheme #3, convergence to a loitering radius smaller than the initial one.

Convergence to a Loitering Radius Greater than the Initial One

Similar to the previous subsection, we assume that the UAV is equipped with a virtual forward-looking camera and a side-looking camera. The scenario proposed here consists of three phases, as shown in Fig. 5.

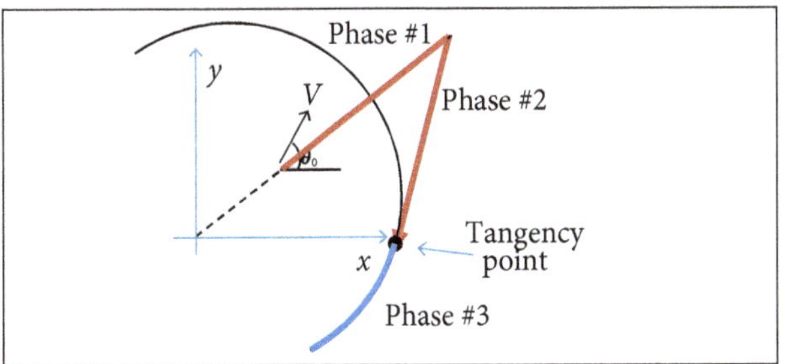

Figure 5. Scheme #3, convergence to a loitering radius greater than the initial one.

In the first phase, based on scheme #1, the UAV flies to a virtual objective position which is on the extension of the line connecting the origin to the initial position of the UAV. Let's denote the UAV distance from the origin by d and the desired loitering radius by R^*. At the end of the first phase, when d is relatively greater than R^*, the UAV finds the tangent lines from its current position to the circle with the radius R^*. With d relatively greater than R^*, one can assume that, in the second phase, based on scheme #1, the UAV flies on the tangent line to reach the tangency point (second virtual objective position). Once at this point, the UAV has reached the reference radius and switches to scheme #3. In the third phase, based on scheme #3, the UAV loiters around the origin with $R_1 = R^*$. The details of feedback linearization, control laws and stability analysis of scheme #1 and scheme #3 were discussed in the previous subsections and are not repeated here. Simulation results of this scheme will be given in "Simulations" section.

SIMULATIONS

In this section, some case studies are developed to verify the feasibility of the proposed schemes. A light fixed-wing UAV is considered, with its characteristics given in Table 1. In the simulations, where applicable, the initial condition of the UAV is assumed to be $[-1{,}800\text{ m} \;\; 2{,}500\text{ m} \;\; 240\text{ deg} \;\; 1\text{ deg/s} \;\; 300\text{ m} \;\; 10\text{ m/s} \;\; 10\text{ deg} \;\; 1\text{ deg/s}]^T$. For the first scheme, the reference signal is assumed to be $[100\text{ m} \;\; -20\text{ m} \;\; 500\text{ m}]^T$. For the described case study, simulations were carried out and the results are shown in Fig. 6.

As it can be readily seen from Fig. 6, after the initial transition, the UAV has aligned its motion such that it flies almost over the objective position on a straight line. Once on this line, the objective position is monitored by merely controlling the angle of the camera (see Fig. 7).

As it was expected, in this scheme, the camera angle will approach zero as the UAV flies toward the objective position. As the UAV flies away from the objective position, the camera angle will approach π, as $t \to \infty$.

Table 1. Characteristics of the light fixed-wing UAV.

Mass (kg)	Moment of inertia — z-axis (kg·m²)	Moment of inertia of the camera (kg·m²)	Forward velocity (m/s)
1	0.01	0.001	10

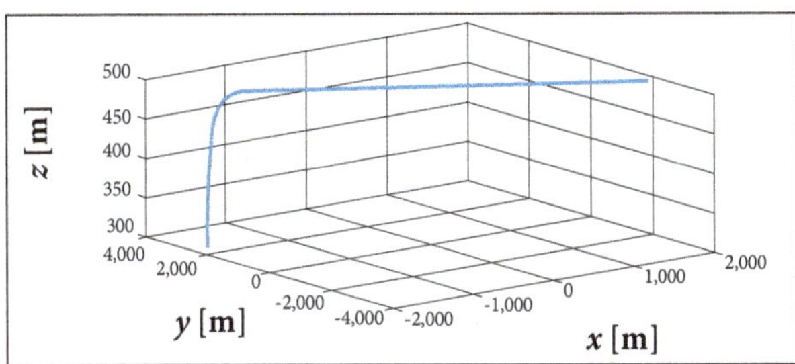

Figure 6. UAV trajectory obtained from scheme #1.

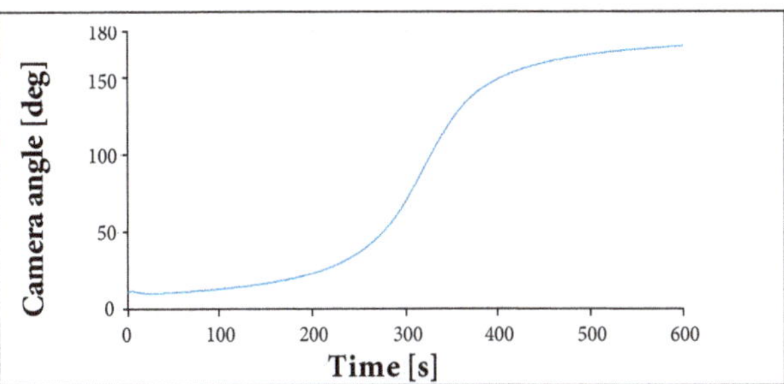

Figure 7. Forward-looking camera angle obtained from scheme #1.

For the second scheme, the reference signal is assumed to be $[-50\text{ m } -50\text{ m } 500\text{ m}]^T$. For the described case study, simulations were carried out and the results are shown in Fig. 8.

As it can be seen from Fig. 8, the UAV has successfully converged to a loitering motion around the objective position. Also, as expected, the camera angle converges to a fixed value in the loitering motion (see Fig. 9).

The initial conditions of the third scheme are assumed to be identical to those of the first scheme. Here, the reference signal is assumed to be $[R_0\text{ 500 m 30 deg}]^T$, where R_0 is the square distance of the UAV from the origin, at the initial time. Similar to the previous scenarios, simulations were carried out and the results are shown in Fig. 10.

Also, in this scheme, it is possible for the UAV to converge to a loitering circle with radius smaller/greater than the initial one. For the loitering radius of 1,500 and 4,500 m, simulations were carried out and the results are shown in Figs. 11 and 12, respectively. It can be seen from Figs. 11 and12 that the UAV has successfully converged to a loitering motion around the origin in both scenarios. Yet, it must be reminded that scheme #3 can fail if the UAV flies in the radial direction. Thus, it is recommended that one employs scheme #4 if a loitering motion is desirable.

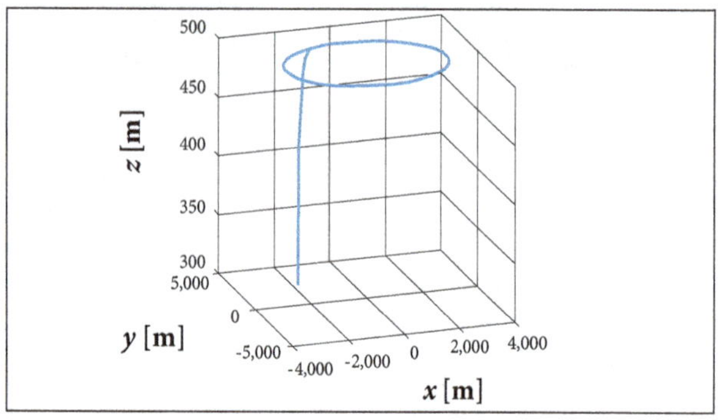

Figure 8. UAV trajectory obtained from scheme #2.

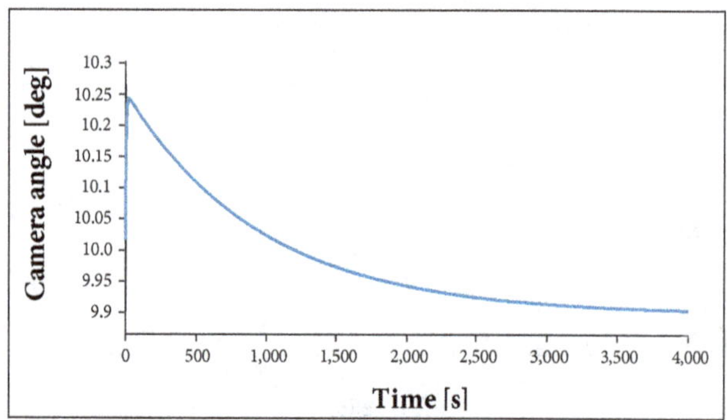

Figure 9. Side-looking camera angle obtained from scheme #2.

It is important to note that scheme #4 includes scenarios where the loitering radius can be smaller/greater than the initial distance of the UAV from the origin.

To verify the feasibility of scheme #4, a case study is developed in which the UAV is at the same initial condition as before. In the first example, let's assume that the UAV is desired to loiter around the origin with a radius of 1,500 m, a value smaller than its initial distance to the origin. For this case study, simulation results are shown in Fig. 13.

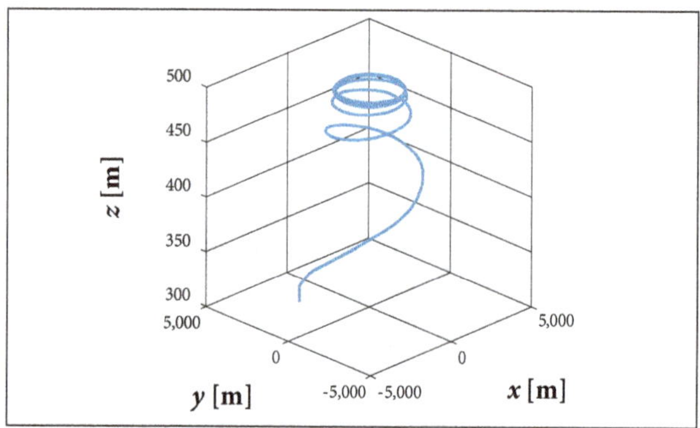

Figure 10. UAV trajectory obtained from scheme #3 — first example.

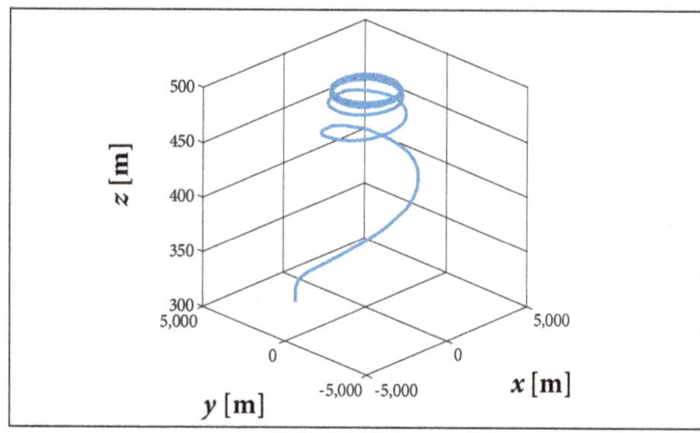

Figure 11. UAV trajectory obtained from scheme #3 — second example.

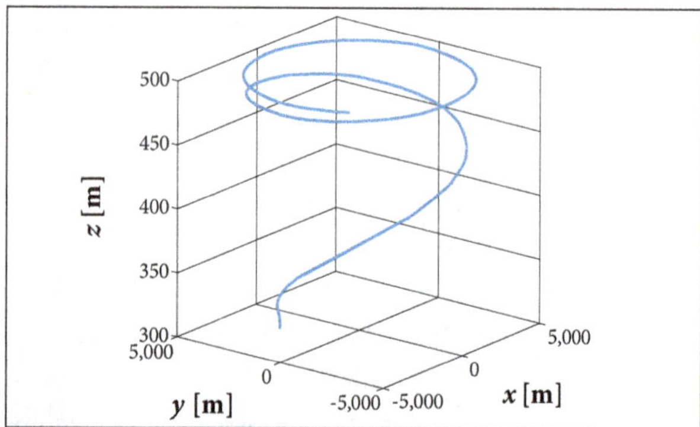

Figure 12. UAV trajectory obtained from scheme #3 — third example.

In Fig. 13, the first and the second phases of the path are shown in red and blue, respectively (see Fig. 4). As it can be seen from the figure, the UAV has successfully converged to the desired reference signal. In the second case study, assume that the UAV is desired to loiter around the origin with a radius of 4,000 m, a value greater than its initial distance to the origin. For this case study, simulation results are shown in Fig. 14, where the first and the second phases of the path are shown in red and the last phase is shown in blue (see Fig. 5). As it can be seen from Fig. 14, the UAV has successfully converged to a loitering motion around the origin with the desired loitering radius.

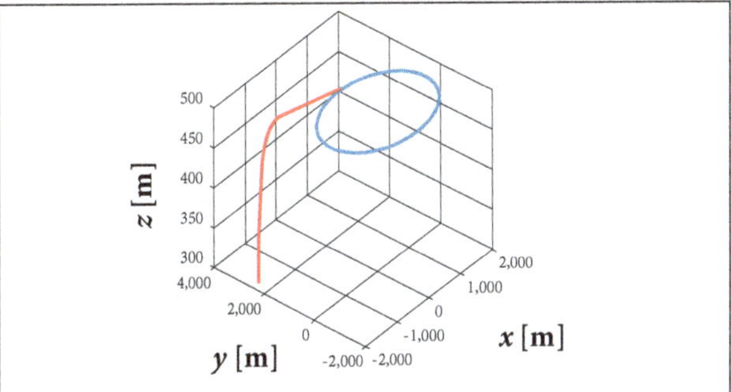

Figure 13. UAV trajectory obtained from scheme #4 — first case study.

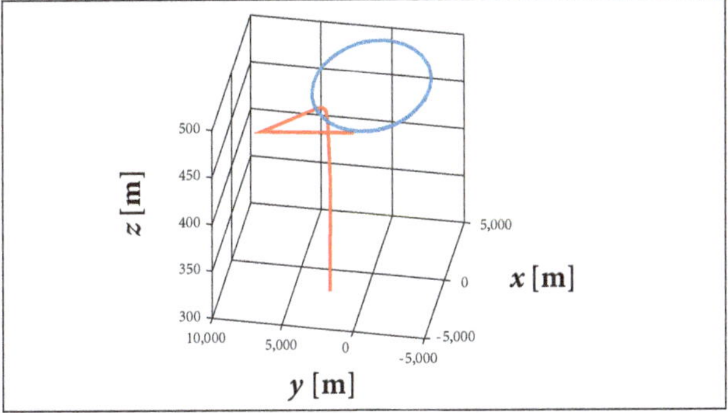

Figure 14. UAV trajectory obtained from scheme #4 — second case study.

CONCLUSION

In this paper, several feedback linearization schemes were studied to make a fixed-wing UAV, with constant forward velocity, (i) fly over or (ii) loiter around a stationary objective position. Throughout the paper, advantages and disadvantages of each scheme were discussed. A main drawback of the proposed schemes is that they do not take account of the maximum angular velocity constraint of fixed-wing UAVs. In scheme #1, the proposed method failed if the UAV was exactly above the objective position. Scheme #2 was disadvantageous in the sense that the loitering radius cannot be explicitly controlled. This was improved in scheme #3. Yet, the method in scheme #3 failed when the UAV had to fly radially inward or radially outward. However, if the heading of the UAV did not fall within this region, the UAV converged to a loitering motion around the origin. In scheme #4, the UAV had to be far enough from the objective circle. In this manner, one could assume that the UAV reaches the tangency point with its heading far from inward-outward direction, and therefore scheme #3 could bring the UAV to loiter around the objective position, with desirable radius. It is important to note that, in scheme #3 and scheme #4, the switching from a control strategy to another was given as an equality-type condition. Therefore, for real-world implementation, thresholds must be defined, and the equalities must be replaced by appropriate inequalities.

ACKNOWLEDGEMENTS

The authors gratefully acknowledge the financial support of Conselho Nacional de Desenvolvimento Científico e Tecnológico (CNPq), Financiadora de Estudos e Projetos (FINEP), Fundação de Amparo à Pesquisa do Estado de Minas Gerais (FAPEMIG) and Coordenação de Aperfeiçoamento de Pessoal de Nível Superior (CAPES), Brazil.

REFERENCES

Almurib HAF, Nathan PT, Kumar TN (2011) Control and path planning of quadrotor aerial vehicles for search and rescue. Proceedings of the IEEE SICE Annual Conference; Tokyo, Japan.

Beard RW, McLain TW, Nelson DB, Kingston D, Johanson D (2006) Decentralized cooperative aerial surveillance using fixed-wing miniature UAVs. Proc IEEE 94(7):1306-1324. doi: 10.1109/JPROC.2006.876930

Bonyan Khamseh H, Pimenta LCA, Tôrres LAB (2014) Decentralized coordination of constrained fixed-wing unmanned aerial vehicles: circular orbits. Proceedings of the IFAC World Congress; Cape Town, South Africa.

Casbeer DW, Kingston DB, Beard RW, McLain TW, Li SM, Mehra R (2006) Cooperative forest fire surveillance using a team of small unmanned air vehicles. Int J Syst Sci 37(6)351-360. doi: 10.1080/00207720500438480

Costa FG, Ueyama J, Braun T, Pessin G, Osório FS, Vargas PA (2012) The use of unmanned aerial vehicles and wireless sensor network in agricultural applications. Proceedings of the IEEE International Geoscience and Remote Sensing Symposium; Munich, Germany.

Fan W, Zhiyong G (2009) An approach to formation maneuvers of multiple nonholonomic agents using passivity techniques. Proceedings of the Chinese Control and Decision Conference; Guilin, China.

Frew E, Lawrence D, Morris S (2008) Coordinated standoff tracking of moving targets using Lyapunov guidance vector fields. J Guid Contr Dynam 31(2):290-306. doi: 10.2514/1.30507

Gonçalves MM, Pimenta LCA, Pereira GAS (2011) Coverage of curves in 3D with swarms of nonholonomic aerial robots. Proceedings of the IFAC World Congress; Milano, Italy.

Gonçalves VM, Pimenta LCA, Maia CA, Dutra BCO, Pereira GAS (2010) Vector fields for robot navigation along time-varying curves in n-dimensions. IEEE Trans Robot26(4):647-659. doi: 10.1109/TRO.2010.2053077

Hafez AT, Marasco AJ, Givigi AN, Beaulieu A, Rabbath CA (2013) Encirclement of multiple targets using model predictive control. Proceedings of the American Control Conference; Washington, USA.

Hsieh MA, Kumar V, Chaimowicz L (2008) Decentralized controllers for shape generation with robotic systems. Robotica 26(5):691-701. doi: 10.1017/S0263574708004323

Hsieh MA, Loizou S, Kumar RV (2007) Stabilization of multiple robots on stable orbits via local sensing. Proceedings of the IEEE International Conference on Robotics and Automation; Rome, Italy.

Kanchanavally S, Ordonez R, Schumacher CJ (2006) Path planning in three dimensional environment using feedback linearization. Proceedings of the American Control Conference; Minneapolis, USA.

Li H, Wang B, Liu L, Tian G, Zheng T, Zhang J (2013) The design and application of SmartCopter: an unmanned helicopter based robot for transmission line inspection. Proceedings of the Chinese Automation Congress; Changsha, China.

Marasco AJ, Givigi SN, Rabbath CA (2012) Model predictive control for the dynamic encirclement of a target. Proceedings of the American Control Conference; Montreal, Canada.

14

Wind Tunnel Testing on a Generic Model of a Hybrid Lifting Hull

Anwar Ul Haque[1], Waqar Asrar[1], Asharf Ali Omar[2], Erwin Suleiman[1]

ABSTRACT: In this research, an experimental investigation was carried out at the International Islamic University Malaysia - Low Speed Wind Tunnel facility on a generic model of a hybrid lifting hull. Based on the historical trends of non-rigid airships, the fineness ratio of the said hull has been selected equal to 4. Free stream velocity was kept at 20 $m \cdot s^{-1}$ and, along with the estimation of aerodynamic parameters, longitudinal and lateral stability characteristics were determined over a range of angles of attack from −8° to +12° and angles of sideslip from −10° to +10°. Zero lift coefficient was obtained at −4.2°, and the corresponding value was found to be greater than that at zero angle of attack. The comparison of the experimental results with the existing analytical relationships of wing has revealed that such an airfoil shaped hull cannot be considered as a wing due to 37% less analytical value of lift coefficient than that obtained by CFD simulations of the said hull. Existing equation of form factor of hull for conventional airships was also revisited, and a correction factor equal to 1.16 in the fundamental drag equation of aircraft's fuselage was also proposed for fineness ratio equal to 4. Trends of the experimental data and comparison of the same with the theoretical calculations and computational results posed some interesting findings. The longitudinal and directional stabilities of a hybrid lifting hull were found to be statically unstable.

KEYWORDS: Aerodynamics, Hybrid airship, Fineness ratio, Static stability, Wind tunnel testing.

INTRODUCTION

The concept of lifting fuselage for aircraft and hybrid lifting hull for hybrid airship is derived from nature as a few marine animals do generate aerodynamic lift from the body (Ul Haque *et al.* 2016a). Vogel (1994) was among the first who noticed that, if one looks at a housefly or fruit fly from the side, the head, thorax and abdomen also seem to form a non-symmetrical airfoil-flatter on the bottom, more rounded on top. Regarding marine animals, there are some cases in which there is a requirement of negative lift (Vogel 2013). For example, ducks have enough air in their plumage so they are awkwardly buoyant and may need negative lift (Prange and Schmidt-Nielsen 1970). Recently, Ul Haque *et al.* (2015a) argued that the lift generated by the body is perhaps free of cost lift and can be utilized if such a marine animal want to swim at constant level/height in the sea. Additional lift is also required for those flight segments like coming out from water, sharp turns etc.

In the field of aviation, it is not a new concept and, as per the review paper by Wood and Bauer (2011), Vincent Burnelli has earlier designed a number aircraft based on the lifting fuselage concept. Perhaps, the lifting body designs usually have greater operational flexibility as the adjustment of dynamic lift can accommodate changes in vehicle's weight due to the burning of fuel. In an earlier reference, Wood (2003) mentioned that, during the period of 1920 to 1955, about 57 aircraft were developed, but these were either single or 2-seat. Other hybrid concepts have aerodynamic lift coming from the wing as well as from the fuselage. This concept was later applied by Santos Dumont (1973) for the design of a winged hybrid airship in which partial

1.International Islamic University Malaysia – Department of Mechanical Engineering – Experimental and Computational Thermofluid Mechanic Research Group – Kuala Lumpur – Malaysia. **2.**University of Tripoli – Department of Aeronautical Engineering – Tripoli – Libya.

Author for correspondence: Waqar Asrar | International Islamic University Malaysia – Department of Mechanical Engineering – Experimental and Computational Thermofluid Mechanic Research Group | Jalan Gombak | 53100 – Kuala Lumpur – Malaysia | Email: waqar@.iium.edu.my

aerodynamic lift is obtained from the aerodynamic contour of hull; he was the first to prove this concept by flight testing.

According to Becker (1958), if we look back in the history, the first lifting-body concepts involved very blunt half cones. Later, the concepts evolved by Saltzman *et al.* (1999) into higher fineness-ratio cones to achieve the capability of an unpowered horizontal landing. Numerous wind tunnel tests were performed on candidate versions of the half cones and shapes having flattened bottom surfaces. In 1962, Reed and Darlene (1977) have provided the details of an unpowered horizontal landing and controllable flight with a miniature lightweight-radio-controlled model of an M_2 half-cone configuration. Unfortunately, none of them has explored the airfoil shaped hull from the aerodynamic and stability point of view.

Several studies can be found in the literature starting from the early 1920s, such as Munk (1924) and Rizzo (1924), and the late 1970s and early 1980s, like DeLaurier and Schenck (1979) and Tischler *et al.* (1981), as well as many others, about the aerodynamic and stability characteristics of airships. However, in the case of hybrid airships, there is limited experimental data available for aerodynamic lift generated by the aerodynamic lifting profile of hull (alone) and its stability characteristics. Most of the experimental data is related to the lifting bodies of hypersonic vehicles. For example, some experimental data can be found in an old NACA report by Gumse (1967), which is based on W-F_2 configuration. As per Ash (1972), this was a modified design of M_2-F_2 with modifications carried out for the afterbody, the control surfaces, and the canopy location. Others include this on NASA HL-20 to develop a preliminary subsonic aerodynamic model for simulation studies of its lifting body Jackson and Christopher (1992). Nevertheless, all of these bodies have fineness ratio (λ) greater than 6 and wings blended with the fuselage.

Hybrid airships are among the potential candidates for tourism industry and transportation of agricultural products. Such vehicles also have the potential to reduce the aviation transport gas emissions and many have come up with promising conceptual designs like Aeroscraft (2013), the airship Sky Freighter (Millennium Air Ship Inc. 2012) and Lockheed's LEMV (Harrison 2010). Unfortunately, the aerodynamics and stability characteristics of airfoil shaped hulls (alone) of airships have not been fully explored yet. One of the probable reasons is due to the non-availability of the experimental data which can reveal its aerodynamic behavior. In the present study, an effort was done to fill this gap by carrying out an experimental study on a generic model of a hybrid lifting hull (HLH). The selection of its λ value was done such that it falls within the known range of non-rigid airships, discussed in detail in the following section.

SELECTION OF FINENESS RATIO

HLH is a type of unconventional hull with voluminous volume for the buoyant lift. It is usually designed to be partially supported by buoyancy lift generated by the buoyant gas while the remaining weight is held up by the aerodynamic lift generated by the aerodynamic contour of the hull (Trenkle 2014). However, the selection of λ is found to be quite trivial, especially in a scenario where there is no guideline/data bank from the historical trends. From the certification point of view, most of the certified airships are flexible in terms of structural anatomy, *i.e.* non-rigid airships. But from an aerodynamic point of view, high value of λ is always desired. However, a low value is desired for optimum structural weight Tanaka *et al.* (2005). If we look at the historical trends of non-rigid airships (Table 1), the lowest value of λ was found in the first winged airship, which was designed and flown by Santos Dumont (1973); DM-20, on the other hand, has the highest value.

Table 1. Fineness ratio of non-rigid airships.

Serial number	Airship title	λ
1	WDL-1	3.793
2	WDL-2	3.89
3	WDL-3	4.0
4	Star	2.507
5	Santos-Dumont	2.561
6	AD-500	3.571
7	B-10	3.864
8	B-12 bis	3.98
9	DM-20	4.05

In the present study, the aerodynamic contour of the generic model of HLH was designed by using RONCZ 1082 airfoil till 70% of the chord length, and the λ value was set at 4. The length and the diameter of the hull were 0.56 and 0.135 m, respectively. Against this λ value, the form factor of airship and aircraft are a bit close to each other. But as per the findings of Tanaka *et al.* (2005), the effects of pressure drag will be there for a hull having $\lambda < 5$.

EXPERIMENTAL SETUP

In order to find the aerodynamic and static stability behavior of a HLH model, tests were conducted at the International Islamic University Malaysia - Low Speed Wind Tunnel (IIUM-LSWT) at Reynolds number (*Re*) equal to 6.3×10^5 against free stream velocity equal to 20 $m \cdot s^{-1}$. IIUM-LSWT is a closed-loop wind tunnel with a test section of dimensions 1.5 × 2.3 × 6 m and a maximum achievable speed equal to 50 $m \cdot s^{-1}$ (Hasim *et al.* 2008). The dynamic pressure in the test section varies from −0.5 to 0.4% from the plane mean value, and the flow angularity holds within ±0.2° (Wiriadidjaja *et al.* 2012).

It is well known that the scaled-down models for the wind tunnel testing are either simplified or have extra bolts and nuts for attachment purposes, so this drawback of small-scale models cannot reproduce a ditto copy of the full-scale model. Therefore, the scaled-down model of HLH of aspect ratio (A_R) equal to 0.25 is manufactured in a single piece and polished to get a gloss finish. Such a wooden model cannot be made hollow from side due to issues related to its attachment with the strut. In order to fulfill the requirement of blockage ratio, defined by Pope and Rae (1984), the size of HLH model is kept small, and its blockage ratio including that of the strut was just 2.5% of the cross-sectional area of the test section. The model is attached with the strut in the test section by manufacturing an adopter made of stainless steel. However, similarly to any other wind tunnel model, this adopter will create an unavoidable cavity in the model. Hence, it is placed inside the hull body to avoid any additional drag due to model-strut attachment. Moreover, due to the requirement of the estimation of yaw stability parameters, the option of half-model testing is not explored.

Sign conventions are used in the present study such that positive axial force is along the positive *x* axis (forward from tail to nose), and positive normal force is along the positive *z* axis (upwards). The lift and drag coefficients (C_L and C_D) are estimated in the wind-axis system, and the results so obtained were plotted without any curve fitting — wing span (*b*) and chord length (*c*) equal to 0.135 and 0.54 m, respectively. The area (*S*) equal to 0.042 m^2 was the reference parameter used for the data reduction. All the moments were obtained about the moment reference center (MRC) of the balance, which is at 0.14 m from the nose of the HLH. In the case of airships, as well as in the hybrid ones, the gondola is usually located at the base of the hull. Therefore, the offset distance in the vertical *z* direction was defined accordingly.

Because the data measured by balance includes the weight components of the model in a special state, it is therefore necessary to deduct the weight components to obtain the real aerodynamic coefficients. The method for deducting the influence was collecting the measurement data without wind, *i.e.* wind-off and model-off (W_0M_0) condition, and then it is deducted in the test matrix. Tests were also conducted to estimate the drag and pitching moment of the strut and fairing, referred here as wind-on and model-off (W_1M_0) condition. As per the guidelines provided by Tucker (1990) to obtain the real aerodynamic coefficients for wind-on and model-on (W_1M_1) condition, results so obtained from wind tunnel testing were subtracted from the W_1M_0 and W_0M_0 conditions. The pictorial views of tests conducted for W_1M_0 and W_1M_1 conditions are shown in Figs. 1a and 1b, respectively, and the major dimensions of the HLH in the side and top views are shown in Figs. 2a and 2b, respectively.

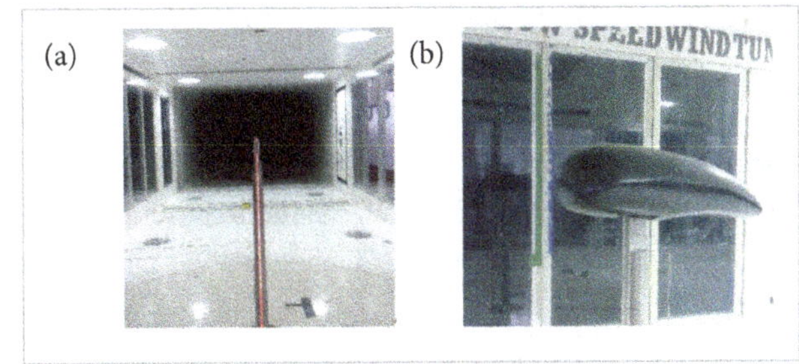

Figure 1. Pictorial view of W_1M_0 and W_1M_1 tests. (a) Strut alone test; (b) Model attached inside the test section.

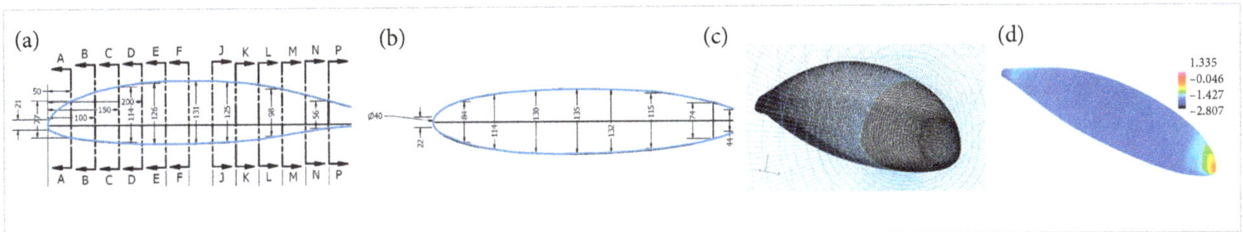

Figure 2. Geometric details of HLH along with its surface grid and distribution of pressure coefficient at $R_e = 6.3 \times 10^5$. (a) Side view; (b) Top view; (c) Surface grid; (d) Distribution of pressure coefficient over the hull's surface.

RESULTS AND DISCUSSION

AERODYNAMICS AND STATIC STABILITY DERIVATIVES

In order to get the 1st-order approximation of aerodynamics and static stability derivatives of the HLH, the steady-state simulations were run by using ANSYS® FLUENT software for which the SIMPLE scheme is employed for pressure velocity coupling along with the *k*-*ω* SST model. The details of the geometry used for the mentioned purpose are shown in Figs. 2a and 2b. Structured mesh is generated for the said purpose with additional focus on grid refinement at the nose (Fig. 2c). No slip condition is used for the wall surface. Since the temperature problem is not of interest, the adiabatic wall conditions with no slip boundary condition are employed. Velocity magnitude, along with its direction components, is specified for the velocity inlet. Outflow boundary is also used, for which all the flow variables are basically extrapolated for incompressible flow. For subsonic flows, it is required to apply the boundary conditions on the computational domain of radius equal to 20 times the length of the HLH. Therefore, the domain extents are kept at 20 times the length of the body to avoid the flow variations near the surface. Coupled explicit solver is used with 2nd-order accuracy. All the simulations were run on an Intel® hex-core dual processor, 3.33-GHz system, with 32 GB of RAM, and the results were overlapped with those obtained from the wind tunnel testing (Fig. 3). A stagnation point is observed at the nose of the HLH, as shown from the contours of the pressure coefficient (Fig. 2d).

For longitudinal tests ($\alpha \neq 0, \beta = 0$, where α is the angle of attack and β is the side-slip angle), α values were changed from −8° to +12°, and, for β sweep test, the model strut assembly remained rested on tunnel turntable and rotated at a range from −10° to +10° with 5°-increments. The graphs in Fig. 3 show the trends of aerodynamic and static stability characteristics of HLH in longitudinal as well as in the lateral direction. The lift curve slope (C_{L_α}) value is obtained by applying the 2nd-order polynomial fit on the curve shown in Fig. 3a. Its value comes out to be equal to 0.0067/deg. This figure also shows a continuous increase in C_L for a defined range of α and a reduction in its value for a negative α. Moreover, there will certainly be a stall value of α beyond 12°; however, if a wing is also attached to it, then the critical value will be that of the wing. Thus, the stalling phenomenon which gives rise to reduction in C_L and dramatic rise in C_D is not observed in these graphs. The angle of attack corresponding to zero-lift condition (α_{OL}) corresponding to C_{L_0} condition is equal to −4.2° and its corresponding C_D value, *i.e.* zero lift drag coefficient (C_{D_0}), is equal to 0.028 (Fig. 3b). It can also be observed from this figure that the C_D at zero angle of attack (0.041) was lower than that at zero-lift condition. Moreover, the induced drag is perhaps responsible for high values of C_D values at a higher angle of attack.

Figure 3c illustrates the pitching moment coefficient for defined values of the angle of attack. Pitching moment coefficient

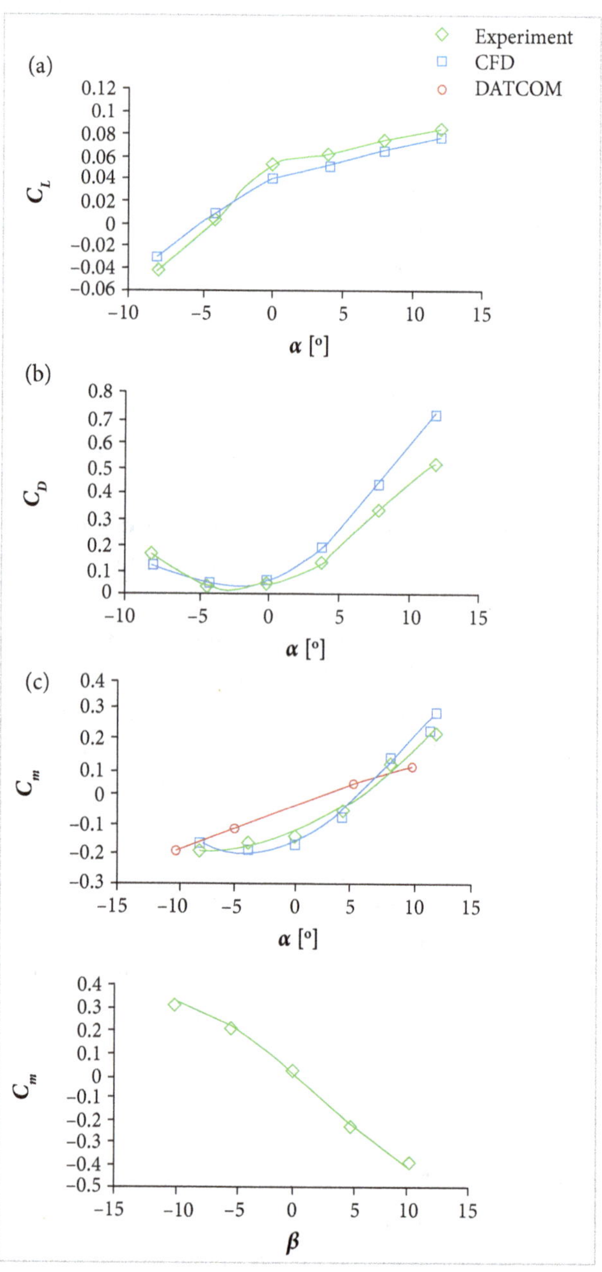

Figure 3. Experimental results of aerodynamic and stability coefficients of HLH. (a) Variation in C_L w.r.t α; (b) Variation in C_D w.r.t α; (c) Variation in C_m w.r.t α; (d) Variation in C_n w.r.t β.

exhibits the same behavior as the drag, with a small difference at negative angles of attack. In addition, the slope of pitching moment (C_{m_α}) is negative for the entire angle of attack range, indicating that it is statically unstable throughout this range. C_{m_α} is positive and its value is 0.016/deg, indicating that the HLH model under study is longitudinally unstable, and the coefficient of pitching moment at zero lift (C_{m_0}) is 0.137 (Fig. 3c). It is an important design parameter as it is always added to the C_{m_0} values of other components, such that the total C_{m_0} can be made available for the fundamental stability equation. Similarly to the C_{L_α} value, the slope of yawing moment coefficient (C_{n_β}) was also obtained by curve fitting and its value is equal to –0.01/deg. An anomaly was seen for the β sweep curve, *i.e.* an unsymmetric pattern in the value of the yawing moment coefficient (C_n) for positive and negative β was observed. One of the probable reasons is due to the asymmetry in the model, made of wood. All these results are presented in Table 1 for quick reference.

Wind tunnel testing always reflects the real flow over defined shape with the help of reliable data. Such results can be employed to develop as well as to check the applicability of known analytical relationships. Therefore, the experimental results are then compared with the existing analytical relationships, specially the slopes.

COMPARISON WITH EXISTING ANALYTICAL RELATIONSHIPS

If we consider HLH as a wing, then C_{L_α} obtained by using Eq. 1 is equal to 0.00422/deg. It is important to note that Eq. 1, taken from Raymer (2012), will provide C_{L_α} in per radian, and, for the purpose of comparison of results, this digit has been converted into per degree.

$$C_{L_\alpha}=\frac{\pi A_R}{2+\sqrt{4+A_R}} \tag{1}$$

The C_{L_α} value predicted by using the above-mentioned relationship is 0.0054/deg. This value is lesser than that predicted by the experiment, which is equal to 0.006/deg. Nevertheless, the experimental value is consistent with that obtained from the computational fluid dynamics (CFD) results. The basic difference is due to the fact that this relationship is for estimation of lift produced by the wing (a lifting surface) and which considers the wing as a flat plate. In our case, the (lifting body) hull is not a flat plate, and a wing with A_R equal to 0.25 is hardly seen in any aircraft. Moreover, the analytical relationships estimated 37% less value of lift coefficient than that obtained by using CFD. Therefore, the use of analytical relationship of the wing for estimation of aerodynamic lift produced by lifting hull will not be suitable. Experimental data of space shuttles are available, but the A_R for such lifting body includes the area of wings attached to it. The C_{L_α} values of M_2-F_2 ($A_R = 0.712$) and space shuttle prototype ($A_R = 2.26$) are 0.02 and 0.0446/deg, respectively (Paulson *et al.* 1960). Hence, considering the HLH as a wing planform cannot be justified for the airfoil shaped hull being investigated for airship.

C_{D_0} is an important parameter required for aerodynamic analysis and design-related studies (Roskam and Lan 1997). Its exact estimation is very important for accurate prediction of performance parameters of any flying vehicle. The analytical expression for C_{D_0} is given in Eq. 2, taken from Nicolai and Carichner (2013). This relationship is basically the function of form factor (FF). For the case of airship's hull, it was defined with the help of Eq. 3, and, for aircraft's fuselage, Eq. 4 is usually used.

$$C_{D_0}=C_f\,FF.Q.\frac{S_{wet}}{S_{ref}} \tag{2}$$

$$FF=1+\frac{1.5}{(FR)^{\frac{3}{2}}}+\frac{7}{(FR)^3} \tag{3}$$

$$FF=1+\frac{60}{(FR)^3}+\frac{FR}{400} \tag{4}$$

where: S_{wet} is the wetted area and S_{ref} is the reference one, both in m^2; C_f is the coefficient of skin friction; Q is the dynamic pressure; FR is the fineness ratio.

Against the calculated value, if C_f is equal to 0.00488, considering the hull of airship as fuselage (Eq. 3), then the value of FF comes out to be 2.35, which is quite high when compared with FF equal to 1.94, obtained from Eq. 4 (Raymer, 2012). For airship's hull of any λ value, this author suggested an adjustment factor of 0.8 in basic drag equation to include scaling effects for the airship's hull. For the value of S_{wet}/S_{ref} equal to 2.62, regarding the geometry under study, the values of C_{D_0} obtained by using Eqs. (3) and (4) are 0.029 and 0.024, respectively. The predicted value of C_{D_0} obtained from the CFD was quite high and equal to 0.048. Although, Eq. 3, which is taken from Nicolai and Carichner (2013), is not really a FF by the same definition as used by Hoerner (1965), but it has provided quite close results to the experimental value. According to a recent review of FF by Ul Haque *et al.* (2016b), Eq. 3 is taken from Hoerner (1965), in which the definition of FF is actually based on a constant frontal area as reference, and the term S_{wet}/S_{ref} was defined to be equal to 3FR (Eq. 5). In fact, it is a simple adjustment which allows the use of S_{ref} for the calculation of C_D by using Eq. 6:

$$\frac{C_{D0}}{C_f}=3FR\left[1+\frac{1.5}{(FR)^{\frac{3}{2}}}+\frac{7}{(FR)^3}\right] \tag{5}$$

$$\frac{C_{D0}}{C_f}=FR\frac{S_{wet}}{S_{ref}} \tag{6}$$

Against the value of FR equal to 4, if we use Eq. 5, then its value is 0.137. This huge digit is due to the fact that Hoerner (1965) had derived the frontal area as reference. Moreover, based on the experimental results and its comparison study, we suggest a correction factor of 1.16 for the FF value obtained by using Eq. 4 and the planform area as reference. It also shows that, in comparison with conventional hull, an airfoil shaped hull will have more drag count.

The comparison of analytical values of C_{m_0} and C_{m_α} and those obtained from the experimental data is perhaps not straight forward, because the existing relationships add the effect of the wing as well in Eqs. 7 and 8. If we closely inspect these 2 equations, shape effects were considered for the value of C_{m_0}, but, surprisingly, it has not catered for estimating the value of the slope of pitching moment of the fuselage ($C_{m_{\alpha\,fus}}$) (Ul Haque *et al.* 2015b).

$$C_{m_{0\,fus}}=\frac{k_2\text{-}k_1}{36.5.S.C_{MGC}}\sum_{x=o}^{x=l_{fus}} w_f^2(\alpha_{ZL_W}+i_f)dx \tag{7}$$

$$C_{m_{\alpha\,fus}}=\frac{k_2\text{-}k_1}{36.5.S.C_{MGC}}\sum_{x=o}^{x=l_{fus}} w_f^2\frac{\partial\beta}{\partial\alpha}dx \tag{8}$$

where: k_2 - k_1 is the correction factor to account for the fuselage slenderness ratio; w_f is the average width of the fuselage section, in m; α_{ZLW} is the wing zero-lift angle relative to the fuselage reference line, in degrees; $C_{m_{\alpha\,fus}}$ is the pitching moment coefficient of the fuselage at zero lift; C_{MGC} is the pitching moment coefficient about the mean geometric chord.

The pitching moment contribution of the fuselage of general aviation aircraft can be approximated by Eq. 9, taken from Raymer (2012). In this relationship as well, the presence of wing can be observed from the factor C and the reference area of the wing (S_w, in m^2). W_f is the maximum width of the fuselage, L_f is the length and K_f is the empirical pitching moment coefficient:

$$C_{m_{\alpha\,fus}}=\frac{K_f W_f^2 L_f}{C.S_w} \tag{9}$$

In this case, we use a low-fidelity tool like Aircraft Digital DATCOM®, then the results so obtained reveal that it has underpredicted C_{m_0} and overpredicted C_{m_α}, when compared with the experimental and computational values. One of the probable reasons is due to the interference of strut with HLH model. On the other side, CFD underpredicted the C_{m_0} value when compared with the experiment, which is equal to −0.187. However, the C_{m_α} value obtained from CFD (0.0204/deg) is higher than that of the experiment, 0.0165/deg. The opposite is true when CFD results are compared with the DATCOM® ones. Similarly to C_{L_α}, C_{n_β} is obtained and is equal to −0.01/deg. Aircraft Digital DATCOM® does not provide the directional stability contribution at individual angles but provides the value of C_{n_β}, which is equal to −0.0065/deg for this case of HLH. The comparison of C_L and C_D is done as Aircraft Digital DATCOM® does not compute lift for an aircraft configuration which has no wing attached to the fuselage. Even if we add a fictitious wing in the hull, based on our experience on similar hybrid lifting fuselage, the lift estimated by CFD is 433.3% higher than that predicted by using the Aircraft Digital DATCOM® , since that produced by the HLH is also responsible for generating the induced drag. Therefore, the comparison of C_D is also out of question. However, the same is not true for the case of pitching moment as the analytical relation derived by Munk-Multhopp (see Eqs. 7 and 8) contains the fuselage camber incidence angle (i_f) to represent the curvature effects of the fuselage (Multhopp 1942).

The wind tunnel test data has provided the aerodynamic as well as static stability characteristics of an HLH. This data can be used in flight performance estimation and system design of hybrid buoyant aerial vehicles. Except for the value of C_D, the CFD results underpredicted the lift and pitching moment coefficients. Moreover, additional combinations of pitch and yaw angles can also be analyzed in the future to be added to the aerodynamic and stability database of such unconventional shaped hull.

CONCLUSION

During α sweep tests, it was observed that the C_L was increased with some initial value. If we consider a hull as a wing, then a poor result of lift curve slope is obtained. The existing formula for airship FF has provided quite reasonable comparison with the experimental value. Overall, the CFD results were in good agreement with the wind tunnel ones. Moreover, the HLH is statically unstable in longitudinal and directional modes in the range of specified angles of attack and side slip angles.

ACKNOWLEDGEMENTS

The support of the Ministry of Science, Technology and Innovation (MOSTI), Malaysia, under the grant 06-01-08-SF0189, is gratefully acknowledged. The authors are thankful to the team of IIUM-LSWT for providing assistance during the experimental phase.

AUTHOR'S CONTRIBUTION

Haque AU conceived the idea, conducted the experiments and completed the literature review. Asrar W post process the raw data of wind tunnel, drafting of the article along with the CFD work for the purpose of comparison of the experimental data of the test article. Omar AA has done the data analysis and interpretation. Suleiman E discussed the findings of the research work and critical revision of the article.

REFERENCES

Aeroscraft (2013) Launching a new century of aviation innovation; [accessed 2013 Apr 2]. http://www.aerosml.com/imagegallery.html

Ash LG (1972) Flight test and wind tunnel performance characteristics of the X-24A lifting body; [accessed 2016 Aug 26]. http://oai.dtic.mil/oai/oai?verb=getRecord&metadataPrefix=html&identifier=AD0901465

Becker JV (1958) Preliminary studies of manned satellites. Winged configurations. Proceedings of the NACA Conference on High-Speed Aerodynamics; Hampton, USA.

DeLaurier J, Schenck D (1979) Airship dynamic stability. Proceedings of the AIAA Lighter-Than-Air Systems Technology Conference; Palo Alto, USA.

Gumse B (1967) Low-speed wind-tunnel tests of a full-scale M2-F2 lifting body model. NASA TM X-1347.

Harrison J (2010) Lockheed's LEMV consolation; [accessed 2012 May 8]. http://edgefighter.com/2010/06/17/lockheeds-lemvconsolation

Hasim F, Rusyadi R, Surya WI, Asrar W, Omar AA, Mohamed Ali JS, Kafafy R (2008) The IIUM low speed wind tunnel. Proceedings of the 2nd Engineering Conference on Sustainable Engineering Infrastructures Development & Management (EnCon2008); Kuching, Malaysia.

Hoerner SF (1965) Fluid-dynamic drag: theoretical, experimental and statistical information. Bakerfield: published by the author.

Jackson EB, Christopher IC (1992) Preliminary subsonic aerodynamic model for simulation studies of the HL-20 lifting body. NASA TM-4302.

Millennium Air Ship Inc. (2012) The airship "Sky Freighter"; [accessed 2012 May 8]. http://www.millenniumairship.com/products.html

Multhopp H (1942) Aerodynamics of the fuselage. NACA-TM-1036.

Munk MM (1924) The aerodynamic forces on airship hulls. NACA-TR-184.

Nicolai LM, Carichner GE (2013) Fundamentals of aircraft and airship design. Vol. 2: airship design and case studies. Reston: American Institute of Aeronautics and Astronautics.

Paulson JW, Shanks RE, Johnson JL (1960) Low-speed flight characteristics of reentry vehicles of the glide-landing type. NASA TM-X-67563.

Pope A, Rae WH (1984) Low-speed wind tunnel testing. 2nd edition. New York: Wiley.

Prange HD, Schmidt-Nielsen K (1970) The metabolic cost of swimming in ducks. J Exp Biol 53:763-777.

Raymer DP (2012) Aircraft design: a conceptual approach. 5th edition. Reston: American Institute of Aeronautics and Astronautics.

Reed R, Darlene L (1977) Wingless flight: the lifting body story. NASA SP-4220; [accessed 2016 Aug 26]. http://history.nasa.gov/SP-4220/contents.htm

Rizzo F (1924) A study of static stability of airships. NACA Technical Note 204.

Roskam J, Lan CTE (1997) Airplane aerodynamics and performance. Lawrence: DAR Corporation.

Saltzman EJ, Wang KC, Iliff KW (1999) Flight-determined subsonic lift and drag characteristics of seven lifting-body and wing-body reentry vehicle configurations with truncated bases. Proceedings of the 37th Aerospace Sciences Meeting and Exhibit; Reno, USA.

Santos Dumont A (1973) My airships; the story of my life. New York: Dover Publications. (Original work published in 1904).

Tanaka K, Hamaguchi Y, Maekawa S, Ishikawa T (2005) Structures and advanced composites for lighter-than-air aircraft. Proceedings of the 56th International Astronautical Congress; Fukuoka, Japan.

Tischler MB, Jex HR, Ringland RF (1981) Simulation of Heavy Lift Airship dynamics over large ranges of incidence and speed. Proceedings of the Lighter-Than-Air Systems Technology Conference; Annapolis, USA.

Trenkle K (2014) Hybrid Certification Criteria (HCC) for transport category hybrid airships document number: 1008D0122. Revision: C; [accessed 2014 Jun 4]. https://www.federalregister.gov

Tucker VA (1990) Measuring aerodynamic interference drag between a bird body and the mounting strut of a drag balance. J Exp Biol 154:439-461.

Ul Haque A, Asrar W, Omar AA, Sulaeman E, Mohamed Ali JS (2015a) Cambered profile of a California sea lion's body. J Exp Biol 218:1270-1271. doi: 10.1242/jeb.117556

Ul Haque A, Asrar W, Omar AA, Sulaeman E, Mohamed Ali JS (2015b) Preliminary aerodynamic and static stability analysis for hybrid buoyant aerial vehicles at low speeds using digital DATCOM. Can Aeronaut Space J 61(3):51-60. doi: 10.5589/q16-001

Ul Haque A, Asrar W, Omar AA, Sulaeman E, Mohamed Ali JS (2016a) Hydrodynamic contour of Steller sea lion — an inspiration for future hybrid buoyant aircraft. Anim Rev 3(1):1-9. doi: 10.18488/journal.ar/2016.3.1/101.1.1.9

Ul Haque A, Asrar W, Omar AA, Sulaeman E, Mohamed Ali JS (2016b) Assessment of engine's power budget for hydrogen powered hybrid buoyant aircraft. Propulsion and Power Research 5(1):34-44. doi: 10.1016/j.jppr.2016.01.008

Vogel S (1994) Life in moving fluids: the physical biology of flow. Princeton: Princeton University Press.

Vogel S (2013) Comparative biomechanics: life's physical world. Princeton: Princeton University Press.

Wiriadidjaja S, Hasim F, Mansor S, Asrar W, Rafie ASM, Abdullah EJ (2012) Subsonic wind tunnels in Malaysia: a Review. Appl Mech Mater 225:566-571. doi: 10.4028/www.scientific.net/AMM.225.566

Wood RM (2003) The contributions of Vincent Justus Burnelli. Proceedings of the 41st AIAA Aerospace Sciences Meeting and Exhibit; Reno: USA.

Wood RM, Bauer SXS (2011) Flying wings/flying fuselages. Proceedings of the 39th Aerospace Sciences Meeting and Exhibit; Reno, USA.

15

Mathematical Modeling And Numerical Simulation of Dropwise Condensation on an Inclined Circular Tube

Hamid Reza Talesh Bahrami[1], Hamid Saffari[1]

ABSTRACT: Dropwise condensation can improve heat transfer process and, consequently, leads to considerable reduction in size and weight of condensers as well as improvement in the dehumidification process in many applications, especially in civil transport aircraft. It can also be used as an efficient cooling tool for electronics and electrical systems in aircraft engineering and aerospace technology. In this paper, the stable dropwise condensation on an inclined tube is mathematically analyzed. To do this, the population of small droplets is estimated by population balance theory while an empirical correlation is used for large droplets. To calculate heat transfer across each droplet, sum of temperature drops due to droplet curvature, phase change at droplet-vapor interface, conduction through the droplet and promoter layer, are equated with surface subcooling. The total heat transfer is calculated with the given droplets population and heat transfer through single droplet. Subsequently, effects of various parameters, including surface subcooling, contact angle and contact angle hysteresis on the growth rate, maximum radius of droplet, droplets population, and total heat transfer rate, are investigated. Results show that growth rate and heat flux of small droplets are much higher than those of the larger ones; hence, surface with small droplets is preferred for dropwise condensation purposes. Droplets with low contact angle and contact angle hysteresis have higher heat transfer rates. Increasing the inclination of tube improves heat transfer process to such an extent that vertical tubes have higher heat transfer rate than the horizontal ones. This fact indicates that vertical tubes must be used for designing condensers with dropwise condensation, which is quite the opposite for condensers designed based on filmwise condensation.

KEYWORDS: Dropwise condensation, Mathematical modeling, Air conditioning, Inclined tubes, Contact angle.

INTRODUCTION

Condensation is one of the most important regimes of heat transfer, which plays a significant role in many industries including aerospace engineering, power plants and refrigeration as well as natural phenomena such as fog or rain formation. Condensation process in industrial applications usually occurs on surfaces, which appears as a liquid film (filmwise condensation), droplets (dropwise condensation), or a combination of both. In dropwise condensation on surfaces, small droplets appear on nucleation sites and grow initially by direct condensation from adjacent vapor. Next, when their size becomes notable so that they can contact neighbor droplets, growing process proceeds by coalescing. Large droplets leave the surface by gravity or other shear forces and sweep other droplets in their way. In stable condition, these processes occur repeatedly and a hierarchical process is formed (Sikarwar *et al.* 2013). Dropwise condensation plays 2 paradox roles in many industries. It shows a negative effect on dew formation on airplane windshield (Fayazbakhsh and Bahrami 2013) and vapor trail around airplanes (Goncalv-ccedil *et al.* 2003; Yamamoto 2003), while it has a positive role in air conditioning of civil transport aircraft. If dropwise condensation is used in the condensers of air conditioning system of an aircraft, it leads to considerable reduction in size and weight of condensers besides improvement in the dehumidification process (Leipertz and Fröba 2008). Dropwise condensation can also be used in the cooling of electronic systems of aerospace industries (O'Callaghan and Babus'Haq 1990). In the following, the positive role of dropwise condensation and enhancement of heat transfer process are addressed.

1.Iran University of Science and Technology – School of Mechanical Engineering – Liquefied Natural Gas Research Laboratory – Tehran/Tehran – Iran.

Author for correspondence: Hamid Saffari | Iran University of Science and Technology – School of Mechanical Engineering – Liquefied Natural Gas Research Laboratory | Narmak | 16846-13114 – Tehran/Tehran – Iran | Email: saffari@iust.ac.ir

Previous studies have shown that dropwise condensation has about one order of magnitude greater heat transfer coefficient than filmwise condensation (Schmidt *et al.* 1930; Rose 2002). Hence, many investigations have experimentally or numerically conducted to the dropwise condensation in recent years (Leipertz and Fröba 2008; Kananeh *et al.* 2010; Sikarwar *et al.* 2012; Reis *et al.* 2016). As an example, McNeil and Burnside (2000) experimentally investigated the performance of both dropwise condensation and filmwise condensation on small tube bundles. Their results show that a durable dropwise condensation can significantly reduce condensers size. Although experimental design of industrial systems is the most reliable approach, it is very expensive especially when many trial and error processes are required. Thus, the development of numerical approaches, which can be used as robust and inexpensive tools for primary or even final design, is important. The earliest theoretical approach for dropwise condensation is proposed by LeFevre and Rose (1966). Consequently, many efforts have been made to improve theoretical approaches (Glicksman and Hunt 1972; Wen and Jer 1976; Maa 1978; Abu-Orabi 1998; Wu *et al.* 2001; Sun *et al.* 2007). Sikarwar *et al.* (2011) conducted an experimental study on the dropwise condensation underneath chemically textured surfaces, while they numerically simulated the same conditions. Battoo *et al.* (2010) presented a numerical simulation for dropwise condensation on inclined surfaces. They investigated the effects of various parameters including contact angle, inclination, contact angle hysteresis, and saturation temperature on dropwise condensation. Wu *et al.* (2001) conducted a numerical simulation on the dropwise condensation on various substrates to discover effects of surface conductivity. They showed that heat transfer of dropwise condensation declines with increasing substrate thermal conductivity. Enright *et al.* (2014) reviewed recent experimental and theoretical studies on the dropwise condensation on micro- and nano-structured surfaces.

The literature shows that very little theoretical studies have been published on dropwise condensation on horizontal or inclined tubes although many experimental ones have been reported (Miljkovic and Wang 2013; Preston 2014). Hosokawa *et al.* (1995) studied single droplets departure heat transfer characteristics in dropwise condensation on an inclined tube. They show that heat transfer coefficient is maximal at inclination angle of 30°. Hu and Tang (2014) presented a theoretical model for dropwise condensation on a horizontal tube. They investigated the effects of various parameters including subcooling temperature and contact angle on both single droplet and overall heat transfer process.

Although numerous research studies have been published on numerical simulation of filmwise condensation inside or outside of tubes (Ji *et al.* 2009; Palen *et al.* 1979; Yun *et al.* 2016), there are very few studies reported in the literature on total heat transfer behavior of dropwise condensation on inclined tubes. According to much higher thermal performance of dropwise condensation compared to filmwise condensation and recent advances in achieving durable dropwise condensation, it is necessary to develop more reliable and robust numerical tools to simulate dropwise condensation.

In this study, a numerical model for dropwise condensation on an inclined tube is developed based on the method presented by LeFevre and Rose (1966). Accordingly, small droplets population is estimated with population balance theory and large droplets population is estimated with the correlation proposed by LeFevre and Rose (1966). Single droplet heat transfer is derived according to contact angle, contact angle hysteresis, tube inclination, nucleation site density, and promoter layer thickness. With the given single droplet heat transfer rate and droplets population, the overall heat transfer rate is calculated. In the following, the effects of various parameters on single droplet behavior or overall heat transfer are investigated.

HEAT TRANSFER MODEL

It is assumed that condensing droplets grow on the nucleation sites (active area) and other sections of the substrate are inactive. Heat transfer through these inactive surfaces is neglected because vapor loses its latent heat at active sites while losing its sensible heat in inactive areas.

The heat flow between the vapor and surface through a single droplet must overcome some thermal resistances, as shown in Fig. 1, where R_d, R_c, R_i and R_{hc} are, respectively, droplet conduction thermal resistance, droplet curvature thermal resistance, liquid-vapor interfacial thermal resistance and hydrophobic coating thermal resistance, all in k/W. Thermal resistances in dropwise condensation modeling are usually expressed as temperature drop, which will be extracted in the following sections.

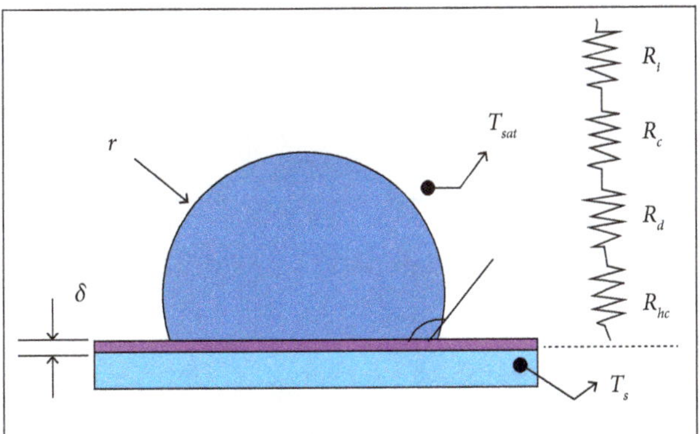

Figure 1. Sketch of necessary parameters and heat transfer resistances between the surface and the vapor through a droplet.

TEMPERATURE DROP OF DROPLET SURFACE CURVATURE

Droplet surface curvature results in higher vapor pressure of the droplet. Temperature drop through the droplet surface (K) is calculated by assuming a continuous Gibbs function across the interface, ideal-gas behavior of the vapor, small and constant specific volume of fluid, and using Clausius-Clapeyron equation along with Young-Laplace equation as follows (Thome 2015):

$$\Delta T_c = \frac{2T_{sat}\sigma}{rh_{fg}\rho_w} \tag{1}$$

where T_{sat} is the saturation temperature (K); σ is the condensate surface tension (N/m); r is the droplet radius (m; Fig. 1); h_{fg} is the latent heat (J/kg); ρ_w is the condensate density (kg/m^3).

The minimum droplet radius (r_{min}; m) for the smallest stable droplet is calculated by considering thermodynamic constraints. Molecule clusters with a size smaller than this limit are unstable and decompose; r_{min} is written as (Carey 2007):

$$r_{min} = \frac{2T_{sat}\sigma}{h_{fg}\rho_w\Delta T} \tag{2}$$

where ΔT is the subcooling temperature.

Combining Eqs. 1 and 2, the temperature drop due to droplet surface curvature can be simplified as:

$$\Delta T_C = \frac{r_{min}}{r}\Delta T \tag{3}$$

According to Eq. 3, smaller droplets have higher ΔT_c.

Temperature drop of vapor-liquid interfacial resistance

Only some fraction of vapor molecules colliding with the droplet surface is absorbed by the liquid phase. This non-ideal process imposes an extra resistance through the heat transfer route as:

$$\Delta T_i = \frac{q_d}{2\pi r^2 h_i(1-\cos\theta)} \tag{4}$$

where q_d is the heat transfer rate through the droplet (W); θ is the contact angle (deg; Fig. 1); *hi* is the interfacial heat transfer coefficient between the vapor and liquid (W/mK), which is derived based on the Kinetic Theory of Gases (Schrage 1953; Rohsenow 1972):

$$h_i = \frac{2\varepsilon}{2-\varepsilon}\frac{1}{\sqrt{2\pi R_g T_s}}\frac{h_{fg}^2}{\upsilon_g T_s} \tag{5}$$

where R_g, h_{fg}, T_s and υ_g are gas constant, latent heat of condensation, saturation temperature and specific volume of the vapor, respectively; ε, which is called condensation coefficient (Wen and Jer 1976) or accommodation coefficient (Liu and Cheng 2015a), is the ratio of molecules absorbed by the liquid phase from the total colliding molecules to the liquid surface; h_i depends on the vapor pressure ranging from 0.383 to 15.7 MW/m^{2}K for pressure from 0.01 to 1.0 atm (Tanasawa 1991).

TEMPERATURE DROP OF CONDUCTION THROUGH DROPLET

The heat flux transferred to the liquid surface must conduct through the droplet to reach the condenser surface, which causes the following temperature drop (Kim and Kim 2011):

$$\Delta T_d = \frac{q_d\theta}{4\pi r k_w \sin\theta} \tag{6}$$

where k_w is the water thermal conductivity (W/mK).

The value of ΔT_d depends highly on the droplet contact angle. When contact angle reaches to 180°, ΔT_d goes up to very large values. This means that surfaces with very large contact angle are not proper for condensation purposes. On the other hand, droplets with larger radius (with $\theta < 180°$) have lower temperature drop with respect to smaller droplets with the same contact angle. This is because larger droplets have greater interface with the condenser surface.

TEMPERATURE DROP OF PROMOTER LAYER

The surfaces of condensers are usually made of industrial metals such as steel, aluminum, or copper, which are hydrophilic and have high surface energy. Therefore, some hydrophobic materials are used as coating to enhance surface hydrophobicity. This coating causes the following temperature drop (Miljkovic *et al.* 2012):

$$\Delta T_{hc} = \frac{q_d \delta}{\pi r^2 k_{coat} (\sin\theta)^2} \tag{7}$$

where δ and k_{coat} are the promoter layer thickness and conductivity, respectively.

This temperature drop adversely depends on the contact angle. Due to smaller base area, droplets with larger contact angle have smaller ΔT_{hc}.

SINGLE DROPLET GROWTH RATE

Equating the sum of all temperature drops, Eqs. 1 to 7, with subcooling temperature, ΔT, the heat transfer through a single droplet is evaluated as follows:

$$q_d = \frac{\Delta T \pi r^2 (1 - \frac{r_{min}}{r})}{\frac{1}{2h_i(1-\cos\theta)} + \frac{r\theta}{4k_w \sin\theta} + \frac{\delta}{k_{coat}(\sin\theta)^2}} \tag{8}$$

The total heat transfer through a single droplet can also be calculated by considering the droplet phase change rate:

$$q_d = \rho_w h_{fg} \frac{dV}{dt} = \frac{\pi}{3} \rho_w h_{fg} \frac{d}{dt} \left\{ (1-\cos\theta)^2 \times (2+\cos\theta) r^3 \right\} \tag{9}$$

where V is the droplet volume; t represents time (s).

Making Eqs. 8 and 9 equivalent, the droplet growth rate (G, m/s), is obtained as:

$$G = \frac{dr}{dt} = \frac{1}{\rho_w h_{fg} (1-\cos\theta)^2 (2+\cos\theta)} \times \frac{\Delta T (1 - \frac{r_{min}}{r})}{\frac{1}{2h_i(1-\cos\theta)} + \frac{r\theta}{4k_w \sin\theta} + \frac{\delta}{k_{coat}(\sin\theta)^2}} \tag{10}$$

Equation 10 can be summarized by introducing the following parameters:

$$A_1 = \frac{\Delta T}{\rho_w h_{fg} (1-\cos\theta)^2 (2+\cos\theta)} \tag{11}$$

$$A_2 = \frac{\theta}{4k_w \sin\theta} \tag{12}$$

$$A_3 = \frac{\delta}{k_{coat}(\sin\theta)^2} + \frac{1}{2h_i(1-\cos\theta)} \tag{13}$$

Then, Eq. 10 is rewritten as:

$$G = A_1 \frac{(1 - \frac{r_{min}}{r})}{A_2 r + A_3} \tag{14}$$

MAXIMUM DROPLET RADIUS

After the birth of a condensing droplet, it first grows with direct condensation and then condenses with neighbor droplets. Along with the growth of the droplet, both surface tension, which adheres it to the surface, and gravity forces (in the absence of other shear forces) increase. When the gravity force overcomes the adhering force, the droplet falls or slides on the surfaces. Surface curvature determines the magnitude of gravity force. A single droplet, which is deposited on an inclined tube and is on the verge of sliding, is schematically shown in Fig. 2.

FIRST METHOD TO DETERMINE THE DROPLET MAXIMUM RADIUS

Initially, it is assumed that the droplet on the verge of sliding is a spherical cap and r_{max} is found by equating gravity and surface tension forces. This method is used by numerous

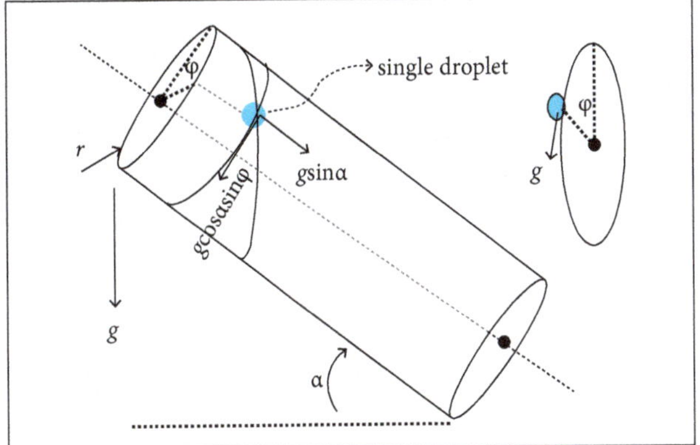

Figure 2. Schematics of gravity forces acting on a droplet on an inclined tube.

researchers in dropwise condensation modeling on inclined surfaces (*e.g.*, Dimitrakopoulos and Higdon 1999; Kim and Kim 2011; Kumar and Yuvaraj 2013; Lee *et al.* 2013; Hu and Tang 2014; Cheng *et al.* 2015; Liu and Cheng 2015b). The maximum radius is evaluated as follows. The gravity force exerted on the droplet is:

$$F_g = \left(2 - 3cos\theta + cos^3\theta\right)\pi r_{max}^3 \frac{\rho g}{3} \times \sqrt{\cos^2\alpha \sin^2\varphi + \sin^2\alpha} \tag{15}$$

where r_{max} is the maximum droplet radius; g is the acceleration of gravity (m/s^2); α is the inclination angle (deg); φ is the peripheral angle (deg).

The surface tension force depending on the droplet shape is (Kim et al. 2002):

$$F_\sigma = 2\sigma r_{max} sin\theta(cos\theta_r - cos\theta_a) \tag{16}$$

where θ_r and θ_a are the receding and the advancing angles, respectively.

When a droplet is deposited on an inclined surface, it alters its shape to produce extra surface tension and resists leaving the surface. This reaction appears in the droplet receding and advancing angles.

Making Eqs. 15 and 16 equivalent, r_{max} is found as:

$$r_{max}(\varphi,\alpha) = \left[\frac{6\sigma sin\theta(cos\theta_r - cos\theta_a)}{\pi(2 - 3cos\theta + cos^3\theta)\rho g\sqrt{\cos^2\alpha \sin^2\varphi + \sin^2\alpha}}\right]^{1/2} \tag{17}$$

SECOND METHOD TO DETERMINE THE DROPLET MAXIMUM RADIUS

Droplets on inclined surfaces deform and take a shape other than spherical cap. For this reason, receding and advancing angles appear. Some researchers have considered this deformation and shown that the assumption of spherical cap cannot properly estimate droplet volume on inclined surfaces. As an example, ElSherbini and Jacobi (2004, 2006) can be mentioned. However, the relation for volume of droplets on the verge of sliding on inclined surfaces proposed by Dussan (1985) is straightforward and can be simply used considering peripheral inclination:

$$\begin{aligned} v_{max}(\varphi,\alpha) = & \left(\frac{\rho g\sqrt{\cos^2\alpha \sin^2\varphi + \sin^2\alpha}}{\sigma}\right)^{-\frac{3}{2}} \left(\frac{96}{\pi}\right)^{\frac{1}{2}} \\ & \times\left(\left(\cos\theta_r - \cos\theta_a\right)^{\frac{3}{2}}\left(1 + \cos\theta_a\right) \right. \\ & \left. \times\left(1 - \frac{3}{2}\cos\theta_a + 1/2\cos^3\theta_a\right)\right) \\ & \times\left(\left(\cos\theta_a + 2\right)^{\frac{3}{2}}\left(1 - \cos\theta_a\right)^{9/4}\right)^{-1} \end{aligned} \tag{18}$$

The equivalent radius can be derived as:

$$r_{max} = \left(\frac{v_{max}(\varphi,\alpha)}{\pi\left(2 - 3cos\theta + cos^3\theta\right)}\right)^{\frac{1}{3}} \tag{19}$$

The results are reported for hysteresis angle up to 10 deg in Dussan (1985). In the next sections, both Eqs. 17 and 19 will be used for simulation.

EFFECTIVE RADIUS

Effective radius (r_e, m) is the half of the average distance between nucleation sites given by Wen and Jer (1976), Abu-Orabi (1998), and Kim and Kim (2011):

$$r_e = \sqrt{\frac{1}{4N_s}} \tag{20}$$

where N_s is the density of nucleation sites on the condensing surface (1/m).

Equation 20 is derived based on the assumption that nucleation sites form a square array. The effective radius is a measure that determines when a growing droplet contacts a neighbor droplet.

DROPLET SIZE DISTRIBUTION

After estimating single droplet thermal behavior, droplets size distribution must be established. This distribution determines how many droplets with a specific size exist on the surface in the stable dropwise condensation at every moment. Population balance theory has been used in the literature (Tanaka 1975; Maa 1978; Abu-Orabi 1998; Vemuri and Kim 2006; Kim and Kim 2011) to derive size distribution of small droplets, which mainly grow by direct condensation. Based on this theory, in the stable dropwise condensation, the number of droplet entering a specific droplet radius range is equal with those leaving the range. Assuming a droplet radius range of $r_1 < r < r_2$, the number of droplets that enters this range in the interval Δt is:

$$AG_1 n_1 \Delta t \tag{21}$$

and the number of the droplets that leaves this range is:

$$AG_2 n_2 \Delta t \tag{22}$$

where n is the population density of small droplets growing by direct condensation (m^3).

The number of small droplet swept with larger sliding droplets can be considered as:

$$S\bar{n}\Delta t \Delta r \tag{23}$$

where S is the rate at which the small droplets are swept by falling droplets; $\bar{n}$ is the average population density in the range $r_1 < r < r_2$, and $\delta r = r_2 - r_1$.

Making Eqs. 21 - 23 equivalent as well as Δt and Δr sufficiently close to 0, the population balance theory is reduced to:

$$\frac{dG}{dt} + \frac{n}{\tau} = 0 \tag{24}$$

where τ is the sweeping period ($\tau = A/S$).

Equation 24 is suitable for small droplets growing mainly by direct condensation. However, when the radii of small droplets exceed the effective radius, they coalesce with neighbor droplets. After this step, droplets mainly grow by coalescence and they are called large droplets. LeFevre and Rose (1966) proposed the following relation for size distribution of large droplets, which have been used in many studies in the literature (e.g., Vemuri and Kim 2006; Hu and Tang 2014; Liu and Cheng 2015a):

$$N(r) = \frac{1}{3\pi r^2 r_{max}(\varphi,\alpha)}\left(\frac{r}{r_{max}(\varphi,\alpha)}\right)^{-2/3} \tag{25}$$

where N is the population density of large droplets (m^3).

Equation 24 is an ordinary differential equation of order one with an exact closed-form solution. Rearranging and integrating Eq. 24, one has:

$$n(r) = \frac{(Gn)_{min}}{G}\exp\left(\frac{A_2}{\tau A_1}\left[\frac{\left(r - r_{min}^2\right)}{2} + 2r_{min}\left(r - r_{min}\right) + r_{min}^2 \ln\left(r - r_{min}\right) + \frac{A_3}{\tau A_1}\left[r - r_{min} + r_{min}\ln\left(r - r_{min}\right)\right]\right]\right) \tag{26}$$

where $(Gn)_{min}$ and τ are unknown terms determined by the following boundary conditions:

$$N(r) = n(r) \qquad at \;\; r = r_e \tag{27}$$

and

$$d\left(\ln n(r)\right)/d(\ln r) = d\left(\ln n(r)\right)/d(\ln r) = -8/3 \tag{28}$$

which means that the value and slop of small and large droplet distributions are equal at effective radius. Applying boundary conditions, one has:

$$n(r) = \frac{1}{3\pi r_e^3 r_{max}}\left(\frac{r_e}{r_{max}}\right)^{-2/3}\frac{r\left(r_e - r_{min}\right)}{r - r_{min}} \times \frac{A_2 r + A_3}{A_2 r_e + A_3}\exp\left(B_1 + B_2\right) \tag{29}$$

where:

$$B_1 = \frac{A_2}{\tau A_1}\left[\frac{r_e^2 - r^2}{2} + r_{min}\left(r - r_{min}\right) - r_{min}^2 \times \ln\left(\frac{r - r_{min}}{r_e - r_{min}}\right)\right] \tag{30}$$

$$B_2 = \frac{A_3}{\tau A_1}\left[r_e - r - r_{min}\ln\left(\frac{r - r_{min}}{r_e - r_{min}}\right)\right] \tag{31}$$

and

$$\tau = \frac{3r_e^2(A_2 r_e + A_3)^2}{A_1\left(11A_2 r_e^2 - 14A_2 r_e r_{min} + 8A_3 r_e - 11A_3 r_{min}\right)} \tag{32}$$

TOTAL HEAT FLUX

With known single droplet heat transfer and droplets distribution, total heat flux can be determined as:

$$q''(\alpha,\theta) = \frac{1}{2\pi}\left[\int_0^{2\pi}\int_{r_{min}}^{r_e} q_d(r,\theta,\varphi,\alpha)n(r,\theta,\varphi,\alpha)drd\varphi + \int_0^{2\pi}\int_{r_e}^{r_{max}(\theta,\varphi,\alpha)} q_d(r,\theta,\varphi,\alpha)N(r,\theta,\varphi,\alpha)drd\varphi\right] \tag{33}$$

where: q^n is the heat flux (W/m^2).

NUMERICAL PROCEDURE VALIDATION

To verify the presented procedure, the calculated heat flux is compared with reported experimental results of Peng *et al.* (2015). Nucleation site numbers reported in the literature are in the range of $10^9 < N_s < 10^{13}$ (Vemuri and Kim 2006). The nucleation site number is considered 10^{12} in this comparative study. According to Peng *et al.* (2015), contact angle, advancing and receding angles are 120, 142, and 102°, respectively. Effects of non-condensable gases and thermal resistance of hydrophobic promoter layer are neglected. In addition, the simulation is performed in atmospheric conditions. The comparative results are illustrated in Fig. 3. As it can be seen, there is a good agreement between the predicted and the experimental results.

As another example, the current study is compared with the reported experimental results of Vemuri and Kim (2006). They experimentally measured the maximum radius of droplet on the surface, r_{max} = 1.5 mm. Therefore, r_{max} is directly put in the code and Eq. 19 is omitted. The simulation is done with nucleation site number of 10^{12}, contact angle of 149° and at atmospheric condition according to Vemuri and Kim (2006). A good agreement can be seen between the current study and that of Vemuri and Kim (2006), as shown in Fig. 4.

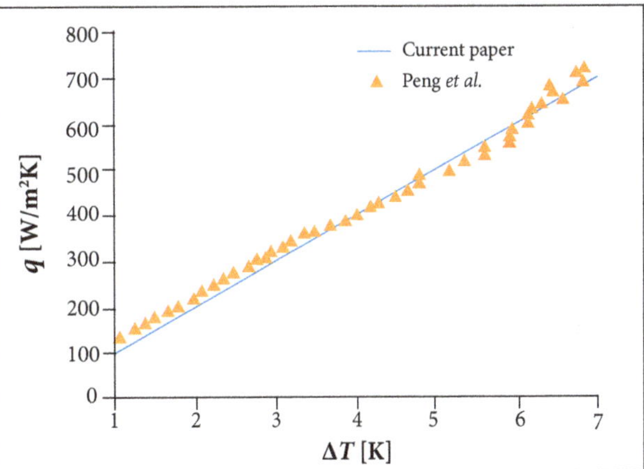

Figure 3. Comparison of the current simulation and that of Peng *et al.* (2015).

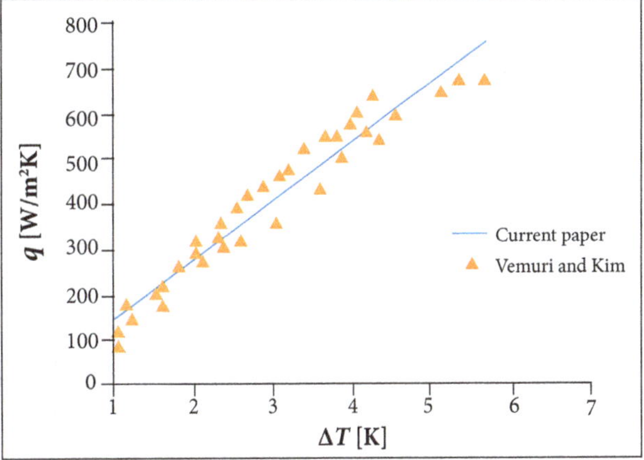

Figure 4. Comparison of the current simulation and that of Vemuri and Kim (2006).

RESULTS AND DISCUSSION

SINGLE DROPLET BEHAVIOR

The variation of different temperature drops is shown in Fig. 5, where θ = 120°, $\Delta\theta$ = 40°, α = 90°, ΔT = 1 K, and δ = 1 μm; $\Delta\theta$ is the contact angle hysteresis (deg). All temperature drops are comparable in very small radii while ΔT_d is the prevalent temperature drop in very large droplets. The variation of single droplet heat flux and heat transfer rate are shown in Fig. 6, where θ = 120°, $\Delta\theta$= 40°, α = 90°, ΔT = 1 K, and δ = 1 μm. Although larger droplets have higher heat transfer rate, they have low heat flux. It means that surfaces with more small droplets have higher heat transfer rate. Hence, for dropwise condensation purpose, it is preferred to have surfaces in which small droplets could slide quickly and easily from the surface to more small droplets form.

Variation of thermal resistances with respect to contact angle (CA) is shown in Fig. 7. All thermal resistances increase with contact angle and the thermal resistance curvature is not considerable with respect to other resistances due to vapor-liquid interface. However, all resistances increase with contact angle. For example, it can be roughly said that all resistances increase by 100 times when contact angle is increased from 100° to 160°.

The variation of single droplet growth rate due to direct condensation (Eq. 14) with respect to radius and subcooling temperature is shown in Fig. 8, where $\theta = 120°$, $\Delta\theta = 40°$, $\alpha = 90°$, and $N_s = 10^{12}$ (1/m²). The growth rate of small droplets is higher at greater subcooling. Also, growth rate of large droplets is very low at all subcooling temperatures.

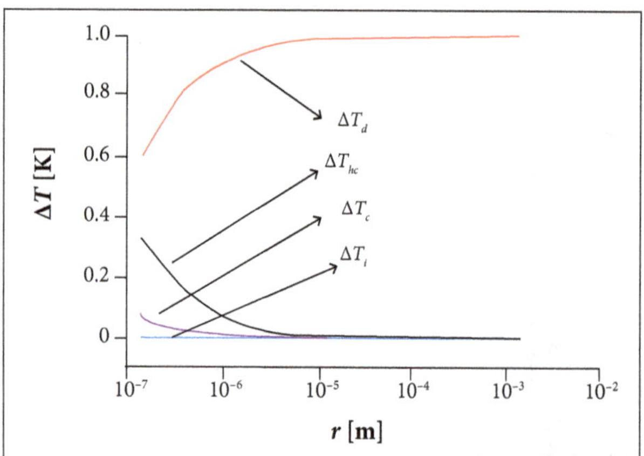

Figure 5. Variation of different temperature drops with respect radius.

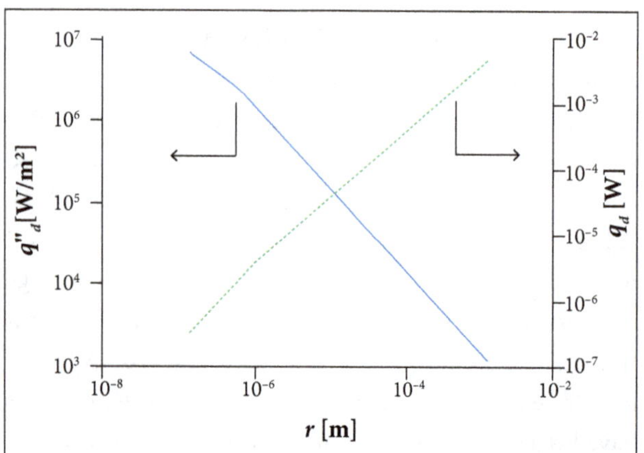

Figure 6. Variation of heat flux and heat transfer rate of a single droplet with respect radius.

The variation of single droplet growth rate with respect to radius and contact angle is illustrated in Fig. 9, where $\Delta T = 5$ K, $\Delta\theta = 40°$, $\alpha = 90°$, and $N_s = 10^{12}$ (1/m²). At the

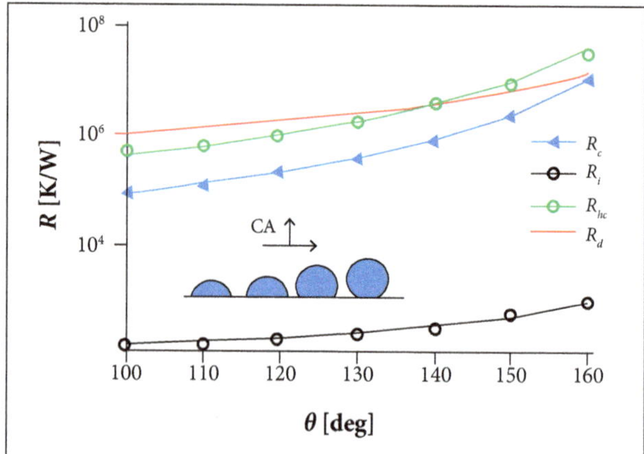

Figure 7. Variation of thermal resistances with contact angle (r = 1 μm).

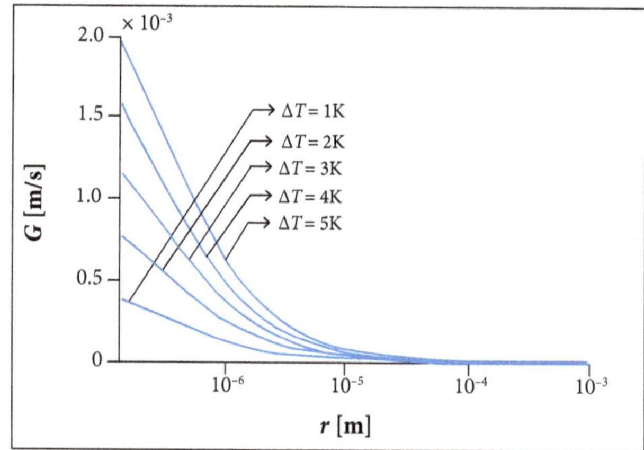

Figure 8. Variation of a droplet growth rate for radius and surface subcooling temperature.

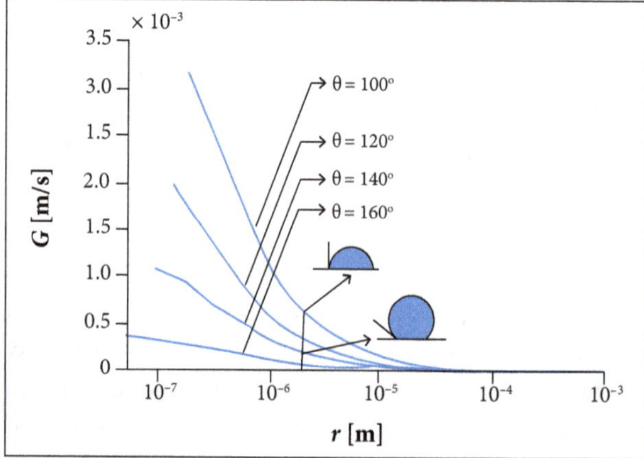

Figure 9. Variation of a droplet growth rate with respect to its radius and contact angle.

same radius, droplets with higher contact angle have lower growth rate. For example, droplets with contact angle of 100° have 6 times higher growth rate than droplets with contact angle of 160° at small radii. This is because droplets with smaller contact angle have larger interface with the surface. On the other hand, droplets with lower contact angle have smaller mobility. Therefore, there is a conflict between droplet mobility and its heat transfer rate.

Previous studies have shown that dropwise condensation is a nucleation phenomenon and that droplets form at the same positions repeatedly (McCormick and Westwater 1965). Assuming that dropwise condensation is a nucleation process, nucleation site number has a significant effect on it. With increasing nucleation sites, the density of small droplets increases and accordingly heat transfer process is improved. Nucleation site number depends on the topography and chemical properties of the surface (Mu *et al.* 2008). However, there is not any comprehensive correlation for estimating nucleation site number on various surfaces. The variation of droplet distribution at various nucleation site numbers is shown in Fig. 10, where $\theta = 120°$, $\Delta T = 10$ K, and $\alpha = 0°$. The nucleation site numbers has a significant effect on small droplets population with radii smaller than effective radius. For radii greater than r_e droplets, the population is independent of nucleation sites number determined from Eq. 25. Very high nucleation site numbers, $N_s > 10^{13}$, lead to high droplets population and small effective radii. These consequences result in droplets coalescence once they nucleate, and filmwise condensation appears. As previous studies have revealed, small nucleation site numbers lead to inaccurate results (Citakoglu and Rose 1969).

Droplet minimum radius (r_{min}) is extracted from thermodynamic minimization. Figure 11 depicts the variation of droplets population with respect to radius at different minimum radii, where $N_s = 10^9$ (1/m^2), $\theta_a = 142°$, $\theta_r = 102°$, $\theta = 120°$, $\Delta T = 10$ K, and $\alpha = 0°$. It can be considered that population distributions (except for $1{,}000 r_{min}$) have a bell-like shape where the maximum population is $n = 1.2 \times 10^{13}$ (1/m^3) locating at effective radius (here, $r_e = 1.5 \times 10^{-5}$ m). The average radius of droplets is $\bar{r} = 6.6 \times 10^{-4}$ m with variance 1.46×10^{-7}. If the minimum radius is exaggeratedly increased ($1{,}000\, r_{min}$), a non-physical condition appears. However, even 100 times increase in minimum radius does not have a considerable effect on droplets population.

The maximum radius variation with respect to hysteresis angle at different inclination angles is depicted in Fig. 12, where $\Delta T = 10$ K and $\theta = 120°$ (Fig. 12a) and $\Delta T = 10$ K, $\theta = 120°$, and $\alpha = 10°$ for the difference of 2 estimates for maximum radius at different hysteresis angles (Fig. 12b). Equation 17 overestimates r_{max} with respect to Eq. 19. Also, the difference of 2 estimations reduces in higher hysteresis angles. In the next sections, Eq. 19 will be used for estimations with hysteresis angle lower than 10°.

The variation of maximum radius with respect to hysteresis angle at different contact angle is depicted in Fig. 13 ($\Delta T = 10$ K). As hysteresis or contact angle increases, the maximum radius increases. The figure shows that if droplet contact angle increases from 2° to 90°, the maximum radius increases 6 times. On the other hand, droplet population distributions are close together in small sliding angles. It should be mentioned that producing surfaces with simulta-

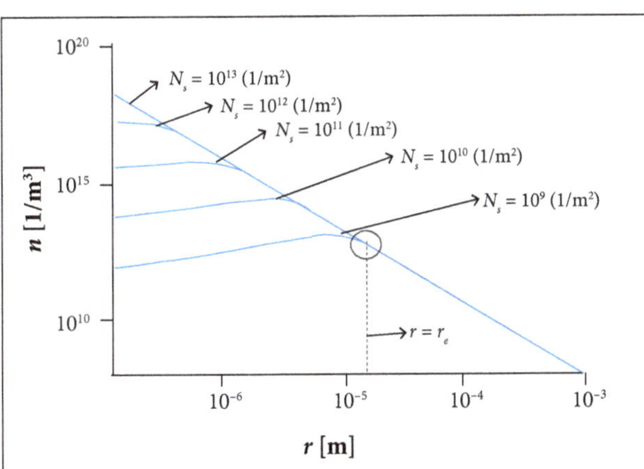

Figure 10. Variation in population of droplets for radius at various numbers of nucleation sites.

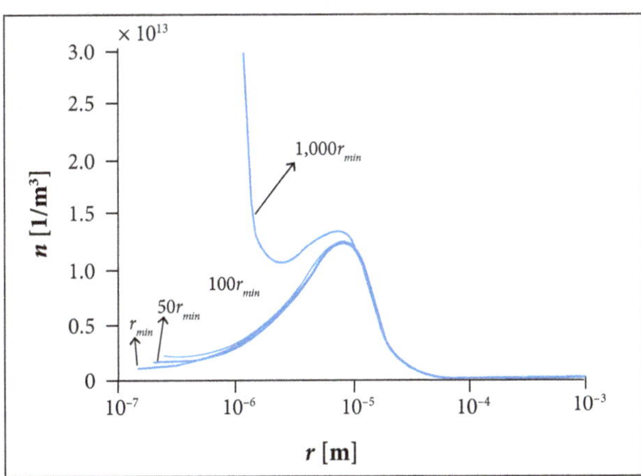

Figure 11. Variation in the number of droplets for radius at various minimum ones.

neous low contact angle and low hysteresis angle is not practically possible (Talesh Bahrami *et al.* 2017). The effect of maximum radius on dropwise condensation will be considered in the next sections.

The variation of maximum droplet radius with respect to peripheral angle at different tube inclinations is shown in Fig. 14, where $\Delta T = 10$ K and $\Delta\theta = 10°$. The maximum radius increases with decrease in inclination. This analysis is not valid for horizontal surfaces where due to the absence of a proper sweep mechanism, droplets continuously coalesce together. Finally, a continuous liquid film covers the surface and dropwise condensation assumption is violated. On the other hand, the maximum radius of droplets on the vertical tube is constant for all peripheral angles and is smaller than other inclination.

The variation of heat flux with respect to subcooling temperature at different inclinations is shown in Fig. 15. The heat flux increases as subcooling or inclination increases. This behavior can be interpreted by considering the variation of the maximum droplet radius. According to Fig. 14, the maximum radius increases as inclination decreases.

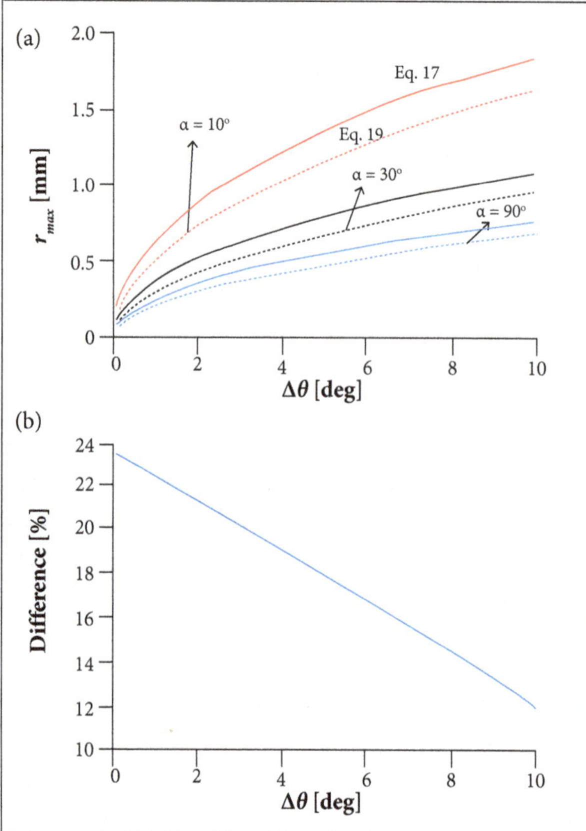

Figure 12. Maximum radius variation for hysteresis angle at different inclination angles.

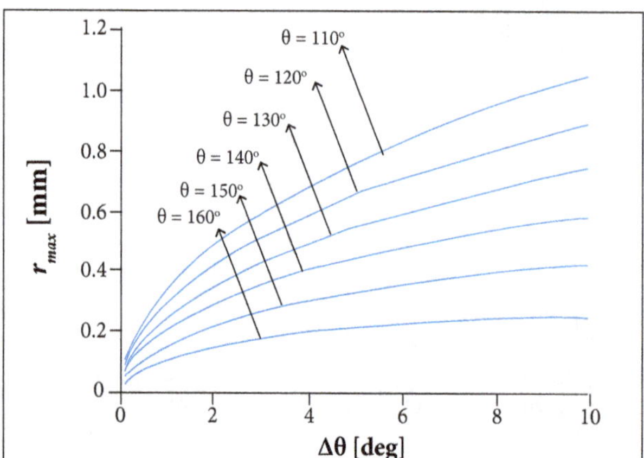

Figure 13. Variation of maximum droplet radius for contact angle hysteresis on a vertical plate.

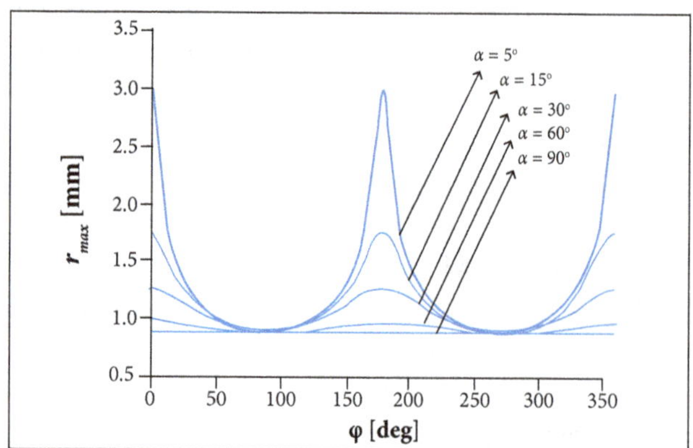

Figure 14. Variation of maximum droplet for peripheral angle at various inclinations.

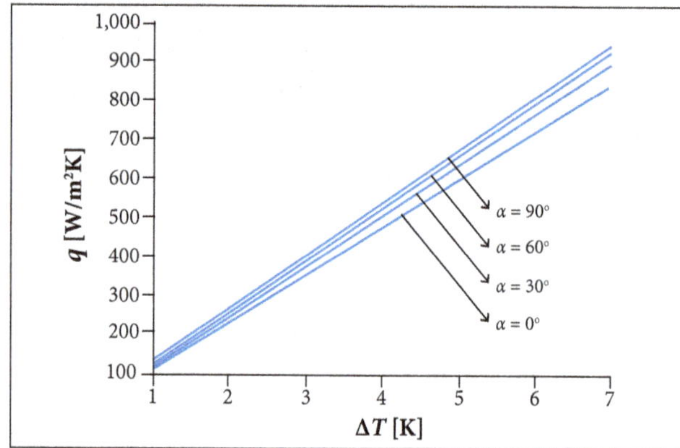

Figure 15. Variation of heat flux for surface subcooling at various inclinations of the tubes.

TOTAL BEHAVIOR OF DROPWISE CONDENSATION

The variation of heat flux with respect to subcooling temperature at various inclination of tube is shown in Fig. 15, where $N_s = 10^{12}\,(1/m^2)$, $\Delta\theta = 10°$, and $\theta = 120°$. Due to better sweep mechanism of vertical tube, the case of $\alpha = 90°$ has the maximum heat flux. This result can be more cleared by comparing filmwise condensation and dropwise condensation behaviors. The heat transfer coefficient in filmwise condensation inversely depends on the condensate thickness, which is directly affected by the surface length. It means that higher condensate thickness leads to lower heat transfer coefficient in the case of filmwise condensation. Hence, horizontal tubes are employed in all shell and tube heat exchanger working in filmwise condensation. On the other hand, heat transfer coefficient of dropwise condensation is independent of the location and is uniform over the surface. These advantages indicate that vertical tubes can be used in heat exchangers working with dropwise condensation without any restriction in length.

The variation of heat flux with respect to contact angle is depicted in Fig. 16, where $N_s = 10^{12}\,(1/m^2)$, $\Delta\theta = 10°$, and $\Delta T = 5$ K. The contact angle has a significant effect on heat flux so that increasing contact angle from 100° to 160° nearly decreases q'' to ⅓. Increasing contact angle results in higher thermal resistances (see Fig. 7), which leads to lower heat flux. It is worth mentioning that high contact angle usually couples with low contact angle hysteresis. Effect of contact angle hysteresis on the heat flux is given in Fig. 17, where $N_s = 10^{12}\,(1/m^2)$, $\theta = 120°$, and $\Delta T = 5$ K. Contact angle hysteresis has a considerable influence on the maximum radius. Increasing contact angle hysteresis leads to rapid decrease in the maximum radius, meaning that droplets with smaller radii slide quickly, off the surface and new droplets nucleate. Consequently, more small droplets population causes higher heat flux (according to Fig. 6).

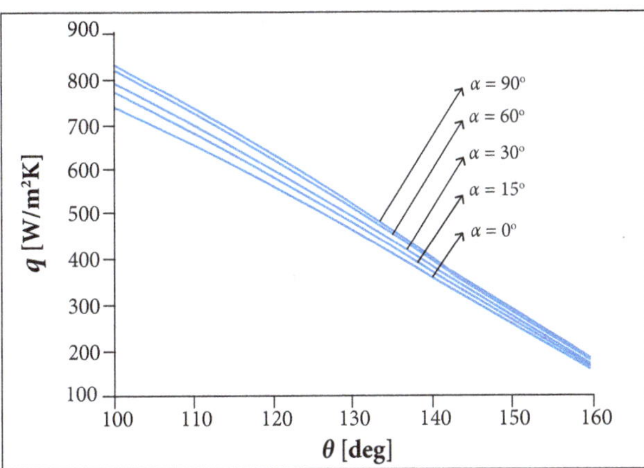

Figure 16. Variation of heat flux for contact angle at various inclinations of the tubes.

Figure 18 shows the effect of nucleation site number on the predicted dropwise condensation heat flux in a horizontal tube, where $\Delta\theta = 10°$, $\theta = 120°$, and $\alpha = 0°$. The dropwise condensation heat flux increases with N_s. This outcome is explained by considering Fig. 10, where the nucleation site number has notable effect on the droplet population. Higher nucleation site number leads to greater droplet population and consequently higher total heat flux.

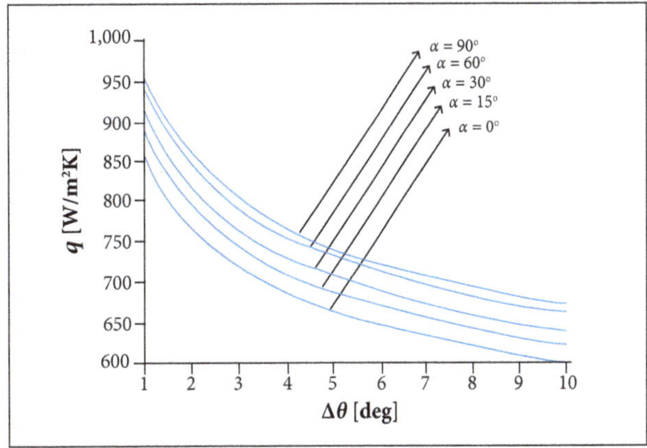

Figure 17. Variation of heat flux for contact angle hysteresis at various inclinations of the tubes.

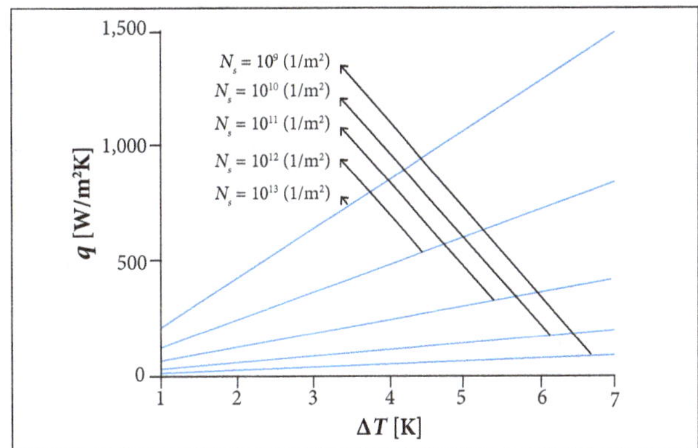

Figure 18. Variation of heat flux for surface subcooling at various nucleation sites.

CONCLUSION

An analytical model for calculation of dropwise condensation heat transfer on an inclined tube is presented. The results showed that heat transfer of vertical tubes is higher than other inclinations. This finding indicates that vertical tube must be used in shell and tube heat exchangers working in dropwise condition. On the other hand, the current study, unlike the one of Hosokawa *et al.* (1995), indicates that increasing inclination enhances heat transfer in dropwise condensation on circular tubes. The reason is that a sliding droplet on more inclined tubes sweeps more droplets in their way. Therefore, further bare surface contacts with the vapor and heat transfer is enhanced.

AUTHOR'S CONTRIBUTION

Conceptualization, Funding Acquisition, Resources, and Supervision, Saffari H; Methodology and Writing – Review & Editing, Saffari H and Bahrami HRT; Investigation, Simulation, and Writing – Original Draft, Bahrami HRT.

REFERENCES

Abu-Orabi M (1998) Modeling of heat transfer in dropwise condensation. Int J Heat Mass Tran 41(1):81-87. doi: 10.1016/S0017-9310(97)00094-X

Battoo NK, Sikarwar BS, Khandekar S, Muralidhar K (2010) Mathematical modeling and simulation of dropwise condensation and inclined surfaces exposed to vapor flux. Proceedings of the 20th National and 9th International ISHMT-ASME Heat and Mass Transfer Conference; Mumbai, India.

Carey VP (2007) Liquid vapor phase change phenomena: an introduction to the thermophysics of vaporization and condensation processes in heat transfer equipment. 2nd edition. Boca Raton: CRC Press.

Cheng K, Kim S, Lee S, Kim KJ (2015) Internal dropwise condensation: modeling and experimental framework for horizontal tube condensers. Int J Heat Mass Tran 83:99-108. doi: 10.1016/j.ijheatmasstransfer.2014.11.084

Citakoglu E, Rose JW (1969) Dropwise condensation — the effect of surface inclination. Int J Heat Mass Tran 12(5):645-650. doi: 10.1016/0017-9310(69)90045-3

Dimitrakopoulos P, Higdon JJL (1999) On the gravitational displacement of three-dimensional fluid droplets from inclined solid surfaces. J Fluid Mech 395:181-209. doi: 10.1017/S0022112099005844

Dussan EB (1985) On the ability of drops or bubbles to stick to non-horizontal surfaces of solids. Part 2. Small drops or bubbles having contact angles of arbitrary size. J Fluid Mech 151:1-20. doi: 10.1017/S0022112085000842

ElSherbini AI, Jacobi AM (2004) Liquid drops on vertical and inclined surfaces: II. A method for approximating drop shapes. J Colloid Interface Sci 273(2):566-575. doi: 10.1016/j.jcis.2003.12.043

ElSherbini AI, Jacobi AM (2006) Retention forces and contact angles for critical liquid drops on non-horizontal surfaces. J Colloid Interface Sci 299(2):841-849. doi: 10.1016/j.jcis.2006.02.018

Enright R, Miljkovic N, Alvarado JL, Kim K, Rose JW (2014) Dropwise condensation on micro- and nanostructured surfaces. Nanoscale and Microscale Thermophysical Engineering 18(3):223-250. doi: 10.1080/15567265.2013.862889

Fayazbakhsh MA, Bahrami M (2013) Analytical modeling of mist condensation by natural convection over inclined flat surfaces. Proceedings of the ASME 2013 Heat Transfer Summer Conference collocated with the ASME 2013 7th International Conference on Energy Sustainability and the ASME 2013 11th International Conference on Fuel Cell Science, Engineering and Technology; Minneapolis, USA.

Glicksman LR, Hunt AW (1972) Numerical simulation of dropwise condensation. Int J Heat Mass Tran 15(11):2251-2269. doi: 10.1016/0017-9310(72)90046-4

Goncalvès E, Houdeville R (2003) Numerical simulation of shock oscillations over airfoil using a wall law approach. AIAA J 41:1829-1832. doi: 10.2514/2.7303

Hosokawa T, Fujiwara Y, Ogami Y, Kawasima Y, Yamasaki Y (1995) Heat transfer characteristic of dropwise condensation on an inclined circular tube. Heat Recovery Systems and CHP 15(1):31-39. doi: 10.1016/0890-4332(95)90035-7

Hu HW, Tang GH (2014) Theoretical investigation of stable dropwise condensation heat transfer on a horizontal tube. Appl Therm Eng 62(2):671-679. doi: 10.1016/j.applthermaleng.2013.10.022

Ji T, Liebenberg L, Meyer JP (2009) Heat transfer enhancement during condensation in smooth tubes with helical wire inserts. Heat Tran Eng 30(5):337-352. doi: 10.1080/01457630802 14466

Kananeh AB, Rausch MH, Leipertz A, Fröba AP (2010) Dropwise condensation heat transfer on plasma-ion-implanted small horizontal tube bundles. Heat Tran Eng 31(10):821-828. doi: 10.1080/01457630903547545

Kim HY, Lee HJ, Kang BH (2002) Sliding of liquid drops down an inclined solid surface. J Colloid Interface Sci 247(2):372-380. doi: 10.1006/jcis.2001.8156

Kim S, Kim KJ (2011) Dropwise condensation modeling suitable for superhydrophobic surfaces. J Heat Transfer 133(8):081502. doi: 10.1115/1.4003742

Kumar DS, Yuvaraj R (2013) Simulation of dropwise condensation on a superhydrophobic inclined substrate. Proceedings of the IEEE International Conference on Energy Efficient Technologies for Sustainability (ICEETS); Nagercoil, India.

Lee S, Yoon HK, Kim KJ, Kim S, Kennedy M, Zhang BJ (2013) A dropwise condensation model using a nano-scale, pin structured surface. Int J Heat Mass Tran 60:664-671. doi: 10.1016/j.ijheatmasstransfer.2013.01.032

LeFevre EJ, Rose JW (1966) A theory of heat transfer by dropwise condensation. In: American Institute of Chemical Engineers. Chemical Engineering Progress Symposium Series. 69 – 74 editions. New York: American Institute of Chemical Engineers.

Leipertz A, Fröba AP (2008) Improvement of condensation heat transfer by surface modifications. Heat Tran Eng 29(4):343-356. doi: 10.1080/01457630701821563

Liu X, Cheng P (2015a) Dropwise condensation theory revisited: Part I. Droplet nucleation radius. Int J Heat Mass Tran 83:833-841. doi: 10.1016/j.ijheatmasstransfer.2014.11.009

Liu X, Cheng P (2015b) Dropwise condensation theory revisited: Part II. Droplet nucleation density and condensation heat flux. Int J Heat Mass Tran 83:842-849. doi: 10.1016/j.ijheatmasstransfer.2014.11.008

Maa JR (1978) Drop size distribution and heat flux of dropwise condensation. Chem Eng J 16(3):171-176. doi: 10.1016/0300-9467(78)85052-7

McCormick JL, Westwater JW (1965) Nucleation sites for dropwise condensation. Chem Eng Sci 20(12):1021-1036. doi: 10.1016/0009-2509(65)80104-X

McNeil DA, Burnside BM, Cuthbertson G (2000) Dropwise condensation of steam on a small tube bundle at turbine condenser conditions. Experimental Heat Transfer 13(2):89-105. doi: 10.1080/089161500269481

Miljkovic N, Enright R, Wang EN (2012) Growth dynamics during dropwise condensation on nanostructured superhydrophobic surfaces. Proceedings of the 3rd International Conference on Micro/Nanoscale Heat and Mass Transfer. American Society of Mechanical Engineers; Atlanta, USA.

Miljkovic N, Wang EN (2013) Condensation heat transfer on superhydrophobic surfaces. MRS Bulletin 38(5):397-406. doi: 10.1557/mrs.2013.103

Mu C, Pang J, Lu Q, Liu T (2008) Effects of surface topography of material on nucleation site density of dropwise condensation. Chem Eng Sci 63(4):874-880. doi: 10.1016/j.ces.2007.10.016

O'Callaghan PW, Babus'Haq RF (1990) Cooling problems facing the electronics and aerospace industries. Aircraft Engineering and Aerospace Technology 62(7):17-19. doi: 10.1108/eb036968

Palen JW, Breber G, Taborek J (1979) Prediction of flow regimes in horizontal tube-side condensation. Heat Tran Eng 1(2):47-57. doi: 10.1080/01457637908939558

Peng B, Ma X, Lan Z, Xu W, Wen R (2015) Experimental investigation on steam condensation heat transfer enhancement with vertically patterned hydrophobic–hydrophilic hybrid surfaces. Int J Heat Mass Tran 83:27-38. doi: 10.1016/j.ijheatmasstransfer.2014.11.069

Preston DJ (2014) Electrostatic charging of jumping droplets on superhydrophobic nanostructured surfaces: fundamental study and applications (Master's thesis). Cambridge: Massachusetts Institute of Technology.

Reis FMM, Lavieille P, Miscevic M (2016) Dropwise condensation enhancement using a wettability gradient. Heat Tran Eng 38(3):377-385. doi: 10.1080/01457632.2016.1189277

Rohsenow WM (1972) Status of and problems in boiling and condensation heat transfer. In: Hetsroni G, Sideman S, Hartnett JP, editors. Progress in heat and mass transfer. Vol. 6: Proceedings of the International Symposium on Two-phase Systems. Oxford: Pergamon Press.

Rose JW (2002) Dropwise condensation theory and experiment: a review. Proc IME J Power Energ 216(2):115-128. doi: 10.1243/09576500260049034

Schmidt E, Schurig W, Sellschopp W (1930) Versuche über die Kondensation von Wasserdampf in Film-und Tropfenform. Technische Mechanik und Thermodynamik 1(2):53-63.

Schrage RW (1953) A theoretical study of interphase mass transfer. New York: Columbia University Press.

Sikarwar BS, Battoo NK, Khandekar S, Muralidhar K (2011) Dropwise condensation underneath chemically textured surfaces: simulation and experiments. J Heat Transfer 133(2):021501. doi: 10.1115/1.4002396

Sikarwar BS, Khandekar S, Agrawal S, Kumar S, Muralidhar K (2012) Dropwise condensation studies on multiple scales. Heat Tran Eng 33(4-5):301-341. doi: 10.1080/01457632.2012.611463

Sikarwar BS, Muralidhar K, Khandekar S (2013) Effect of drop shape on heat transfer during dropwise condensation underneath inclined surfaces. Interfacial Phenomena and Heat Transfer 1(4):339-356. doi: 10.1615/InterfacPhenomHeatTransfer.v1.i4.30

Sun FZ, Gao M, Lei SH, Zhao YB, Wang K, Shi YT, Wang NH (2007) The fractal dimension of the fractal model of dropwise condensation and its experimental study. International Journal of Nonlinear Sciences and Numerical Simulation 8(2):211-222. doi: 10.1515/IJNSNS.2007.8.2.211

Talesh Bahrami HR, Ahmadi B, Saffari H (2017) Optimal condition for fabricating superhydrophobic copper surfaces with controlled oxidation and modification processes. Mater Lett 189:62-65. doi: 10.1016/j.matlet.2016.11.076

Tanaka H (1975) A theoretical study of dropwise condensation. J Heat Transfer 97(1):72-78. doi: 10.1115/1.3450291

Tanasawa I (1991) Advances in condensation heat transfer. Advances in Heat Transfer 21:55-139. doi: 10.1016/S0065-2717(08)70334-4

Thome JR (2015) Encyclopedia of two-phase heat transfer and flow I: fundamentals and methods. Vol. 2. New Jersey: World Scientific.

Vemuri S, Kim KJ (2006) An experimental and theoretical study on the concept of dropwise condensation. Int J Heat Mass Tran 49(3-4):649-657. doi: 10.1016/j.ijheatmasstransfer.2005.08.016

Wen HW, Jer RM (1976) On the heat transfer in dropwise condensation. The chemical engineering journal 12(3):225-231. doi: 10.1016/0300-9467(76)87016-5

Wu YT, Yang CX, Yuan XG (2001) Drop distributions and numerical simulation of dropwise condensation heat transfer. Int J Heat Mass Tran 44(23):4455-4464. doi: 10.1016/S0017-9310(01)00085-0

Yamamoto S (2003) Onset of condensation in vortical flow over sharp-edged delta wing. AIAA J 41(9):1832-1835. doi: 10.2514/2.7304

Yun BY, No HC, Shin CW (2016) Modeling of high pressure steam condensation in inclined horizontal tubes of PAFS in APR+. J Nucl Sci Tech 53(9):1353-1365. doi: 10.1080/00223131.2015.1110505

16

Application of a Greedy Algorithm to Military Aircraft Fleet Retirements

Jeffrey Newcamp[1], Wim Verhagen[1], Heiko Udluft[1], Richard Curran[1]

ABSTRACT: This article presents a retirement analysis model for aircraft fleets. By employing a greedy algorithm, the presented solution is capable of identifying individually weak assets in a fleet of aircraft with inhomogeneous historical utilization. The model forecasts future retirement scenarios employing user-defined decision periods, informed by a cost function, a utility function and demographic inputs to the model. The model satisfies first-order necessary conditions and uses cost minimization, utility maximization or a combination of the 2 as the objective function. This study creates a methodology for applying a greedy algorithm to a military fleet retirement scenario and then uses the United States Air Force A-10 Thunderbolt II fleet for model validation. It is shown that this methodology provides fleet managers with valid retirement options and shows that early retirement decisions substantially impact future fleet cost and utility.

KEYWORDS: Aircraft retirement, Fleet manager, Aircraft cost, Retirement model.

INTRODUCTION

Military aircraft fleet managers are responsible for providing strategic capability to their owning command. Thus, aircraft are based around the globe to perform various roles under a variety of operating conditions. As these individual aircraft are flown over time, each one develops a historical utilization profile that is related to its fatigue life expended (Molent *et al.* 2012). When a fleet of individual assets nears projected end-of-life, it is imperative that the fleet manager plan for retirement so that operational demand can be satisfied. Retirement planning varies greatly across military services and within service fleets (Garcia 2001; AFSB 2011). It can be proactive and data-driven but at times it has been reactionary, driven by changing budgetary conditions or critical aircraft failures. As the average age of aircraft fleets is increasing, retirement planning tools and methodology are necessary to aid fleet managers through the retirement decision process (Carpenter and White 2001).

The objective of this research was to develop a tool to provide fleet managers with a list of aircraft serial numbers that should be considered for retirement, sorted by precedence and timing. This tool is called the Fleet and Aircraft Retirement Model (FARM). It provides a list of aircraft indicating which one should be retired first and when this should happen. To improve the applicability of the tool, its interface is simplistic, the greedy algorithm implementation is clear and the inputs are accepted in a variety of formats. FARM was built for the spectrum of fleet managers including those who seek to minimize lifecycle cost, to maximize aircraft utility and to maximize the fleet's utility to cost ratio. The methodology also supports a fleet manager who wishes to use his own objective function that might be based on a variety of weighted metrics.

1.Delft University of Technology – Faculty of Aerospace Engineering – Air Transport and Operations Section – Delft/South Holland – Netherlands.

Author for correspondence: Jeffrey Newcamp | Delft University of Technology – Faculty of Aerospace Engineering – Air Transport and Operations Section | Kluyverweg 1 | 2629 HS – Delft/South Holland – Netherlands | Email: jnewcamp@gmail.com

Prior to discussing retirement, a fleet manager must understand the fleet's demands and historical utilization (Jin and Kite-Powell 2000). A previous study analyzed this opportunity using operational data from the United States Air Force (USAF) A-10 Thunderbolt II fleet (Newcamp 2016). The next step in retirement thinking is to develop replacement policy for a fleet utilizing the operation research methodologies contained in the study of replacement theory (Peters 1956).

Unfortunately, current fleet retirement schemes are primarily based, after an initial objective screening, on subjective means because economic life calculations are exceedingly complex (Tang 2013; Lincoln and Melliere 1999; Unger 2008). For example, the USAF gathers maintenance and logistics experts to decide which aircraft can get retired; however, the decision is very complex, and the decision-makers lack suitable tools (Marx 2016). Aircraft can be identified for retirement based on flight hours, repairs that limit usability, limit exceedances, corrosion, owning unit capabilities, among many other factors. While the bulk of replacement theory literature discusses the replacement of current (defender) assets with more modern (challenger) assets, this study ignores the latter because their acquisition does not directly hasten defender retirements (Robbert *et al.* 2013). Also, the authors treat military aircraft as parallel assets that independently contribute to supply (Stuivenberg *et al.* 2013), which allows for the specificity of individual serial numbers in the fleet.

Military aircraft fleet's assets do not continually operate at maximum capacity. Since retirement schedules depend on utilization, a fleet manager may alter utilization patterns leading to a more optimal retirement schedule. Testing various retirement schedules with an objective tool is necessary to quantify the net present value of each scheme. This paper contributes with a methodology that answers this need and enables fleet managers to make utilization decisions now that will affect future fleet statuses.

The novel contribution of the FARM methodology is the use of individual serial number utilization histories and cost data as a basis for future year predictions. Traditional replacement models have used fleet-wide utilization averaging or ignored asset utilization altogether, which has led to non-optimal solutions (Hartman 1999). To overcome the limitation of basing forecasts on outdated information, fleet managers can periodically use FARM to update their fleet retirement forecasts, including updated cost and utility data for each iteration. This approach also allows fleet managers to alter their utilization levels across a fleet to optimize their retirement scheduling.

The remainder of this article will discuss the methodology employed in the FARM software. The background section contains relevant literature on asset retirement plus a discussion of capital asset replacement theory. In the methodology section, the greedy algorithm approach to the retirement problem and the mathematical formulation for FARM are described. Then the results section shows data from a simulation run using FARM for a virtual fleet. The discussion section highlights the usefulness of a serial number specific retirement tool and shows validation of FARM using the real USAF A-10 fleet. Lastly, the conclusions section emphasizes the major findings from this study.

BACKGROUND

LITERATURE REVIEW

A military aircraft fleet retirement methodology must connect the domains of replacement theory, capital asset economics and military operational analysis. Relevant studies concerning asset replacement include Jones *et al.* (1991), Rajagopalan (1998) and Bethuyne (1998) and the thorough treatment of capital equipment replacement in Jardine and Tsang (2013). While insights can be gained from other domains, 2 considerations are important to aircraft replacements. First, aircraft lifecycles and planning/construction timelines are much greater than some other asset categories. Second, upgrades and overhauls significantly alter the capability and lifetime projection (Tang 2013).

Tang (2013), in a study on replacement schedules, discussed a time-space network approach for helicopters. The study concluded that cost parameters like fixed and variable operating costs can be simplified for benefit of the model's approach. The author assumed all helicopters were homogenous regardless of age and utilization history and excluded variable staff costs from the model. The present research advances this assumption by accommodating variable staff costs in the variable cost function and allows an inhomogeneous fleet input. Hartman's complementary study on replacement schedules showed that these are highly dependent on asset utilization through time (Hartman 2004). Hartman's integer programming method used a cost-minimization technique for asset replacement over a finite horizon (Hartman 1999). His paper suggested that future research should address fleet management and fleet sizing options.

Jin and Kite-Powell (2000) relied on system utilization and replacement decisions to meet the demands of a profit-maximizing manager. The authors looked at operating cost trends and the cost of replacement as factors for the retirement decision for ships. The primary contribution of Jin and Kite-Powell (2000) is the conclusion that an asset should be retired if its net benefit in a fleet is less than the salvage value.

Evans (1989) studied ship replacement theory basing his approach on costs rather than profits and concluding that replacement should occur when it becomes cheaper to purchase a replacement than to continue operating an aging system. The paper has many similarities to aircraft fleet replacement study, mainly that replacement should only be affected by costs in real terms. Additionally, this author posited that replacement decisions should focus on the existing fleet and not on the costs or capabilities of the replacement assets. The present study uses the same approach, suggesting that retirement is based on the current operating costs of the fleet. Since ship replacement requires years for contracting, construction and testing, ships are more similar to aircraft than assets in the motor vehicle, farm machinery and locomotive industries. As Evans posited, ships are often replaced with like replacements. However, aircraft are commonly replaced with newer assets with greater capability (Boness and Schwartz 1969).

Malcomson (1979) determined replacement rules for capital equipment and concluded that an iterative approach was the most efficient. Like in this paper, Malcomson also assumed that the replacement trigger point must be when the operating cost of aging assets is greater than operating new equipment. Further, the author noted that finite answers to the replacement problem are more desirable than approximate answers, and given modern computing power, finite solutions are attainable at very low cost.

Landry (2000) analyzed multiple courses of action for maintaining the aging fleet of Canadian CF188 (F-18) and CP140 (P-3) aircraft. His study treated the problem as a business case analysis with the aim of providing a fleet manager with objective data for a retirement decision. His Airframe Life Extension Program (ALEX) software used fatigue test control point data to forecast early retirement dates.

Lu and Anderson-Cook (2015) concluded that future reliability estimations can be improved when assets of the same age are not treated homogeneously, but are rather based on historical usage. The authors used an automobile example to illustrate that 2 cars of the same age do not possess the same reliability. Understanding mileage and usage conditions can improve maintenance and replacement decisions, just as understanding aircraft demographics can improve retirement decisions.

REPLACEMENT THEORY

Replacement theory is a decision-making process from operations research dealing with substitute system selection conducted by an agent. For a group of assets, the formulation becomes a parallel replacement problem. If the goal is to minimize lifecycle cost, replacement theory can help to determine a capital asset's optimum life. As capital assets age, increasing maintenance costs and reduced utility draw attention to the necessity for replacement (Bethuyne 1998; Lu and Anderson-Cook 2015). Retiring assets is half of the parallel replacement puzzle and the subject of this research. It is assumed that the selection of replacement equipment occurs outside the scope of this methodology.

Generally, new equipment with better capability replaces older equipment (Nair and Hopp 1992). For aircraft, replacement theory might suggest 2 courses of action: upgrades/overhauls or retirement. As Landry's research concluded, the crux is deciding whether it is more fiscally responsible to upgrade aircraft structure or to replace the aircraft altogether (Landry 2000). This paper only addresses the retirement course of action, which is termed the replacement model. It is believed that providing a fleet manager with the best replacement model will yield the most sensible economic replacement policy.

A parallel replacement problem, by its nature, addresses a set of assets. Unlike the single asset case, assets under consideration for parallel replacement can have their utilization levels adjusted to prolong or accelerate deterioration (Bethuyne 1998). This can be an invaluable approach for fleet managers trying to meet operational requirements or retirement mandates.

METHODOLOGY

FRAMING THE PROBLEM

To determine the optimal aircraft to retire at a point in the future, managers could use previous aircraft information as the best predictor for residual aircraft life (Hsu *et al.* 2011; Hawkes and White III 2007). However, analyzing the current fleet and each smaller fleet size was not computationally feasible for fleet sizes greater than approximately 15 assets, so a greedy algorithm was implemented. Calculating every permutation was not necessary since a greedy algorithm provides the same global optimum if the problem is appropriately bounded and local optima are avoided through logic (Cormen *et al.* 2009). This model consisted of a fleet of n aircraft with each subsequent fleet size, $n - 1$, dependent on the previous reduction. This methodology was grounded in the assumption that a fleet manager desiring to retire 2 or more aircraft would always choose the worst asset to retire at each iteration. Therefore, all smaller fleet size problems became $n - 1$ easier until $n - (n - 1)$, when the single remaining aircraft was the least desirable option. This iterative approach resulted in a Pareto front of fleet cost, fleet utility or the ratio of fleet utility to cost. Changing from a minimization model to a maximization model, a second Pareto front could be found. The space between the Pareto fronts indicates the relative goodness or inferiority of retirement choices.

FLEET AND AIRCRAFT RETIREMENT MODEL

FARM uses a greedy algorithm to determine which aircraft in an inhomogeneous fleet should be retired and in what order. For each smaller fleet size, the algorithm chooses the current optimal solution before analyzing the next smaller fleet size. FARM's methodology is outlined in Fig. 1. The multi-year outlook makes retirement decisions using projected asset cost and utility. The model is valid for any initial and final fleet sizes. FARM operates with user inputs (decision periods, minimum/maximum aircraft ages and rate of yearly budget increase) and 3 user functions (fixed cost, variable cost and utility). The fixed cost is distributed evenly across assets while the variable cost and utility are both functions of aircraft age. Costs are modeled as equivalent costflow. Inflation and the effects of various methods for cost reporting were removed from the model by using maintenance man-hours as a proxy in the variable cost calculations. Utility is analogous to aircraft availability, is a number between 0 and 1 and is computed as the number of available days out of 31. However, individual FARM users may alter the format of input functions as necessary.

The methodology underlying FARM is useful for modeling a real fleet of aircraft as well as a virtual fleet of aircraft. Virtual fleet modeling follows the conventions found in literature: aircraft operations and support (O&S) costs are high in the first few years of operation, then decrease sharply as the fleet matures and finally the costs increase at approximately 3% per year of age into the future (Dixon and Project Air Force (U.S.) 2006). Utility begins low for a new aircraft, then quickly peaks, followed by a decrease with age. An example of the cost and utility models used for FARM's development are shown in Fig. 2. Step functions in utility levels and costs that occur due

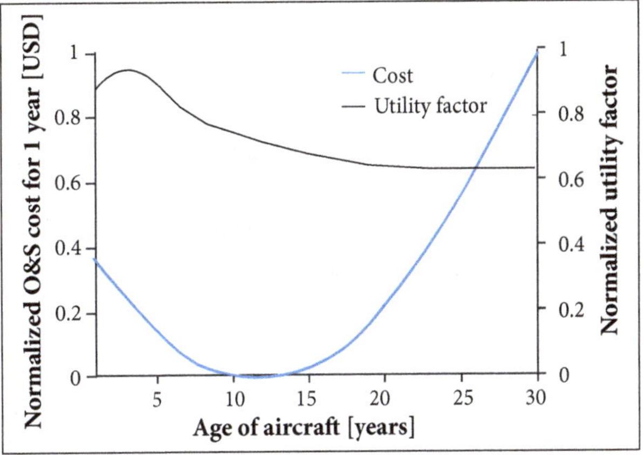

Figure 2. Representation of cost and utility models in FARM.

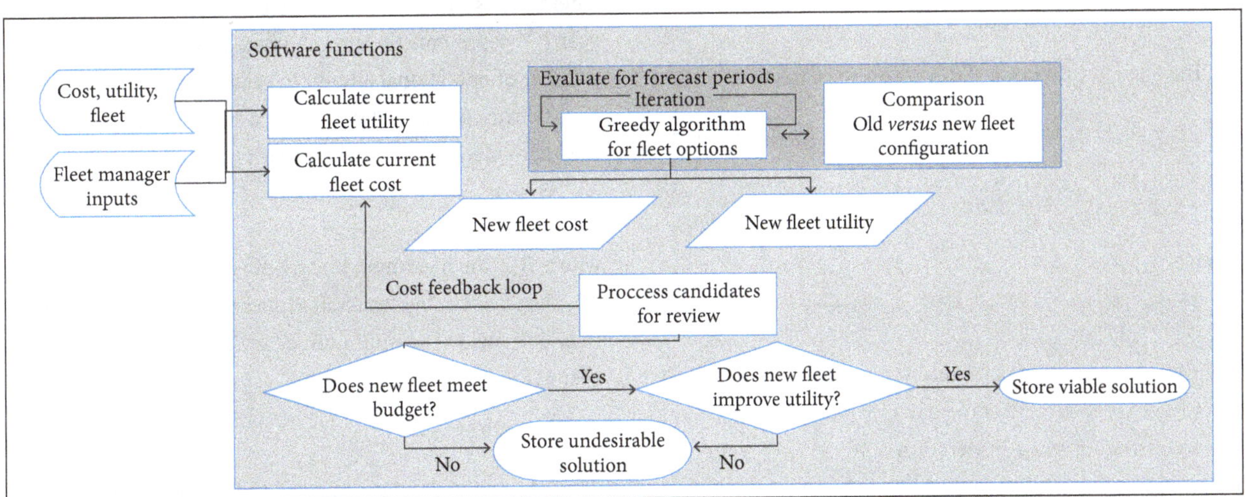

Figure 1. Flow chart for methodology steps.

to major overhaul or repairs were not added to the model. Real fleets were modeled with actual cost and utility functions, which in general were found to follow the published conventions. To forecast future fleet conditions, the most recent cost and utility were extrapolated through time. Otherwise, depending on the age distribution of the fleet, FARM would suggest retiring very young aircraft with high cost and low utility.

For each decision period, FARM outputs the recommended serial numbers to retain for all fleet size options with associated metrics for each option. Fleet managers may use these data to identify their ideal fleet size and makeup. Fleet changes with time can then be evaluated. The limitations of this methodology and associated software model are few but important. The methodology is only valid for 1 mission design series. For example, a mixed fleet of KC-135s and F-15s cannot be evaluated. Second, the methodology does not allow for subjective valuations or weighting factors for the aircraft. Lastly, FARM does not provide a time-sequence of retirement decisions. Rather, FARM forecasts future asset cost and utility to support a retirement decision forecast.

MATHEMATICAL FORMULATION

This section presents the optimization model that the greedy algorithm solves in each of its iterations for a given year of interest. Lastly, the calculation equations and problem constraints are presented.

The decision variables are:

$$X_{ta}^{i} = \begin{cases} 1, operating, \\ 0, not\ operating. \end{cases}$$

$$R_{ta}^{i} = \begin{cases} 1, retired, \\ 0, not\ retired. \end{cases}$$

The objective function (Eq. 1) seeks to maximize:

$$Z = -W_c \int_0^a \int_0^t C_{ta} X_{ta}^i dtda + W_u \int_0^a \int_0^t U_{ta} X_{ta}^i dtda + + W_r \int_0^a \int_0^t \frac{U_{ta} X_{ta}^i}{C_{ta} X_{ta}^i} dtda \quad (1)$$

where:

$$W_c, W_u, W_r \in \{0,1\}, W_c + W_u + W_r = 1 \quad (2)$$

a represents the aircraft of interest; t means year of interest; C_{ta} is the cost of aircraft a in year t; X^{i}_{ta} means that aircraft a is operating in year t in iteration i; W_c, W_u, W_r represent weighting — cost, utility, and utility/cost ratio, respectively; U_{ta} is the utility of aircraft a in year t.

The objective function contains 3 terms. The first is the cost calculation, a combination of all fixed and variable costs for operations and sustainment. The second term is the utility calculation, measured as wished by the fleet manager. The third term is the utility per cost ratio, a way to balance the cost associated with changes to utility. It is assumed that only 1 term can be optimized at a time in the model. That is, 1 and only 1 of the weights is equal to 1 each time the optimization model is solved, as shown in Eq. 2. The following equations are required to evaluate the objective function.

The cost of an aircraft a in year t is the integration of aircraft cost from simulation start until the year of interest, assuming that the integration increment is small enough to yield small error (Eq. 3):

$$C_{ta} = \int_0^t C_a dt \quad (3)$$

where C_a is the annualized cost function of aircraft a.

The utility of an aircraft a in year t is the integration of aircraft utility from simulation start until the year of interest, assuming that the integration increment is small enough to yield small error (Eq. 4):

$$U_{ta} = \int_0^t U_a dt \quad (4)$$

where U_a is the annualized utility function of aircraft a.

The equations are subjected to several constraints. The sum of aircraft a in year t must be between the bounds of operational aircraft in year t (Eq. 5):

$$\underline{NA_t} \leq \sum_{a \in A} X_{ta}^i \leq \overline{NA_t} \quad (5)$$

where $\underline{NA_t}$ is the minimum number of operational aircraft in year t; A represents the aircraft type, a; $\overline{NA_t}$ is the maximum number of operational aircraft in year t.

The sum of the cost of aircraft a times inventory must be less than or equal to budget in year t (Eq. 6):

$$\sum_{a \in A} C_{ta} X_{ta}^i \leq \overline{B_t} \quad (6)$$

where $\overline{B_t}$ is the maximum budget in year t.

The sum of utility of aircraft a times inventory must be greater than or equal to the minimum acceptable utility threshold in year t (Eq. 7):

$$\sum_{a \in A} U_{ta} X_{ta}^i \geq \underline{U_t} \quad (7)$$

where $\underline{U_t}$ represents the minimum utility threshold of the fleet in year t.

The opportunity to retire an aircraft *a* in year *t* is contingent upon the existence of aircraft *a* in the fleet in the previous year (Eq. 8):

$$R_{ta}^{i} \leq X^{i}{}_{(t-1)a}, \forall\, a \in A \quad (8)$$

where R^{i}_{ta} means that the aircraft *a* is retired in year *t* in iteration *i*.

The presence of an aircraft *a* in year *t*, given the knowledge of previous years of interest and the decision made in year *t*, is represented in Eq. 9:

$$\left(X_{(t-1)a}^{i} - R_{ta}^{i}\right), \forall\, a \in A \quad (9)$$

where, upon initialization, all aircraft are operational (Eq. 10):

$$X_{0a}^{i} = 1, \forall\, a \in A \quad (10)$$

The fleet size in year *t*, Eq. 11, is the summation of the operating aircraft:

$$F_{t}^{i} = \textstyle\sum_{a \in A} X_{ta}^{i} \quad (11)$$

where F^{i}_{ta} is the fleet size in year *t* in iteration *i* and must be 1 smaller at each iteration (Eq. 12):

$$F_{t}^{i} = F_{t}^{i-1} - 1 \quad (12)$$

and the initial fleet size, Eq. 13, is the summation of the operating aircraft in the initial year:

$$F_{0a}^{i} = \textstyle\sum_{a \in A} X_{0a}^{i} \quad (13)$$

RESULTS

This section presents results from the FARM program. A virtual fleet is used for simulation and simplified output plots show representative results. Then, to validate the methodology, A-10 case study FARM results are shown with plots showing detail to the tail number level.

To evaluate FARM, this discussion uses a simulated aircraft fleet of size, $n = 100$, over a period of 5 years with cost and utility data similar to those represented in Fig. 2. Aircraft ages were drawn from a uniform distribution. Budget was set at the current budget plus a 1% yearly budget increase to mimic the defense budgeting process. Minimum acceptable utility was set to 45% of the existing utility. Three objective functions are used: cost minimization, utility maximization and utility per cost maximization.

Figure 3 shows simplified simulation cost results for a sample fleet in year 5 for fleet size options from 1:*n*. The 2 lines represent the feasible solutions, which include only those results meeting budget and utility requirements. The bottom curve represents the cost-minimization solutions. These solutions show the cost of the fleet for *n* aircraft, *n* - 1 aircraft, etc. The top curve shows fleet cost for cost maximization or worst case retirement choices made for each fleet size. The vertical gap between the curves is the cost delta that can be saved by making the cost-minimization serial number retirement decisions. The curves are cutoff at the both ends, caused by budget and utility constraints.

Figure 4 is an expanded view of a small portion of the lines in Fig. 3. This expanded view shows that the lines in Fig. 3 are composed of many discreet points. At each fleet size, *n*, FARM calculates all of the possible options. These are shown in Fig. 4 between the most expensive and the least expensive options. Knowing the range of options is useful because it is not always practical for a fleet manager to retire the optimum aircraft.

Figure 5 shows the simplified simulation results for the same scenario, but with a utility-centered management focus.

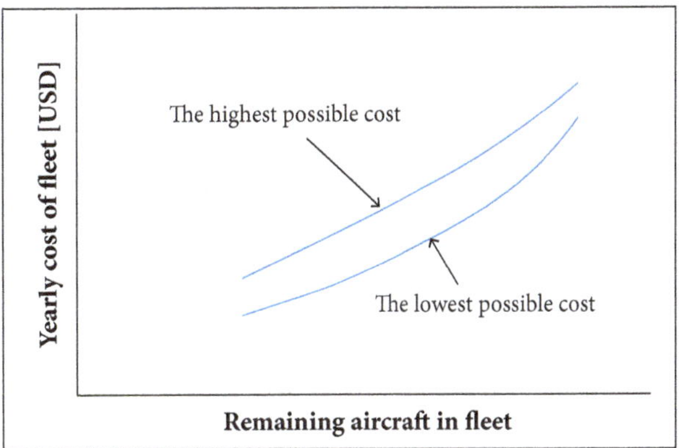

Figure 3. High and low cost choices for fleet of various fleet size options.

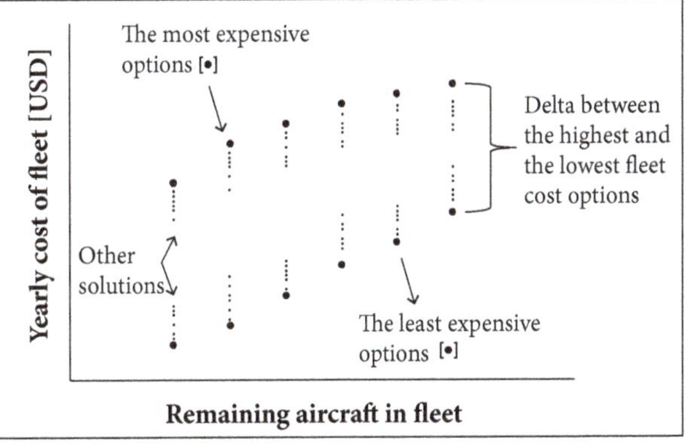

Figure 4. Expanded view of cost options showing all solutions.

These results inform the fleet manager which serial numbers to retire if the fleet goal was to maximize the utility factor, which for this scenario is the sum of aircraft days available per month for the existing fleet. The expanded view shows that, for each fleet size, there are $n - 1$ utility outcomes. The shapes of the curves shown in Fig. 3 to Fig. 5 are the manifestation of the cost and utility input data.

The curves in Fig. 6 show the Pareto fronts for the utility per cost ratio calculations for the sample fleet. As aircraft are retired from the fleet (right to left), the curves diverge, showing that a fleet manager can make poor retirement decisions that impact the fleet's utility per cost ratio. As the fleet size shrinks, the shape of the Pareto curves shifts which is due to the fixed cost distribution function. Maintaining a constant fixed cost distribution function but varying the fleet retirement scenarios always results in local maxima (optimality condition). This result is valuable to fleet managers because it recommends a minimum practical fleet sizing solution. For example, this simulation shows a maximum utility per cost ratio that can be achieved for a fleet size of 30 aircraft.

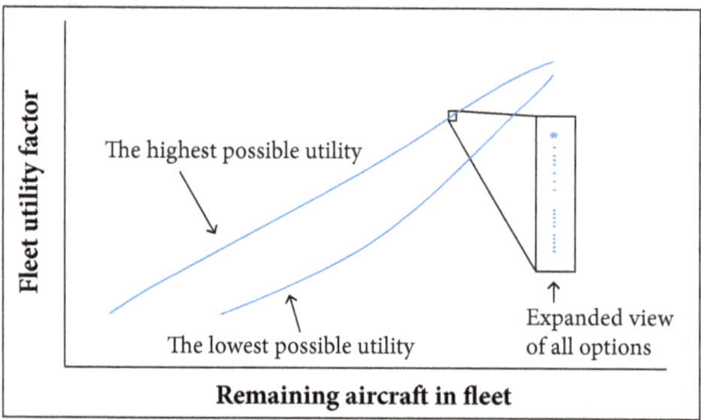

Figure 5. High and low utility choices for various fleet size options with expanded view of all possibilities.

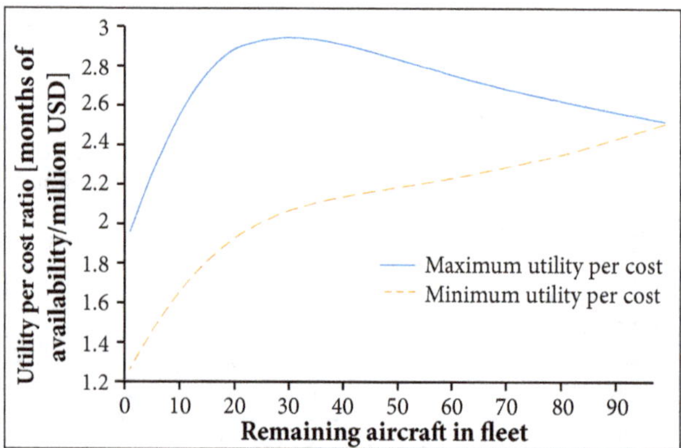

Figure 6. High and low utility per cost of Pareto fronts for various valid fleet size options.

A-10 CASE STUDY

A realistic retirement scenario for the USAF A-10 fleet (2016 active fleet) sought to reduce the fleet size to simulate the closure of a base. Right-censored A-10 data were provided by the USAF and were used as demographic data for FARM. Maintenance man-hour data were provided for each active tail number for each month for fiscal years 1995 to 2015 (66,172 total observations). Figure 7 shows 2 different percentile categories for the distribution of man-hours and the median line of the aircraft in the set. For example, the median number of maintenance man-hours for a 14 year-old A-10 was approximately 100 h per month. The dashed line is a 3% growth prediction, which validates the relationship between aircraft age and maintenance burden for agile aircraft investigated by Dixon and the Project Air Force (U.S.) (2006). The A-10 maintenance man-hour data increased at a rate of approximately 3% per year. A 1-way ANOVA confirmed this age effect (factor: aircraft age; dependent variable: maintenance man-hours; p-factor = 0.014). A 159 USD labor cost rate derived from USAF depot cost data was applied to the man-hour data for illustrative purposes in the case study. Fixed cost and variable cost values were derived from the USAF's Total Ownership Cost Tool (Robbert *et al.* 2013).

The USAF also provided mission capable rates as a utility measure for use in FARM simulations. These data were recorded monthly for each active tail number for the years 2009 – 2015 (2,792 observations). The mission capable rate was a reasonable utility metric to use for the A-10 because it is a function of failure frequency, which represents asset reliability (Balaban *et al.* 2000). The mission capable rate data did fluctuate in response to funding changes, upgrades and operational conditions. During the data collection period, for example, the A-10 fleet underwent a system life extension program that altered the

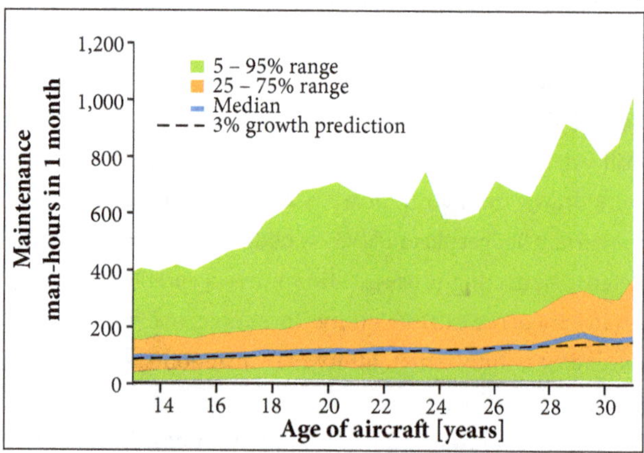

Figure 7. A-10 maintenance man-hours historical look.

mission capable rates of the fleet. These fluctuations in the data were useful for testing the software.

The data from maintenance man-hour (cost) and mission capable rate (utility) were input functions to FARM. Given that information, simulations were run to determine which aircraft would be chosen for retirement. For the active fleet of 349 A-10 aircraft, FARM produced the cost minimization output (Fig. 8) and the utility maximization output (Fig. 9) for the decision period of 5 years. Although not shown here, the accompanying outputs list the serial numbers that should be retired for each desired end-strength fleet size.

The cost-minimization objective function results (Fig. 3 and Fig. 8) exhibit different shapes. This is due to the variance in the cost data inputs ($\sigma_{A\text{-}10} > \sigma_{model}$) and emphasizes the potential advantage to this method's approach in identifying weak assets in a capital equipment fleet. Also, the expanded view in Fig. 9 highlights the inhomogeneity of utility factors in the actual A-10 fleet. The groupings of solutions occur in the expanded view result because the utility input data possess groups of aircraft

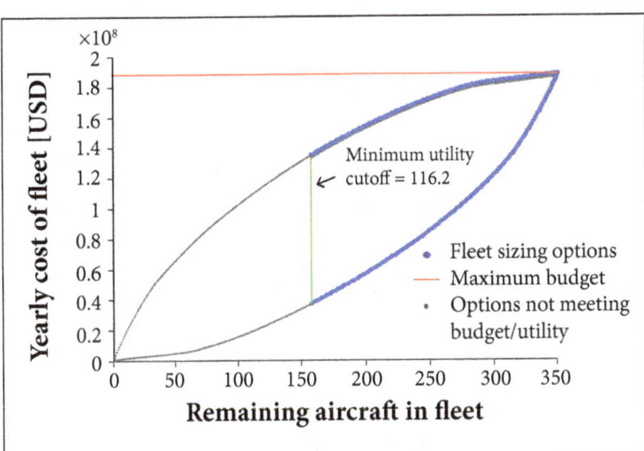

Figure 8. A-10 cost of fleet for various valid fleet size options.

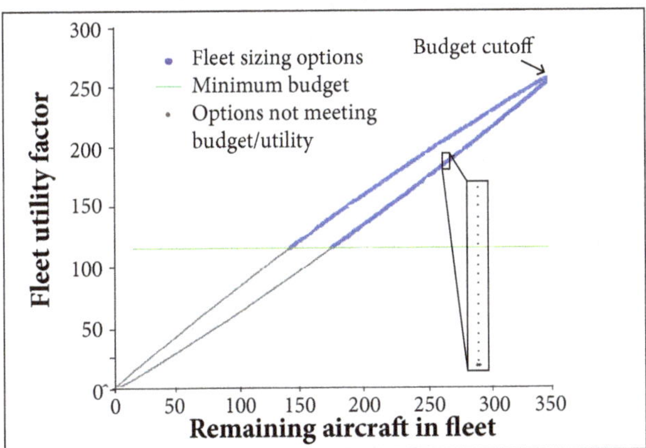

Figure 9. A-10 utility of fleet for various valid fleet size options.

with low factors, probably due to major corrective maintenance on some serial numbers during the data collection period. Fleet managers must be aware that a low utility factor may be the result of corrective maintenance or upgrades, which may make an asset less desirable in the interim but more desirable in the future. FARM allows managers to cater the utility function to reflect this, and recently improved aircraft are not identified for retirement.

DISCUSSION

FARM experiments revealed several tenets important for retirement policy analysis, namely that the inputs drive the results, uncertainty dramatically reduces the model accuracy and the earlier retirement decisions have the greatest impact on lifetime fleet cost and utility. Further, using the greedy algorithm enabled a computationally fast asset retirement model so that each of these tenets could be explored.

The shapes of the input functions directly impact the results. For example, if aircraft cost linearly increases as a function of age, then the oldest aircraft (the most costly) are indicated by the greedy algorithm for retirement first. However, real fleets exhibit more complex input functions so FARM's value increases as the fleet complexity increases.

Once uncertainty is entered the retirement model framework, a fleet manager must be careful about forecasting aircraft that would be candidates for retirement in future years. In year 1, the retirement suggestion is a direct representation of the initial cost and utility inputs. In the following years, uncertainty in cost and utility forecasts grows, therefore making future year retirement decisions mere predictions, worsening with time. Cost uncertainty is shown in Fig. 10. One facet of this uncertainty is the effect of short production runs. For a wide distribution of aircraft ages, FARM results show a finite solution. As the aircraft production timespan decreases, the retirement prediction confidence decreases. This occurs because the cost differences between individual capital assets decrease, thus making assets less distinguishable, particularly with confidence intervals. Retirement planning should be updated yearly with more recent cost and utility functions to lessen the uncertainty.

FARM shows that it is more important to make the right retirement choices from the start. Retirement policy errors propagate through time, making the initial net present value decision an assumption of future net present value. Retiring an asset with more future potential than a neighboring asset will affect the cost baseline in each subsequent year.

For generic fleets, FARM shows that the costliest aircraft possessing the lowest utility should be retired first. Actual fleet data show that the oldest serial numbers sometimes are not the costliest, least useful aircraft because of usage variation. This is the most basic reason for using a methodology like the one developed for FARM in retirement analysis.

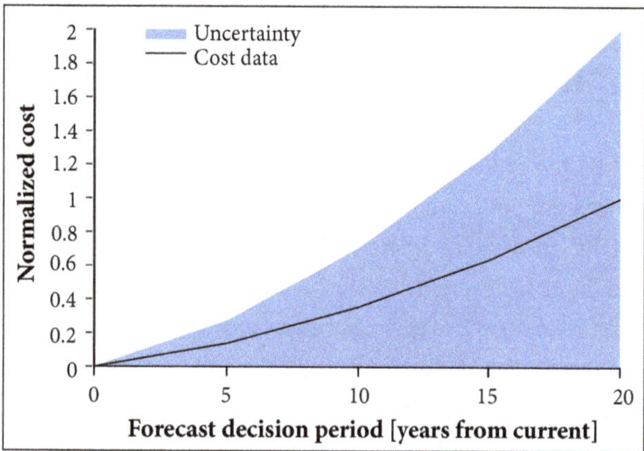

Figure 10. Uncertainty growth for FARM decision periods.

VALIDATION

Sensitivity analysis showed accurate model response to a wide range of reasonable variable and function inputs. FARM calculated fleet retirement options for both very large and very small fleets but the results were most valuable to real-world fleet sizes in the tens to hundreds of aircraft. Computation time for all scenarios described in this article was below 60 s, and the principal component affecting run time was the fleet size. A summary of run times for relevant USAF fleet sizes is shown in Table 1. The model's big O notation is: $O(n^2)$.

The model was developed using assumed values from previous studies but was validated using data from the United

Table 1. Model run times for sample fleet sizes.

Fleet size	Run time (s)*
16	3.2
100	4.2
160	5.5
320	11.3
500	22.5
1,000	95.2
2,000	567.6

*Intel Core 2 Duo, 3 GHz, 16 GB RAM.

States Air Force's Logistics, Installations and Mission Support Enterprise View repository. F-16 Fighting Falcon and A-10 Thunderbolt II data validated the general forms of the cost and utility models. One necessary step for validating the model was to catalog and analyze the aircraft serial numbers recommended for retirement to ensure the model accurately identified the weak assets. The model was found to produce repeatable results, recommending the same serial numbers for retirement given static input conditions. Likewise, whether the fleet manager wanted to retire n aircraft or some multiple of n, the sequence of retired serial numbers remained the same.

To determine model efficacy for an actual retirement scenario, the fiscal year 2013 retirement of 41 A-10s was analyzed. More aircraft were retired during this wave, but this validation effort focused on the 41 aircraft sent to retirement and ignored those aircraft reassigned as maintenance and egress trainers. The decision process to retire the 41 aircraft began in December 2011 and continued until early 2013. The FARM model was fed with cost, utility and demographic data about the fleet in the years preceding and including 2012. Using the utility per cost ratio metric and allowing FARM to choose 41 aircraft for retirement, 19 (46%) FARM choices matched the USAF ones. Using just the cost metric resulted in 17 matches (41%) and just the utility metric resulted in 15 matches (37%). These validation results do not necessarily suggest that the choice of aircraft in the 2013 retirement wave was based on a utility-per-cost metric. The stakeholders involved in the retirement used a risk-based analytical process followed by other metrics and subjective determinations to select aircraft (Thomsen *et al.* 2011).

A second A-10 retirement population was evaluated to test the model. However, the 2011 retirement wave only consisted of 9 serial numbers. Of that group, 7 were reassigned to non-flying duties allowing only 2 serial numbers for model validation. The model would have retired 1 of those 2 aircraft, but the small population size limits the value of the finding. Due to the lack of additional aircraft fleet retirement data, no further validation analyses could be conducted. Retirement decisions are complex, with many subjective factors; but a simple tool that can provide decision-makers with a starting point for choosing serial numbers shows the value of this methodology. In the case of the 2013 retirement wave, FARM would have provided an initial list that was nearly 50% accurate when compared to the final one.

A fleet manager could employ any of the 3 retirement strategies (cost minimization, utility maximization or utility per cost maximization) used in this study. To show validity, each strategy was compared to the others for both the A-10 case study and for a virtual fleet. In each case and as expected, the named strategy outperformed the remaining ones. Figure 11 shows how the 3 strategies for the A-10 fleet compare with each other for the utility-per-cost maximization strategy. The similarity between the utility-per-cost maximization and cost-minimization strategies (Fig. 11) evidences why the 2013 retirement data match well for those 2 strategies.

Other validation plots show greater stratification between the 3 strategies. This shows the value of giving the fleet manager multiple objective function options.

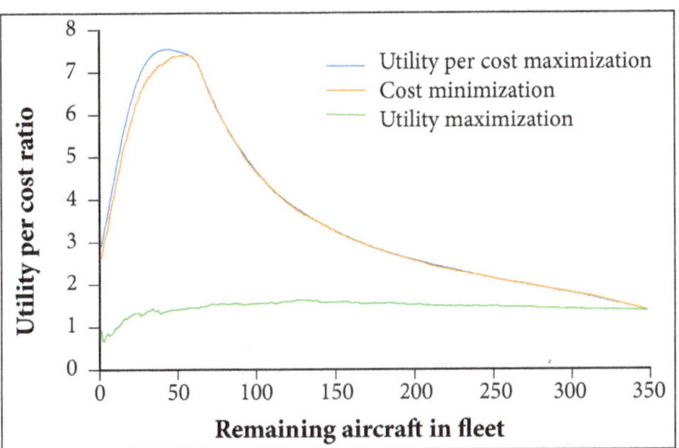

Figure 11. Comparison of retirement strategies for utility per cost ratio.

CONCLUSIONS

This study applied a greedy algorithm to an aircraft fleet retirement decision. It answered the question of which individual aircraft serial numbers should be retired and in what order. The hallmarks of this study were the use of inhomogeneous utilization histories for parallel assets and decision period forecasting. The methodology developed herein showed applicability to a virtual fleet as well as to the current USAF A-10 fleet. It was found that the correlation between usage history and retirement susceptibility could be better understood by fleet managers. The managers can control utilization levels of their assets to prolong or accelerate deterioration, which ultimately impacts the retirement schedule. Because fleet planning is a multi-year forecast, using a tool like FARM to make forecasts and periodically update them is more useful than one with a limited or finite horizon. Since suboptimal early retirement decisions cannot be remedied, a robust retirement policy is necessary.

This methodology can inspire future research in several ways. First, the methods may be extended to similar fields where parallel assets have unique usage histories. Though the objective function may change and the greedy algorithm may not present the globally optimal solution, this approach may fit into other domains. Further, other domains may also wish to study the retirement problem with non-like assets. Second, this methodology did not accommodate decision-makers with complex needs. Only cost minimization, utility maximization and utility-per-cost ratio maximization were considered. An amalgamation of weighted fleet priorities could be applied to this methodology, which can better satisfy some fleet managers. Lastly, future research might expand the scope of this methodology to include multiple aircraft mission designs in the retirement analysis. The F35A Joint Strike Fighter, for example, was designed to replace both the USAF's F-16 and A-10 aircraft. Fleet managers may be interested in evaluating which mission design should be retired first and in what quantities.

AUTHOR'S CONTRIBUTION

Newcamp J and Verhagen W conceived the idea for the study; Udluft H contributed to the methodology section and assisted with code generation; Curran R edited the text and provided the scope for the research. All authors discussed the results and commented on the manuscript.

REFERENCES

Air Force Studies Board (2011) Examination of the U.S. Air Force's aircraft sustainment needs in the future and its strategy to meet those needs. Washington: Air Force Studies Board; The National Academies Press.

Balaban HS, Brigantic RT, Wright SA, Papatyi AF (2000) A simulation approach to estimating aircraft mission capable rates for the United States Air Force. Proceedings of the 2000 Winter Simulation Conference; Orlando, USA.

Bethuyne G (1998) Optimal replacement under variable intensity of utilization and technological progress. Eng Economist 43(2):85-105. doi: 10.1080/00137919808903191

Boness AJ, Schwartz AN (1969) A cost-benefit analysis of military aircraft replacement policies. Nav Res Logist 16(2):237-257. doi: 10.1002/nav.3800160208

Carpenter M, White J (2001) Setting up a strategic architecture for the life cycle management of USAF aging aircraft. Proceedings of the RTO AVT Specialists' Meeting on Life Management Techniques for Aging Air Vehicles; Manchester, United Kingdom.

Cormen TH, Leiserson CE, Rivest RL, Stein C (2009) Introduction to algorithms. 3rd edition. Cambridge: MIT Press.

Dixon MC; Project Air Force (U.S.) (2006) The maintenance costs of aging aircraft: insights from commercial aviation. Santa Monica: RAND Corporation Air Force.

Evans J (1989) Replacement, obsolescence and modifications of ships. Marit Pol Manag 16(3):223-231. doi: 10.1080/03088838900000061

Garcia RM (2001) Optimized procurement and retirement planning of Navy ships and aircraft; [accessed 2017 May 18]. http://calhoun.nps.edu/bitstream/handle/10945/6009/01Dec_GarciaR.pdf?sequence=1

Hartman JC (1999) A general procedure for incorporating asset utilization decisions into replacement analysis. Eng Economist 44(3):217-238. doi: 10.1080/00137919908967521

Hartman JC (2004) Multiple asset replacement analysis under variable utilization and stochastic demand. European Journal of Operational Research 159(1):145-165. doi: 10.1016/S0377-2217(03)00397-7

Hawkes EM, White III ED (2007) Predicting the cost per flying hour for the F-16 using programmatic and operational data. The Journal of Cost Analysis & Management 9(1):15-27. doi: 10.1080/15411656.2007.10462260

Hsu CI, Li HC, Liu SM, Chao CC (2011) Aircraft replacement scheduling: a dynamic programming approach. Transport Res E Logist Transport Rev 47(1):41-60. doi: 10.1016/j.tre.2010.07.006

Jardine AK, Tsang AH (2013) Maintenance, replacement, and reliability: theory and applications. Boca Raton: CRC Press.

Jin D, Kite-Powell HL (2000) Optimal fleet utilization and replacement. Transport Res E Logist Transport Rev 36(1):3-20. doi: 10.1016/S1366-5545(99)00021-6

Jones PC, Zydiak JL, Hopp WJ (1991) Parallel machine replacement. Naval Research Logistics 38(3):351-365. doi: 10.1002/1520-6750(199106)38:3<351::AID-NAV3220380306>3.0.CO;2-U

Landry N (2000) The Canadian Air Force Experience: selecting aircraft life extension as the most economical solution; [accessed 2017 May 18]. http://www.dtic.mil/docs/citations/ADP010316

Lincoln JW, Melliere RA (1999) Economic life determination for a military aircraft. J Aircraft 36(5):737-742. doi: 10.2514/2.2512

Lu L, Anderson-Cook CM (2015) Improving reliability understanding through estimation and prediction with usage information. Qual Eng 27(3):304-316. doi: 10.1080/08982112.2014.990033

Malcomson JM (1979) Optimal replacement policy and approximate replacement rules. Applied Economics 11(4):405-414. doi: 10.1080/758538855

Marx J (2016) A-10 Maintenance Officer and Retirement Specialist. Telephone interview by Jeffrey Newcamp.

Molent L, Barter S, Foster W (2012) Verification of an individual aircraft fatigue monitoring system. Int J Fatig 43:128-133. doi: 10.1016/j.ijfatigue.2012.03.003

Nair SK, Hopp WJ (1992) A model for equipment replacement due to technological obsolescence. Eur J Oper Res 63(2):207-221. doi: 10.1016/0377-2217(92)90026-6

Newcamp J, Verhagen W, Curran R (2016) Correlation of mission type to cyclic loading as a basis for agile military aircraft asset management. J Aero Sci Tech 55:111-119. doi: 10.1016/j.ast.2016.05.022

Peters W (1956) Notes on the Theory of Replacement. The Manchester School 24(3):270-288. doi: 10.1111/j.1467-9957.1956.tb00987.x

Rajagopalan S (1998) Capacity expansion and equipment replacement: a unified approach. Oper Res 46(6):846-857. doi: 10.1287/opre.46.6.846

Robbert AA, Project Air Force (U.S.), RAND Corporation (2013) Costs of flying units in Air Force active and reserve components. Santa Monica: RAND Corporation.

Stuivenberg T, Ghobbar AA, Tinga T, Curran R (2013) Towards a usage driven maintenance concept: improving maintenance value. In: Stjepandić J, Rock G, Bil C, International Society for Productivity Enhancement, editors. Concurrent engineering approaches for sustainable product development in a multi-disciplinary environment. London: Springer. p. 355-365.

Tang CH (2013) A replacement schedule model for airborne service helicopters. Comput Ind Eng 64(4):1061-1073. doi: 10.1016/j.cie.2013.02.001

Thomsen M, Whitman Z, Pilarczyk R, Clark P (2011) Development of a quantitative risk based maintenance prioritization concept. Proceedings of the Aircraft Airworthiness and Sustainment Conference; San Diego, USA.

Unger EJ (2008) An examination of the relationship between usage and operating-and-support costs of U.S. Air Force Aircraft; [accessed 2017 May 18]. http://www.rand.org/content/dam/rand/pubs/technical_reports/2009/RAND_TR594.pdf

17

The Control of Asymmetric Rolling Missiles Based on Improved Trajectory Linearization Control Method

Huadong Sun[1], Jianqiao Yu[1], Siyu Zhang[1]

ABSTRACT: According to motion characteristic of an asymmetric rolling missile with damage fin, a three-channel controlled model is established. The controller which is used to realize non-linear tracking and decoupling control of the roll and angle motion is introduced based on an improved trajector y linearization control method. The improved method is composed of the classic trajectory linearization control method and a compensation control law. The classic trajectory linearization control method is implemented in the time-scale separation principle. The Lipschitz non-linear state observer systematically obtained by solving the linear matrix inequality approach is provided to estimate state variables and unknown parameters, and then the compensation control law utilizing the estimated unknown parameters improves the TLC method. Simulation experiments show that the adaptive decoupling control ensure tracking performance, and the robustness and accuracy of missile attitude control are ensured under the condition of the system parameters uncertainty, random observation noise and external disturbance caused by damage fin.

KEYWORDS: Asymmetric, Rolling missiles, Control, Improved TLC, Lipschitz adaptive observer.

INTRODUCTION

The structure or the aerodynamic asymmetric phenomenon is common for many rolling missiles.

Such unintended asymmetric phenomenon is often caused by two reasons: machining or assembling misalignment and body or fin structural damage by large external forces during the launch or the flight.

Because of uncertainty and random asymmetric factors, the asymmetric rolling missile system is a complex non-linear system with uncertainty parameters. The research on dynamic modeling and control of asymmetric rolling missile is an important problem.

Scholars carried out in-depth research in the dynamic and modeling of asymmetric aircraft. Asymmetric aerodynamic characteristics were the first to be of concern, and wing bending and impact damage were studied by the use of wind tunnel experiments (Render *et al.* 2007; Djellal and Ouibrahim 2008; Render *et al.* 2009). The dynamic problems were also the focus of the study. For an asymmetric rolling missile, when the roll rate nears to the natural frequency of pitch or yaw motion, the roll rate of the missile may be locked and maintained in the natural frequency, and the phenomenon is named lock-in. If the angle of attack of the missile becomes bigger and bigger, the catastrophic yaw happens. Since the lock-in mechanism and the phenomenon of catastrophic yaw were revealed (Murphy 1989), the research about asymmetric rolling missile motion model and dynamic behaviors are widely investigated. By the use of coupling angular motion and roll motion of 5-degrees-of-freedom equations, different dynamic behaviors such as limit

1. Beijing Institute of Technology – School of Aerospace Engineering – Beijing – China.

Author for correspondence: Huadong Sun | Beijing Institute of Technology – School of Aerospace Engineering | Tiyu N Rd, Haidian | Beijing, 10081 – China | Email: huadongsun@163.com

and chaos of asymmetric rolling missile were studied (Murphy 1989; Ananthkrishnan and Raisinghani 1992; Mikhail 1998; Tanrkulu 1999; Sun *et al.* 2015; Morote 2007; Morote *et al.* 2013). Bifurcation analysis was introduced to investigate the evolutionary process of dynamic behaviors such as lock-in and limit circle in quantitatively and qualitatively ways (Sun *et al.* 2015).

The control method of investigation and controller design are important things for an asymmetric rolling missile. Two types of fin damage were studied in the modeling and missile guidance law designing, and the classical proportional-integral-derivative (PID) control method was used (Harris and Slegers 2009). Research on non-linear control for the uncertainty parameters aircrafts was quite extensive, and it has been a hot issue of scholar's attention. Non-linear controls, such as robust adaptive (Rajagopal *et al.* 2010), sliding mode (Yang *et al.* 2012), and dynamic inversion (Nguyen *et al.* 2006), were applied in the presence of asymmetry of aircraft and spacecraft with uncertainties or other factors. Among these methods, trajectory linearization control (TLC) is a simple but effective gain scheduling means to solve non-linear and uncertainty system. TLC has been successfully applied in missiles (Mickle and Zhu 2001), robots (Liu *et al.* 2003), aircrafts (Zhu and Huizenga 2004), and other objects (Bevacqua *et al.* 2004; Su *et al.* 2013).

However, the control performance of TLC method can significantly be reduced or even infeasible in the presence of serious uncertainties (Zhu and Huizenga 2004). Besides, for most physical missile systems, another major difficulty for TLC are strong external disturbances and model uncertainties due to either constant or sudden changes. TLC faces a big challenge to deal with difficulties of cross coupling, modeling errors, external disturbances, and sensor noise effectively. In addition, the complex dynamic behaviors, such as lock-in, limit circle, and even chaos phenomenon, increase the control difficulty, and furthermore the complexity is exacerbated because of strong cross yaw-roll dynamical coupling caused by the missile rotation. Improving TLC algorithm is an issue of great significance.

This paper aims at designing a good performance control system for asymmetric rolling missiles and developing an improved method for TLC algorithm. Firstly, considering the external force caused by damage fin, a three-channel controlled model for asymmetric rolling missiles is established by the time-scale separation principle. Secondly, control law is presented using improved TLC method in which an adaptive compensation control law is added based on Lipschitz observer. Lastly, simulation experiments are carried out, and the results show that the performance of three-channel attitude control is well-exhibited. The control effectiveness of the proposed improved method is more robust then TLC.

MOTION MODEL

For an aerodynamic asymmetric cruciform finned missile with fixed rolling rate, the moment equations expressed by the complex angle of attack ξ and the complex angular velocity μ can be given in the aeroballistic axes (Murphy 1963), illustrated in Fig. 1. For a controlled missile with air rudders, compared with the aerodynamic force produced by the body, the air rudder force is a small term. Neglecting the small force produced by the rudder but taking the moment into consideration, the motions can be transformed to the body fix axes and provided as:

$$\xi' = \left[-C^*_{L\alpha} - \mathrm{i}\phi'\left(C^*_{Np\alpha}+1\right)\right]\xi + \mathrm{i}\gamma\mu \tag{1}$$

$$\begin{aligned}\mu' &= k_t^{-2}\left(\phi' C^*_{Mp\alpha} - \mathrm{i}C^*_{M\alpha}\right)\xi - \mathrm{i}\phi'\tau\mu + \\ &+\left(k_t^{-2}C^*_{Mq} + C^*_D\right)\mu + k_t^{-2}C^*_{M0}e^{\mathrm{i}\phi_{M0}} + k_t^{-2}C^*_{M\delta}\delta\end{aligned} \tag{2}$$

where: $C_{L\alpha}$ is the lift force coefficient; ϕ is the roll angle; $C_{Np\alpha}$ is the Magnus force coefficient; $\gamma = \cos(\sqrt{\alpha^2} + \beta)$, being α and β angles of attack and side-slip; k_t is the transverse radius of gyration; $C_{Mp\alpha}$ is the Magnus moment coefficient; $C_{M\alpha}$ is the normal moment slope coefficient; $\tau = 1 - I_x/I_y$, being I_x and I_y axial and transversal moment of inertia; C_{Mq} is the damping moment coefficient; C_D is the drag force coefficient; $C_{M0}e^{\phi M0}$ is the asymmetric moment coefficient, being C_{M0} the amplitude and ϕ_{M0} the phase; $C_{M\delta}$ is the control moment coefficient; the superscript * means a multiplication by $\rho Sd/(2m)$, being S reference area , ρ air density, and m the mass.

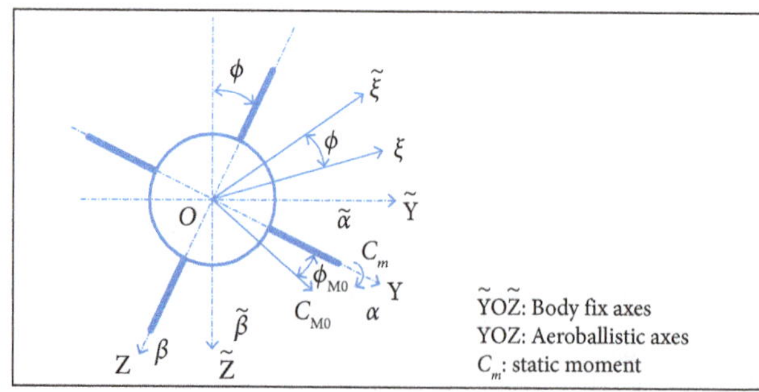

Figure 1. Axes of missile motion.

In Eqs. 1 and 2, ξ' and μ' are the derivatives of ξ and μ with respect to the independent variable l, which has the form $l = d^{-1}\int_0^t V\,dt$, being V the velocity of the missile, d the reference length and t the time; $\xi = \beta + i\alpha$ is the complex angle of attack in the body fix axes, and $\mu = q + ir$ is the complex angle velocity. $k_t^{-2} C^*_{M0} e^{i\phi M0}$ performs the uncertainty provided by the small asymmetric term. $\delta = \delta + i\delta_y$ is the rudder deflection angle in the yaw and pitch channels.

For rolling missiles, canted fins causing a constant roll moment K_δ are usually used to generate a design steady-state roll rate. Induced roll moment must be taken into account in the rolling motion besides roll moment and roll damping moment. The induced roll moment can be expressed in a simply form varying with α. The roll motion then has the form:

$$\phi'' = -K_p\phi' + K_\delta + K_n\alpha + k_a^{-2}C^*_{M\delta_r}\delta_r \tag{3}$$

where: K_p equals to $-(C^*_D + k_a^{-2}\,C^*_{lp})$, being k_a the axial radius of gyration and C_{lp} the roll damping moment; K_δ is the roll moment by canted fins; K_n is the induced roll moment coefficient; $C_{M\delta r}$ is the rolling control moment coefficient; δ_r is the rudder deflection angle in the roll channel.

When the asymmetric uncertainties are severe, they cannot be simply expressed in a constant. As shown in Fig. 2, when a fin surface is seriously damaged, the uncertainty interference caused by the lost lift dealt as an external force can be approximated as a function of the angle of attack α. Equations 2 and 3 are rewritten into the following forms, respectively:

$$\left(\phi' C^*_{Mp\alpha} - iC^*_{M\alpha}\right)\xi - i\phi'\tau\mu + \left(k_t^{-2}C^*_{Mq} + C^*_D\right)\mu + k_t^{-2}C^*_{M0}e^{i\phi_{M0}} + k_t^{-2}C^*_{M\delta}\delta + iF_1^*\alpha \tag{4}$$

$$\phi'' = -K_p\phi' + K_\delta + K_n\alpha + k_a^{-2}C^*_{M\delta_r}\delta_r + F_2^*\alpha \tag{5}$$

where: F_1 and F_2 are the uncertainty force in the angle and roll motion caused by damage.

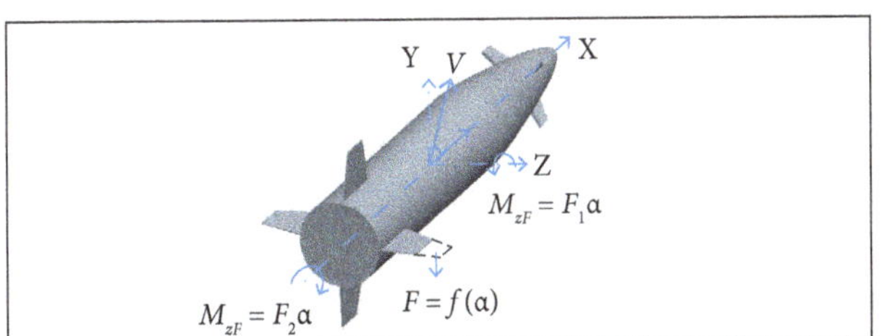

Figure 2. Structural damage schematic diagram.

Thus, Eqs. 1, 4 and 5 constitute the asymmetric rolling missile motion model in three-channel control. Slow loop variables $\Omega = (\beta, \alpha, \phi)^T$ and fast loop variables $\omega = (q, r, p)^T$ are defined, respectively, being q, r, and p yaw, pitch, and roll rates in body fixed axes. Ω responding slowly is the Euler angle vector, and ω responding fast is the angle velocity vector.

According to the time scale separation principle, Eqs. 1, 4 and 5 are rewritten in the forms:

$$\begin{pmatrix}\beta\\ \alpha\\ \phi\end{pmatrix}' = \begin{pmatrix}a_{11} & 0 & 0\\ 0 & a_{12} & 0\\ 0 & 0 & a_{13}\end{pmatrix}\begin{pmatrix}\beta\\ \alpha\\ \phi\end{pmatrix} + \begin{pmatrix}0 & a_{14} & a_{15}\alpha\\ a_{16} & 0 & a_{17}\beta\\ 0 & 0 & a_{18}\end{pmatrix}\begin{pmatrix}q\\ r\\ p\end{pmatrix} = \begin{pmatrix}f_{11}\\ f_{12}\\ f_{13}\end{pmatrix} + \begin{pmatrix}0 & a_{14} & a_{15}\alpha\\ a_{16} & 0 & a_{17}\beta\\ 0 & 0 & a_{18}\end{pmatrix}\begin{pmatrix}q\\ r\\ p\end{pmatrix} \tag{6}$$

$$\begin{pmatrix}q\\ r\\ p\end{pmatrix}' = \begin{pmatrix}a_{21}\alpha + a_{23} + a_{25}q + a_{26}pr + a_{28}p\beta\\ -a_{21}\beta + a_{24} + a_{25}r - a_{26}pq + a_{27}\alpha + a_{28}p\alpha\\ a_{41}p + a_{42} + a_{43}\alpha + a_{44}\alpha\end{pmatrix} + \begin{pmatrix}a_{22} & 0 & 0\\ 0 & a_{22} & 0\\ 0 & 0 & a_{31}\end{pmatrix}\begin{pmatrix}\delta_y\\ \delta_z\\ \delta_r\end{pmatrix} = \begin{pmatrix}f_{21}\\ f_{22}\\ f_{23}\end{pmatrix} + \begin{pmatrix}a_{22} & 0 & 0\\ 0 & a_{22} & 0\\ 0 & 0 & a_{31}\end{pmatrix}\begin{pmatrix}\delta_y\\ \delta_z\\ \delta_r\end{pmatrix} \tag{7}$$

where: $a_{11} = -C^*_{L\alpha}$, $a_{12} = -C^*_{L\alpha}$, $a_{13} = 0$, $a_{14} = -1$, $a_{15} = -(C^*_{Np\alpha+}1)$, $a_{16} = 1$, $a_{17} = -a_{15}$, $a_{18} = 1$, $a_{21} = k_t^{-2}C^*_{M\alpha}$, $a_{22} = k_t^{-2}C^*_{M\delta}$, $a_{23} = k_t^{-2}C^*_{M0}\cos(\phi_{M0})$, $a_{24} = k_t^{-2}C^*_{M0}\sin(\phi_M)$, $a_{25} = k_t^{-2}C^*_{Mq} + C^*_D)$, $a_{26} = \tau, -1 - \tau$, $a_{27} = F_1$, $a_{28} = k_t^{-2}C^*_{Mp\alpha}$, $a_{31} = k_t^{-2}C^*_{M\delta}$, $a_{41} = -K_p$ $a_{42} = -K_\delta$, $a_{43} = -K_{n,}$ $a_{44} = F_2$.

TLC PRINCIPLE

As shown in Fig. 3, TLC design method is consisted of two parts. One is forward loop designed by the use of non-linear dynamic inverse method, which changes the trajectory tracking problem into error adjustment problems. Another is state feedback loop designed by the use of linear varying system parallel-differential (PD) spectral theory, which ensures the robustness of the system with model errors. Control model can be represented in two parts as slow and fast loop. The slow one is missile attitude angles loop, and the fast one is angular velocity loop. The core issue for TLC is the design of gain scheduling control law.

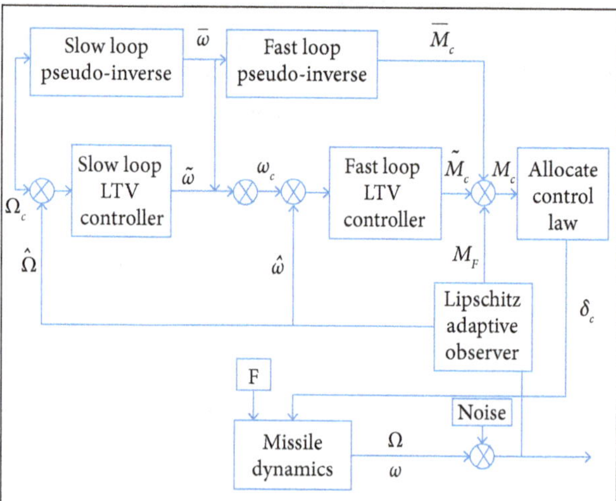

Figure 3. Control system configuration.

NOMINAL CONTROL COMMAND COMPUTING

Let us consider:

$$g_1 = \begin{pmatrix} 0 & a_{14} & a_{15}\alpha \\ a_{16} & 0 & a_{17}\beta \\ 0 & 0 & a_{18} \end{pmatrix}, g_2 = \begin{pmatrix} a_{22} & 0 & 0 \\ 0 & a_{22} & 0 \\ 0 & 0 & a_{31} \end{pmatrix}.$$

Nominal $\bar{\Omega} = \Omega_c$ command of slow loop is the expected control command of missile. Because g_1 is invertible, the nominal command of slow loop is given by:

$$\bar{\omega} = g_1^{-1}\left(\dot{\bar{\Omega}} - \bar{f}_1\right) \tag{8}$$

where: $\bar{\omega}$ is also the nominal command of fast loop.

Then the nominal control moment is represented as follows:

$$\bar{M}_c = \bar{g}_2^{-1}\left(\dot{\bar{\omega}} - \bar{f}_2\right) \tag{9}$$

Derivatives $\dot{\bar{\Omega}}$ and $\dot{\bar{\omega}}$ are computed from Ω and ω using a pseudo-differentiator represented by the transfer function:

$$G_{i,diff} = \frac{\omega_{i,diff}\, s}{s + \omega_{i,diff}}, \quad i = 1,2 \tag{10}$$

Equation 10 is also a low-pass filter and not only passes through input signal but also avoids output saturation by high-frequency noises.

SLOW LOOP CONTROLLER DESIGN

According to TLC method, linear time-varying proportional-integral (PI) regulator is usually designed to track the augmented vector error. The augmented vector can be expressed as:

$$X_1 = \left[\int \beta dl, \int \alpha dl, \int \phi dl, \beta, \alpha, \phi\right]^{\mathrm{T}}$$

Meanwhile, the slow loop in Eq. 6 can be augmented as:

$$X_1' = \tilde{f}_1(X_1) + \tilde{g}_1(X_1)\omega \tag{11}$$

where:, $\tilde{f}_1 = [\beta, \alpha, \phi, f_{11}, f_{12}, f_{13}]^{\mathrm{T}}$ and $\tilde{g}_1 = [O_3\ g_1^{\mathrm{T}}]^{\mathrm{T}}$, and O_3 is a zero matrix 3 × 3.

Equation 11 is linearized in $(\bar{X}_1, \bar{\omega})$. The state and input matrices of the linearized Eq. 11 are represented as:

$$A_1 = \left(\frac{\partial \tilde{f}_1}{\partial X_1} + \frac{\partial \tilde{g}_1}{\partial X_1}\omega\right)\Bigg|_{\bar{X}_1, \bar{\omega}} \tag{12}$$

$$B_1 = \tilde{g}_1\big|_{\bar{X}_1, \bar{\omega}} \tag{13}$$

The expected error dynamics characteristic of the slow closed loop is represented as:

$$A_{c1} = \begin{bmatrix} 0 & 0 & 0 & 1 & 0 & 0 \\ 0 & 0 & 0 & 0 & 1 & 0 \\ 0 & 0 & 0 & 0 & 0 & 1 \\ -\alpha_{111} & 0 & 0 & -\alpha_{112} & 0 & 0 \\ 0 & -\alpha_{121} & 0 & 0 & -\alpha_{122} & 0 \\ 0 & 0 & -\alpha_{131} & 0 & 0 & -\alpha_{132} \end{bmatrix} \tag{14}$$

where: $\alpha_{1,jk}, j = 1, 2, 3, k = 1, 2$, varying with time, are obtained from the closed-loop quadratic PD eigenvalues.

The state feedback matrix $K_1(t)$ is deduced from:

$$A_{c1} = A_1(t) + B_1(t)K_1(t) \tag{15}$$

Dynamic error augmented vectors can be defined as:

$$e_\Omega = \left[\int(\beta - \bar{\beta})dl \quad \int(\alpha - \bar{\alpha})dl \quad \int(\phi - \bar{\phi})dl \quad \int(\phi - \bar{\phi})dl \quad \beta - \bar{\beta} \quad \alpha - \bar{\alpha} \quad \phi - \bar{\phi}\right]^{\mathrm{T}}$$

and slow loop control input can be expressed as:

$$\omega_c = \bar{\omega} + K_1(t)e_\Omega \quad (16)$$

FAST LOOP CONTROLLER DESIGN

Following the same method to define fast loop dynamics augmented vector error,

$$e_\omega = \Big[\int(q-\bar{q})\mathrm{d}l \quad \int(r-\bar{r})\mathrm{d}l \quad \int(p-\bar{p})\mathrm{d}l$$
$$\int(p-\bar{p})\mathrm{d}l \quad q-\bar{q} \quad r-\bar{r} \quad p-\bar{p}\Big]^{\mathrm{T}}$$

augmented equation of fast loop has the form:

$$X_2' = \tilde{f}_2(X_2) + \tilde{g}_2(X_2)u \quad (17)$$

The fast loop linearization state matrix and input matrix are provided as:

$$A_2 = \left.\left(\frac{\partial \tilde{f}_2}{\partial X_2} + \frac{\partial \tilde{g}_2}{\partial X_2}u\right)\right|_{\bar{X}_2,\bar{u}} \quad (18)$$

$$\tilde{g}_2 = \begin{bmatrix} O_3 & g_2^{\mathrm{T}} \end{bmatrix}^{\mathrm{T}} \quad (19)$$

where: $u = [\delta_y \ \ \delta_z \ \ \delta_r]^{\mathrm{T}}$.

Expected error dynamics characteristic matrix of the fast closed loop can be expressed as follows:

$$A_{c2} = \begin{bmatrix} 0 & 0 & 0 & 1 & 0 & 0 \\ 0 & 0 & 0 & 0 & 1 & 0 \\ 0 & 0 & 0 & 0 & 0 & 1 \\ -\alpha_{211} & 0 & 0 & -\alpha_{212} & 0 & 0 \\ 0 & -\alpha_{221} & 0 & 0 & -\alpha_{222} & 0 \\ 0 & 0 & -\alpha_{231} & 0 & 0 & -\alpha_{232} \end{bmatrix} \quad (20)$$

where: $\alpha_{2,jk}$, $j = 1, 2, 3$, $k = 1, 2$, can be given according to PD spectral theory similarly.

State feedback matrix $K_2(t)$ is deduced from the equation:

$$A_{c2} = A_2(t) + B_2(t)K_2(t) \quad (21)$$

The fast loop control input can be expressed as:

$$M_c = \bar{M}_c + K_2(t)e_\omega = \bar{M}_c + \tilde{M}_c \quad (22)$$

LIPSCHITZ ADAPTIVE OBSERVER DESIGN

State observer design is an essential process for a control system, and the compensation control law design is based on the estimations of state variables and unknown parameters. Lipschitz observer is a common non-linear system state observer and still has a good observation performance for strongly non-linear systems with noise disturbances. Specific design process of Lipschitz observer is characterized as follows (Rajamani 1998; Zemouche and Boutayeb 2013; Pourgholi and Majd 2011).

For a classical non-linear system with unknown parameters:

$$\begin{cases} \dot{x} = Ax + \Phi(x,u) + \Psi(x,u)F \\ y = Cx \end{cases} \quad (23)$$

where: A and C are linear matrices; $x \in \mathbb{R}^n$ is the state vector; $u \in \mathbb{R}^m$ is the control vector; $y \in \mathbb{R}^p$ is the output vector; $F \in \mathbb{R}^l$ is an unknown steady bounded parameter; and $|F|| \le \gamma_1$. For all (x,y) and all u the pair (C,A) is observable.

For Eq. 23, make the following three hypotheses (Rajamani 1998; Zemouche and Boutayeb 2013):

Hypothesis (1): non-linear functions $\Phi\,(x,u)$ and $\Psi\,(x,u)$ are both uniform boundedness, and $\forall x \in \mathbb{R}^n$ and $\forall u \in \mathbb{R}^m$, Lipschitz condition is satisfied as follows:

$$\begin{cases} \|\Phi(x_1,u) - \Phi(x_2,u)\|_2 \le \gamma_2 \|x_1 - x_2\|_2 \\ \|\Psi(x_1,u) - \Psi(x_2,u)\|_2 \le \gamma_3 \|x_1 - x_2\|_2 \end{cases} \quad (24)$$

where: $\gamma_2 > 0$ and $\gamma_3 > 0$ are Lipschitz constants.

Hypothesis (2): there exist a gain matrix and a positive number ε making algebraic Riccati equation:

$$(A - LC)^{\mathrm{T}} P + P(A - LC) + \gamma PP + (1+\varepsilon)I = 0 \quad (25)$$

have a positive definite solution P, where $\gamma = \gamma_2 + \gamma_1\gamma_3$.

Hypothesis (3): there exists a vector function $h(x,u)$ making the positive definite solution P satisfy:

$$P\Psi(x,u) = C^{\mathrm{T}}h(x,u) \quad (26)$$

If the Hypotheses (1) to (3) conditions are satisfied, then the observer of the Eq. 23 is given as follows:

$$\begin{cases} \dot{\hat{x}} = A\hat{x} + \Phi(\hat{x},u) + \Psi(\hat{x},u)\hat{F} - L(y - C\hat{x}) \\ \dot{\hat{F}} = \rho h^{\mathrm{T}}(\hat{x},u)(y - C\hat{x}) \end{cases} \quad (27)$$

where: ρ is a constant parameter to adjust the estimation error and $\hat{F} = [\hat{F}_1, \hat{F}_2]$. L is the gain matrix of the observer.

L is obtained by transforming Riccati equation into the linear matrix inequality (LMI) problem (Pourgholi and Majd 2011).

In this paper, according to the Lyapunov stability conditions, the design of observer is changed to the process of solving LMI group.

LMI problem is equivalent to find a definite solution $P > 0$ and a positive number $\eta > 0$ satisfies the inequality equation:

$$\begin{bmatrix} A^{T}P + PA - 2\eta C^{T}C + I & P \\ P & -\frac{1}{\gamma^{2}}I \end{bmatrix} < 0 \tag{28}$$

By solving the above inequality (Eq. 28), L can be obtained as:

$$L = \eta P^{-1}C^{T} \tag{29}$$

COMPENSATION CONTROL LAW DESIGN

Compensation control law is given according to the estimated parameter $\hat{F}$. The control law can compensate for the interference generated by the F in the pitch and roll channels, and then TLC control performance is improved. Furthermore, in order to improve the yaw channel performance, the state feedback stabilization is increased. Compensation control law is shown as follows:

$$M_F = \left[-K_\beta \hat{\beta} / a_{22} \quad -\hat{F}_1 \hat{\alpha} / a_{22} \quad -\hat{F}_2 \hat{\alpha} / a_{31} \right]^{T} \tag{30}$$

where: K_β is the adjusting gain.

The improved TLC control law based on Lipschitz adaptive compensation is then proposed as:

$$M_c = \bar{M}_c + \tilde{M}_c + M_F \tag{31}$$

SIMULATION

Simulation analysis for an asymmetric rolling missile is performed, and system parameters are:

$a_{11} = -2.3\times10^{-4}, a_{12} = -2.3\times10^{-4}, a_{13} = 1, a_{15} = 1, a_{21} = -1.4\times10^{-5}, a_{22} = 1\times10^{-5}, a_{23} = -2.4947\times10^{-7}, a_{24} = 2.4947\times10^{-7}, a_{25} = -1.5\times10^{-4}, a_{26} = 0.99, a_{28} = -1\times10^{-5}, a_{31} = 9\times10^{-4}, a_{41} = -1.3\times10^{-3}, a_{42} = 5\times10^{-6}, a_{43} = 5.2\times10^{-4}, F_1 = 0, F_2 = 0.$

SIMULATION 1

Equations 6 and 7 can be easily transformed into a standard non-linear form with unknown parameter as Eq. 23. In order to verify the state observation and the noise suppressing performance of Lipschitz state observer, external forces F_1 and F_2 are taken to 0. In addition, system state observations often have some uncertainty. For example, the output values of the measurement system are usually superimposed with white noise. In order to represent the sensor noise, the output values are superimposed on white noise in simulation. Meanwhile, the actual value of the aerodynamic parameters increases by 20% compared with the estimated one.

The expected tracking states "command" in yaw, pitch and roll channels are β_c, α_c and p_c, respectively. It means that the motion states β, α and p of the missile are expected to change as the commands β_c, α_c and p_c require, as shown in Fig. 3. The specific command for β_c in yaw channel is a square wave; the specific command for α_c in pitch channel is a step response; and the specific command for p_c in roll channel is a constant. These three commands are all passed through a low-pass filter $5s/(s + 5)$.

According to Eq. 29, let us take the gain matrix L as:

$$L = \begin{pmatrix} 4.25 & 0 & 0 & 0 & -1.90 & 0 \\ 0 & 4.31 & -0.02 & 1.70 & 0 & -2.15 \\ 0 & -0.02 & 5.49 & -0.46 & 0 & -2.33 \\ 0 & 1.70 & -0.46 & 548.21 & 0 & -504.29 \\ -1.90 & 0 & 0 & 0 & 589.88 & 0 \\ 0 & -2.15 & -2.33 & -504.29 & 0 & 483.77 \end{pmatrix}$$

The simulation results are shown in Figs. 4 to 6. The expected motion states in the three channels of β, α and p are meant to track the commands β_c, α_c and p_c which are marked in black line. To track the same commands β_c, α_c and p_c, different control effectiveness in TLC method and "TLC + Lispchitz observer" method is compared. The control effectiveness of TLC method is marked in blue dash line, and the control effectiveness of "TLC + Lispchitz observer" is marked in red line. As shown in Figs. 4 to 6, in the case of aerodynamic parameters deviation and output noise conditions, control accuracy of only TLC method is poor, and there exist large errors. When Lipschitz state

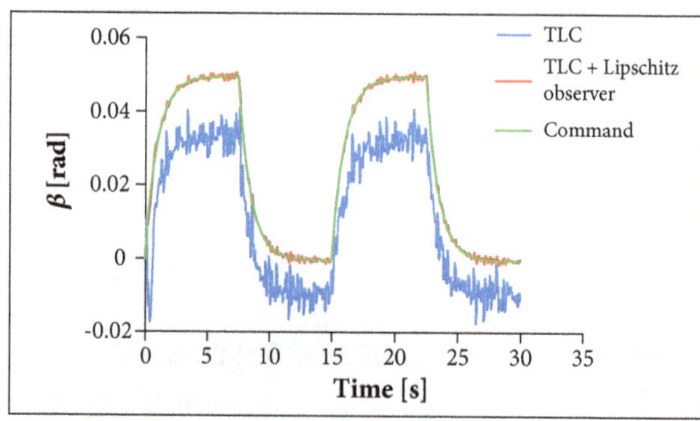

Figure 4. Effectiveness of yaw control.

observer is used, noise is effectively suppressed. It is apparent that the controller has good tracking performance even with output noise interference.

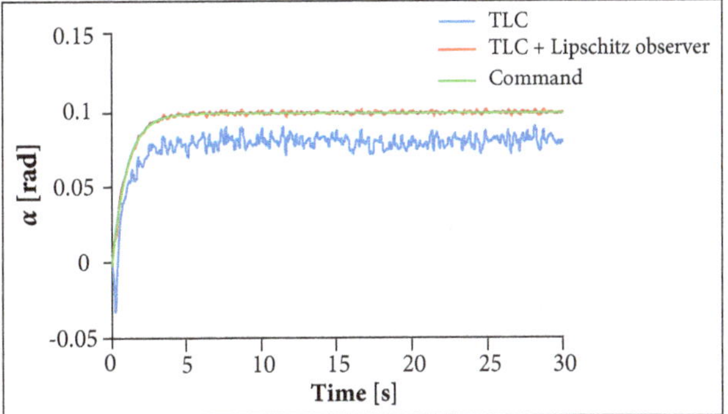

Figure 5. Effectiveness of pitch control.

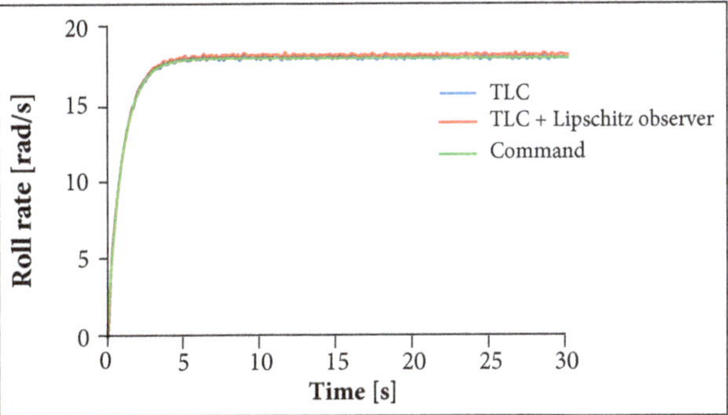

Figure 6. Effectiveness of roll control.

SIMULATION 2

To verify the estimate effectiveness of Lipschitz adaptive observer, let us consider $F_1 = 100$ and $F_2 = 200$; the state variables for the missile in the free movement are estimated, as shown in Figs. 7 to 10. Simulation results show that state observer can estimate the state variables and unknown parameters quickly and accurately.

Another simulation is performed to verify the performance of the improved TLC method. In addition, structural damage for missiles often happens suddenly, and it is supposed that a fin of the missile is suddenly damaged on 5*s* after the beginning of the simulation. The damage effectiveness is $F_1 = 100$ and $F_2 = 200$. At the same time, the actual value of aerodynamic parameters is reduced 25% compared to the estimated value, and the simulation results are shown in Figs. 11 to 14.

The specific command for β_c in yaw channel and α_c in pitch channel is a step response signal, and the specific command for p_c in roll channel is a constant. These three commands are all passed through a low-pass filter $5s/(s + 5)$. In the figures, the expected attitudes which are named "command" are marked in black line. Two different methods "TLC" and improved TLC which is named "TLC + adaptive compensation" are separately applied to control the missile attitudes to track the "command".

The figures show that an only algorithm using TLC has certain robustness in dealing with the uncertainties of aerodynamic parameters. However, the control effectiveness is not ideal in response to a sudden but strong interference, and even the system is going to diverge. "TLC" cannot match "command" in a good performance. The figures exhibit a perfect match between "command" and "TLC + adaptive compensation". It means that TLC combined with Lipschitz adaptive compensation control law improves the

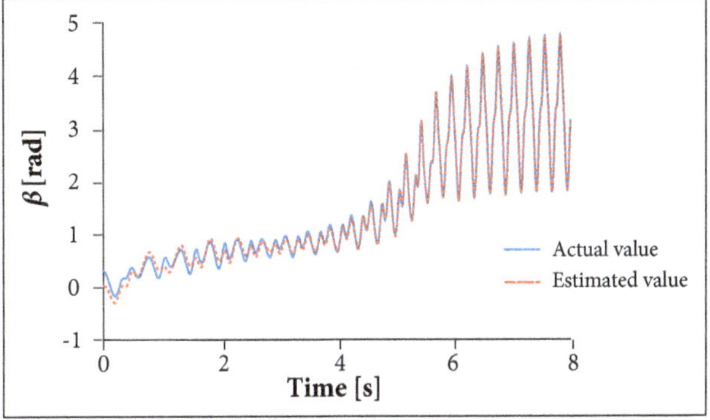

Figure 7. Actual and estimated β.

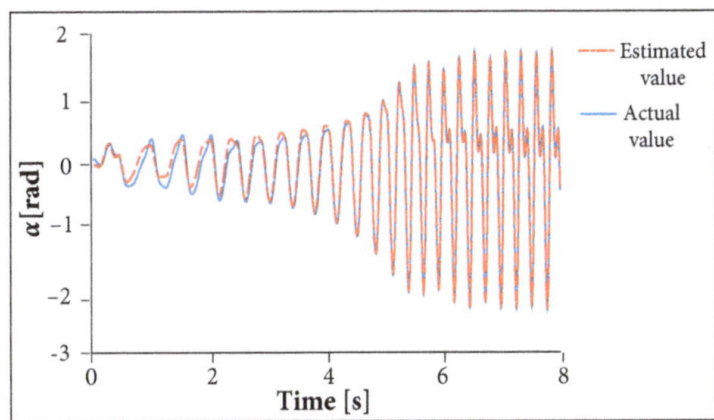

Figure 8. Actual and estimated α.

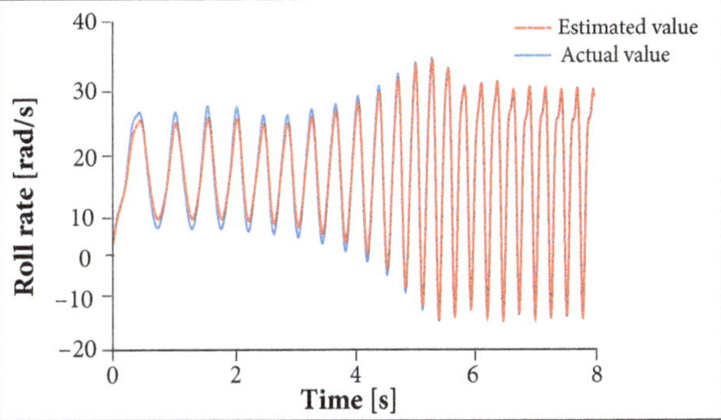

Figure 9. Actual and estimated p.

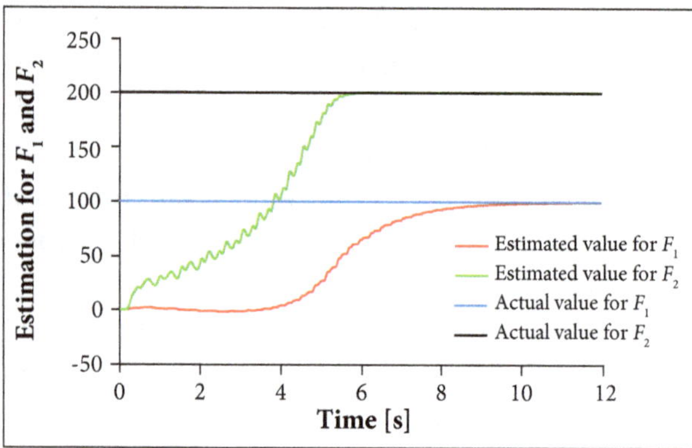

Figure 10. Actual and estimated *F*.

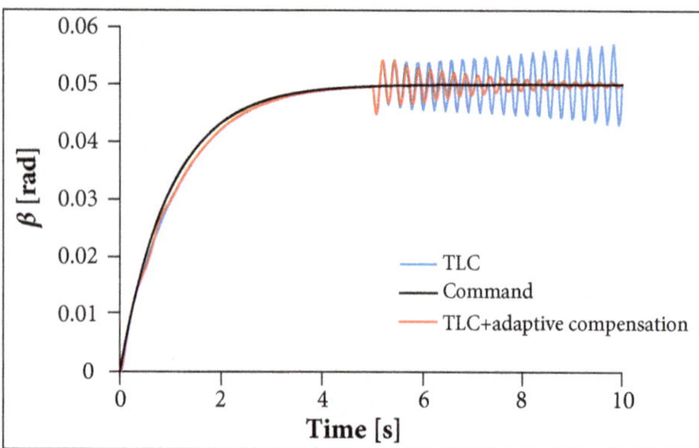

Figure 11. Effectiveness of yaw control.

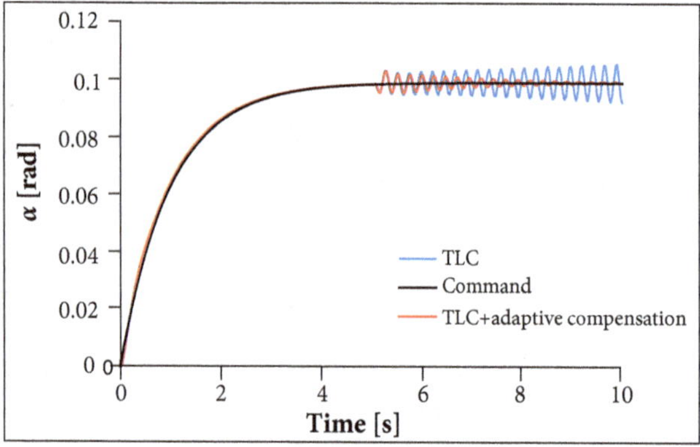

Figure 12. Effectiveness of pitch control.

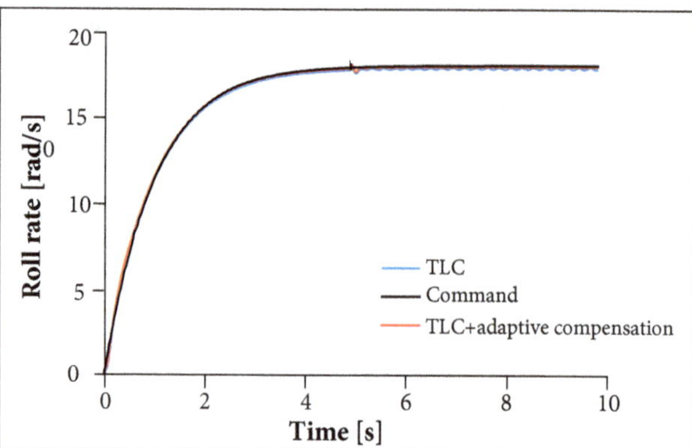

Figure 13. Effectiveness of roll control.

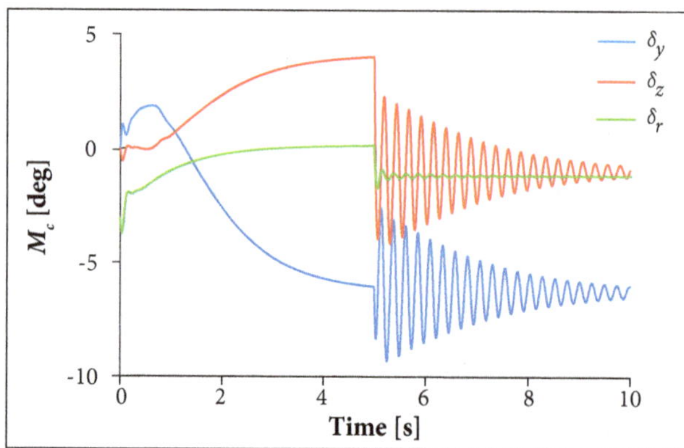

Figure 14. Rudder deflection angles for three-channel control in "TLC + adaptive compensation".

control performance in three channels and enhances the robustness of the system. The missile system can converge to the expected state of motion more accurately and quickly with strong external interference. The adaptive decoupling control solves the cross coupling in three channels. The rudder deflection angles of TLC method are lead to saturation when control system diverges, and then TLC method becomes invalid. Compared to TLC, the rudder deflection angles of improved TLC are easy to achieve without saturation phenomenon, and this means that the improved TLC method is physically realizable.

CONCLUSIONS

In this paper, the non-linear control method for asymmetric rolling missiles is investigated. Firstly, an asymmetric rolling missile controlled motion model is established considering inherent asymmetric aerodynamic force and damage fin uncertain external interference. Then, a preliminary controller is presented by the TLC method in the time-scale separation principle. Thirdly, an improved TLC method with adaptive compensation control law based on Lipschitz observer is proposed. In the improved method, the classic TLC method is complementary by an adaptive compensation control law, and this is designed according to the estimated unknown parameters and state variables from Lipschitz observer. Finally, control effectiveness comparison of TLC and improved TLC is carried out by simulation, and adaptive decoupling control for three channels is achieved.

Simulation results demonstrate that improved TLC is more effective than classic TLC method, and the proposed improved TLC method exhibits a good performance in the track ability, robustness and adaptability.

ACKNOWLEDGEMENT

This study was supported by the National Natural Science Foundation of China (No. 11532002).

REFERENCES

Ananthkrishnan N, Raisinghani SC (1992) Steady and quasisteady resonant lock-in of finned projectiles. J Spacecraft Rockets 29(5):692-696. doi: 10.2514/3.11512

Bevacqua T, Best E, Huizenga A, Cooper D, Zhu JJ (2004) Improved trajectory linearization flight controller for reusable launch vehicles. Proceedings of the 42nd AIAA Aerospace Sciences Meeting and Exhibit; Reno, USA.

Djellal S, Ouibrahim A (2008) Aerodynamic performances of battle-damaged and repaired wings of an aircraft model. J Aircraft 45(6):2009-2023.

Harris J, Slegers N (2009) Performance of a fire-and-forget anti-tank missile with a damaged wing. Math Comput Model 50(1-2):292-305. doi: 10.1016/j.mcm.2009.02.009

Liu Y, Wu X, Zhu JJ, Lew J (2003) Omni-directional mobile robot controller design by trajectory linearization. Proceedings of the 2003 American Control Conference; Denver, USA.

Mickle MC, Zhu JJ (2001) Skid to turn control of the APKWS missile using trajectory linearization technique. Proceedings of the 2001 American Control Conference; Arlington, USA.

Mikhail AG (1998) Fin damage and mass offset for kinetic energy projectile spin/pitch lock-in. J Spacecraft Rockets 35(3):287-295. doi: 10.2514/2.3353

Morote J (2007) Analytic model of catastrophic yaw. J Spacecraft Rockets 44(5):1029-1037. doi: 10.2514/1.29858

Morote J, García-Ybarra PL, Castillo JL (2013) Large amplitude oscillations of cruciform tailed missiles. Part 1: catastrophic yaw fundamental analysis. Aero Sci Tech 25(1):145-151. doi: 10.1016/j.ast.2012.01.002

Murphy CH (1963) Free flight motion of symmetric missiles. BRL Report No. 1216. Aberdeen Proving Ground: U. S. Army Ballistic Research Laboratory. AD 442757.

Murphy CH (1989) Some special cases of spin-yaw lock-in. J Guid Control Dynam 12(6):771-776. doi: 10.2514/3.20480

Nguyen N, Krishnakumar K, Kaneshige J (2006) Dynamics and adaptive control for stability recovery of damaged asymmetric aircraft. Proceedings of the AIAA Guidance, Navigation, and Control Conference and Exhibit; Keystone, USA.

Pourgholi M, Majd VJ (2011) A nonlinear adaptive resilient observer design for a class of Lipschitz systems using LMI. Circ Syst Signal Process 30(6):1401-1415. doi: 10.1007/s00034-011-9320-y

Rajagopal K, Balakrishnan SN, Nguyen N, Krishnakumar K (2010) Robust adaptive control of a structurally damaged aircraft. Proceedings of the AIAA Guidance, Navigation, and Control Conference; Toronto, Canada.

Rajamani R (1998) Observers for Lipschitz nonlinear systems. IEEE Trans Automat Contr 43(3):397-401. doi: 10.1109/9.661604

Render PM, De Silva S, Walton A, Mani M (2007) Experimental investigation into the aerodynamics of battle damaged airfoils. J Aircraft 44(2):539-549. doi: 10.2514/1.24144

Render PM, Samaad-Suhaeb M, Yang Z, Mani M (2009) Aerodynamics of battle-damaged finite-aspect-ratio wings. J Aircraft 46(3):997-1004. doi: 10.2514/1.39839

Su XL, Yu JQ, Wang YF, Wang LL (2013) Moving mass actuated reentry vehicle control based on trajectory linearization. International Journal of Aeronautical and Space Sciences 14(3):247-255. doi: 10.5139/IJASS.2013.14.3.247

Sun H, Yu J, Zhang S (2015) Bifurcation analysis of the asymmetric rolling missiles. Proc IME G J Aero Eng 26(9):1-2. doi: 10.1177/0954410015606941

Tanrkulu O (1999) Limit cycle and chaotic behavior in persistent resonance of unguided missiles. J Spacecraft Rockets 36(6):859-865. doi: 10.2514/2.3504

Yang I, Kim D, Lee D (2012) A flight control strategy using robust dynamic inversion based on sliding mode control. Proceedings of the AIAA Guidance, Navigation, and Control Conference; Minneapolis, USA.

Zemouche A, Boutayeb M (2013) On LMI conditions to design observers for Lipschitz nonlinear systems. Automatica 49(2):585-591. doi: 10.1016/j.automatica.2012.11.029

Zhu JJ, Huizenga AB (2004) A type two linearization controller for a resuable launch vehicle — a singular perturbation approach. Proceedings of the AIAA Atmospheric Flight Mechanics Conference and Exhibit; Providence, USA.

18

Aerothermodynamic Optimization of Aerospace Plane Airfoil Leading Edge

Chen Zhou[1], Zhijin Wang[1], Jiaoyang Zhi[1], Anatolii Kretov[1]

ABSTRACT: Aiming to mitigate the aerodynamic heating during hypersonic re-entry, the aerothermodynamic optimization of aerospace plane airfoil leading edge is conducted. Lift-to-drag ratio at landing condition is taken as a constraint to ensure the landing aerodynamic performance. First, airfoil profile is parametrically described to be more advantageous during the optimization process, and the Hicks-Henne type function is improved considering its application on the airfoil leading edge. Computational Fluid Dynamics models at hypersonic as well as landing conditions are then established and discussed. Design of Experiment technique is utilized to establish the surrogate model. Afterwards, the previously mentioned surrogate model is employed in combination with the Multi-Island Genetic Algorithm to perform the optimization procedure. NACA 0012 is taken as the baseline airfoil for case study. The results show that the peak heat flux of the optimal airfoil during hypersonic flight is reduced by 7.61% at the stagnation point, while the lift-to-drag remains almost unchanged under landing condition.

KEYWORDS: Airfoil optimization, Aerodynamic heating, Hicks-Henne type function, Airfoil parameterization, Surrogate model.

INTRODUCTION

Aerospace planes (ASP) encounter severe aerodynamic heating during the hypersonic phase of atmospheric re-entry. Its reusable nature dictates that it should be able to shield the underlying structure from excessive temperatures during hypersonic flight and still have good aerodynamic performance at the landing speed. Regarding the Space Shuttle, a double-delta wing configuration was adopted to optimize the hypersonic flight as well as to obtain a good lift-to-drag ratio for landing (Launius and Jenkins 2012).

Since Computational Fluid Dynamics (CFD) plays a critical role in the aerospace industry, airfoil optimization has been widely studied during the design process of a winged vehicle. Buckley and Zingg (2013) developed a weighted-integral objective function to perform multipoint aerodynamic shape optimization in which a range of operating conditions were involved. A 2-step approach was introduced to conduct the aerodynamic and structural optimization of the adaptive wing leading edge (Sun *et al.* 2013). Various algorithms were employed for the aerodynamic optimization. A novel global optimization algorithm based on the particle swarm one was developed and applied to a low-velocity airfoil optimization (Yang *et al.* 2015). Koziel and Leifsson (2014) proposed an approach utilizing the multi-objective evolutionary algorithm together with surrogate model to obtain the Pareto front of a transonic airfoil. Li *et al.* (2012) developed an efficient method using the response surface model and genetic algorithm to optimize the transonic airfoil. Xia and Chen (2015) performed the aerothermodynamic optimization of a hypersonic wing profile to decrease the maximum heat flux. However, research on the hypersonic

1. Nanjing University of Aeronautics and Astronautics – College of Aerospace Engineering – Minister Key Discipline Laboratory of Advanced Design Technology of Aircraft – Nanjing/Jiangsu – China.

Author for correspondence: Chen Zhou | Nanjing University of Aeronautics and Astronautics – College of Aerospace Engineering – Minister Key Discipline Laboratory of Advanced Design Technology of Aircraft | No. 29 Yudao Street – Qinhuai District | 210016 – Nanjing/Jiangsu – China | Email: zhouchen@nuaa.edu.cn

aerothermodynamic optimization for ASP considering the landing aerodynamic performance, is still limited.

In this study, based on the Space Shuttle re-entry case, the aerothermodynamic optimization of an airfoil leading edge was carried out to alleviate the severe aerodynamic heating during hypersonic re-entry, while aerodynamic characteristics under landing condition were simultaneously considered. First, linear superposition method of analytic function with a modified Hicks-Henne type function was used for parametric modelling. CFD models for both hypersonic and landing conditions are described here. Then, the adopted optimization approach is presented, followed by a discussion on the optimized results.

SPACE SHUTTLE RE-ENTRY DESCRIPTION

The atmospheric re-entry is particularly challenging, and the Space Shuttle is designed for a predefined schedule to survive the extreme environment (Launius and Jenkins 2012; Powers 1986). Figure 1 illustrates the nominal re-entry flight corridor for Space Shuttle (Sellers 2004). It was controlled to fly at a 40° angle of attack, producing high drag, not only to slow it down to landing speed, but also to reduce re-entry heating. The large amount of potential and kinetic energy is dissipated as heat as the Space Shuttle enters the atmosphere. A large detached bow shock wave carries away most of the heat, with the rest transferred to the vehicle through convection and radiation (Allen and Eggers Jr 1958; Tetzman 2010).

The aerodynamic heating is the severest at the stagnation point. It is assumed that the stagnation zone was in chemical equilibrium. Although the gas behind the shock is most likely in a non-equilibrium state, the approximation of chemical equilibrium boundary layer is reasonable in the stagnation zone. The heat flux at the stagnation point, which is the maximum heat flux of the leading-edge, can be estimated by (Bian and Zhong 1986):

$$q_{ws} \approx \frac{c}{\sqrt{R_s}}\left(\frac{\rho}{\rho_0}\right)^{1/2}\left(\frac{V}{V_0}\right)^m \quad (1)$$

where: c and m are constants; ρ refers to the local air density; ρ_0 is the air density at sea level; R_s represents the radius of curvature at the stagnation point; V is the vehicle's velocity; $V_0 = 7.9$ km/s is the first cosmic velocity.

The peak temperature of the outer surface is always close to the radiation equilibrium temperature. The heat flux arising from the aerodynamic braking should not cause the temperature on the ASP surface (T_w) to exceed the maximum permissible values for materials placed on the outer surface. As for the stagnation point,

$$\sigma\varepsilon T_w^{\ 4} = q_{ws} \quad (2)$$

where: $\sigma = 5.67 \times 10^{-8}$ W/(m^{2}K^4) stands for Stefan-Boltzmann constant; ε is the emissivity of the surface, depending on the material processing and surface temperature.

After a series of steep S-shaped banking turns, the vehicle lowered its nose into a shallow dive and began its approach to the landing site. Then the nose was pulled up to finally slow down the vehicle to approximately 100 m/s at touch-down. Unlike commercial airliners, the Space Shuttle glides to runway with no power and has relative low lift-to-drag ratio, so it needs a big angle of attack to maintain the longitudinal flying quality (Powers 1986).

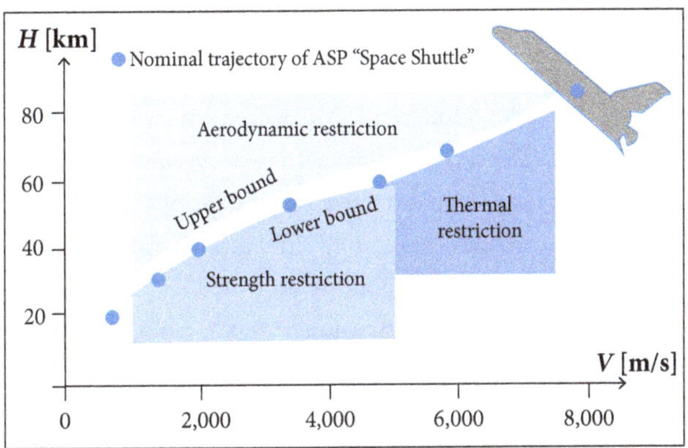

Figure 1. Re-entry corridor for Space Shuttle.

AIRFOIL PARAMETERIZATION

The airfoil profile is regenerated through altering the value of the control points during the optimal design of 2-D airfoil, while parametric description tends to be more advantageous. This paper uses the linear superposition method of analytic function to fit the airfoil profile, which is defined from the baseline one, type function and corresponding coefficients, as:

$$\overline{y}(\overline{x}) = \overline{y}_b(\overline{x}) + \sum_{k=1}^{n} a_k f_k(\overline{x}) \quad (3)$$

where: $\bar{y}_b(\bar{x})$ represents the baseline airfoil profile; n and a_k are the number of control points and the coefficients, respectively; $f_k(\bar{x})$ denotes the type function; $a_k f_k(\bar{x})$ is the perturbation of the baseline airfoil.

In this paper, NACA 0012 is chosen as the baseline airfoil, which is similar to the airfoil for Space Shuttle Orbiter Columbia NACA 0012-64 (Rochelle *et al.* 1973). These airfoils have the same leading-edge radius, and the only difference is the location of the maximum thickness. NACA 0012 airfoil can be described as:

$$\bar{y}_b(\bar{x}) = \pm 0.12 / 0.2 (0.29690 \sqrt{\bar{x}} - 0.12600\bar{x} - 0.35160\bar{x}^2 + 0.28430\bar{x}^3 - 0.10150\bar{x}^4) \quad (4)$$

Since only the leading-edge is considered in this study, a modified Hicks-Henne type function (Hicks and Henne 1978; Zhou *et al.* 2014) is employed to control the first quarter of the airfoil profile, which is expressed as:

$$\begin{cases} f_1(\bar{x}) = 0.25(1-\bar{x})^{0.25}(1-4\bar{x})e^{-20\bar{x}} \\ f_k(\bar{x}) = 0.25\sin^3[\pi(4\bar{x})^{e(k)}],\ k>1 \end{cases} \quad (5)$$

where: $e(k) = \ln 0.5 / \ln x_k$, $0 \le x_k \le 1$, $x_k = 4\bar{x}$, x_k $(k = 2, 3, 4, 5)$.

In this paper, coefficients $a_1 \sim a_5$ and $a_6 \sim a_{10}$ are used to control the upper and lower surfaces of the airfoil, respectively; x_k $(k = 2, 3, 4, 5)$ are set to be [0.04, 0.1, 0.3, 0.6]. The constraint $a_1 = a_6$ is applied to maintain the continuity of the airfoil leading edge. The previously mentioned modified type function with parameters setting is illustrated in Fig. 2.

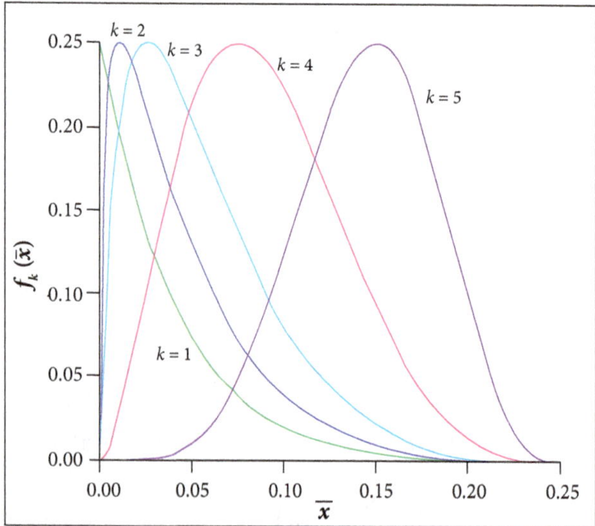

Figure 2. Improved Hicks-Henne type function.

NUMERICAL MODEL

COMPUTATIONAL FLUID DYNAMICS MODEL DESCRIPTION

The flow around an airfoil is numerically simulated by solving compressible 2-D Navier-Stokes equations. The airfoil chord length is set as 5 m. C-type structured grids around the airfoil are generated by using the commercial software CFD-GEOM. Grid convergence studies are conducted, and approximately 4×10^5 cells are distributed in the domain. A close view of the mesh distribution is illustrated in Fig. 3. Then, the commercially available CFD-FASTRAN is employed to calculate both the heat flux at the hypersonic condition and the aerodynamic coefficients at the landing condition.

As for the hypersonic case in this paper, laminar flow model is adopted since the air is thin and the Reynolds number is small at that altitude. Radiative wall boundary condition is used while the emissivity of the airfoil wall is assumed to be 0.8. It allows for radiation heat flux at the wall according to the Stefan-Boltzmann Law (see Eq. 2). Thus, a balance is formed for heat flux at the wall between conduction to the wall and radiation from it. Regarding the landing condition, k-ε turbulent model with wall function is used to solve the problem.

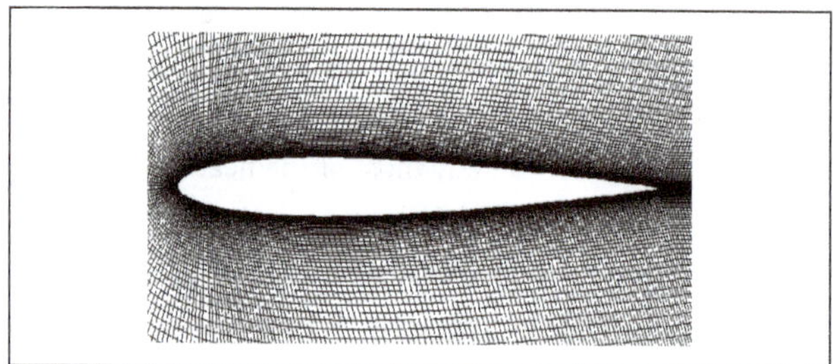

Figure 3. Close view of mesh around the airfoil.

NUMERICAL MODEL VALIDATION

Hypersonic Condition

According to a typical re-entry trajectory of Space Shuttle (Sellers 2004), its altitude and velocity profile is shown in Fig. 4. A series of hypersonic flow simulations were conducted from 7,300 to 3,050 m/s through the re-entry stage, where the angle of attack was maintained at 40°. The heat flux variation with velocity at the stagnation point is shown in Fig. 5. Normally, the velocity for the maximum heat flux is about 80 – 85% of the re-entry velocity (Bian and Zhong 1986; Sellers 2004). In this study, the first cosmic velocity is taken as the re-entry velocity. As shown in Fig. 5, the maximum heat flux occurs when the velocity decreases to approximately 6,700 m/s, *i.e.*, roughly 84.8% of the re-entry velocity. Thus, the CFD model used to calculate the heating results at the stagnation point under hypersonic condition is considered to be viable.

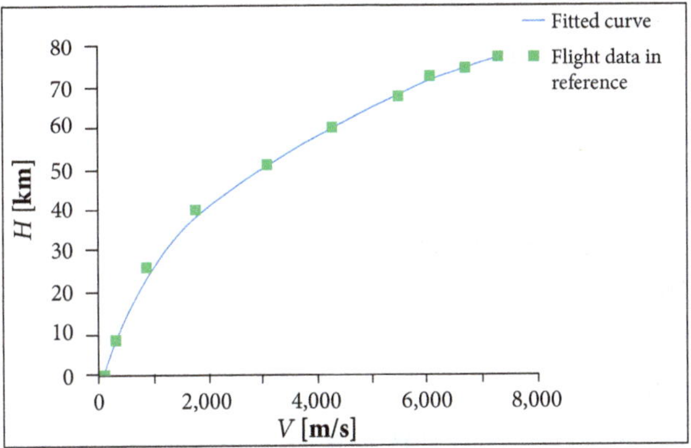

Figure 4. Space Shuttle's altitude *versus* velocity for a typical re-entry.

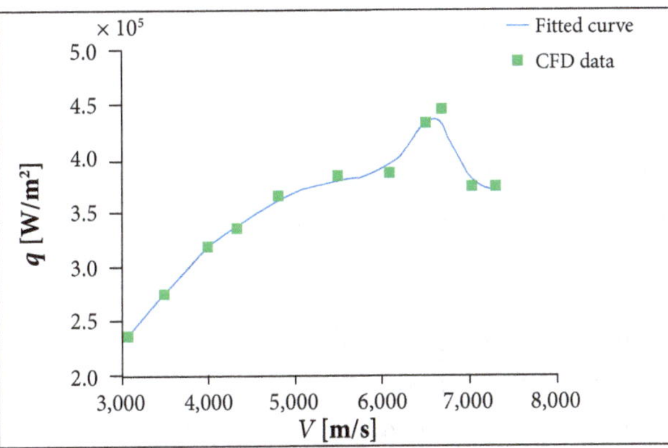

Figure 5. Heat flux at stagnation *versus* velocity.

Landing Condition

Simulations of aerodynamic characteristics of the NACA 0012 airfoil were carried out at $Re = 3.4 \times 10^7$ using the previously described computational model. Angles of attack α ranging from 0° to 16° were considered. Figure 6 shows the lift coefficient variation with angle of attack. The slope of the lift curve for an airfoil at high Re can be estimated by the empirical formula (Lu 2009):

$$C_l^a = 1.8\pi\,(1 + 0.8\bar{c}) \tag{6}$$

where: $\bar{c}$ stands for relative thickness of the airfoil.

According to Fig. 6 and Eq. 6, the relative error of the lift curve slope between the CFD results and the empirical formula result is calculated to be less than 2%. The numerical result is very close to the theoretical one. In addition, experimental data (Ladson 1988) for NACA 0012 at high Reynolds numbers show that the airfoil will normally stall around $\alpha = 16°$, which is consistent with the numerical results obtained. The comparisons indicate that the numerical model is suitable for the landing condition problem.

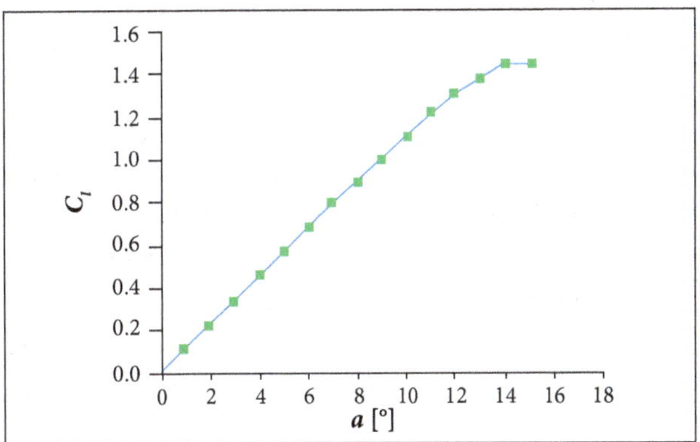

Figure 6. Lift coefficient variation with angle of attack.

OPTIMIZATION DESIGN

OPTIMIZATION APPROACH

The whole optimization process is shown in Fig. 7. All parts were integrated using the Isight framework (Dassault Systèmes Simulia Corp. 2012). Latin Hypercube Sampling (LHS; McKay

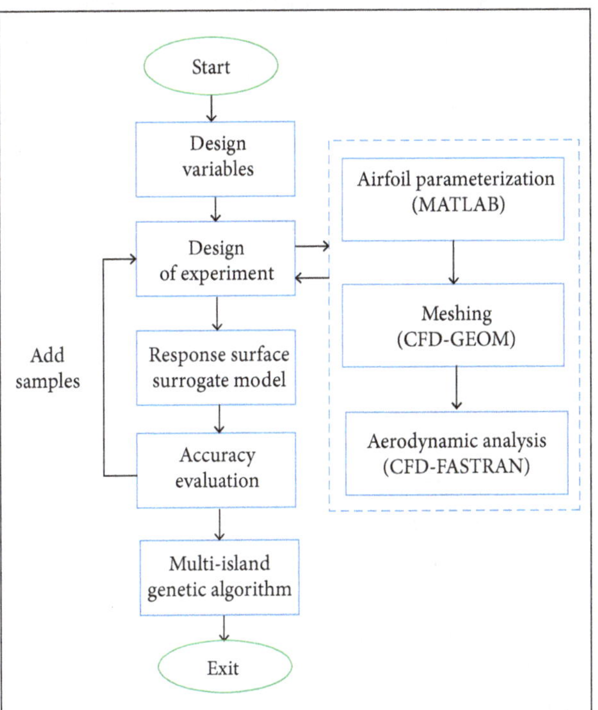

Figure 7. Designing process of optimization.

et al. 1979) was adopted for the design of experiment (DOE). Variables are normally referred to as factors in a DOE study, while the values are known as levels. With the LHS technique, the design space for each factor is uniformly divided, and then these levels are randomly combined to specify sample points defining the design matrix. It provides an efficient method for generating random sample points, which are uniformly distributed over the entire design space. For each sample, MATLAB® was used to automatically generate the corresponding airfoil profile database. Then, CFD-GEOM and CFD-FASTRAN were employed for meshing and flow field calculation, respectively; hypersonic heating environment, as well as landing performance, were also obtained.

Afterwards, surrogate models (Liem *et al.* 2015) are established based on DOE results. Specifically in this study, Response Surface Method (RSM; Park *et al.* 2009) surrogate models were used. The RSM is a statistical technique to explore the relation between design variables and responses. Low-order polynomials are usually applied to approximate the response of an actual analysis. A number of simulations, accomplished at the previous DOE stage, are required initially to construct a model. Then it can be used in optimization with a small computational cost, as only polynomial calculation is involved. For the current optimization problem, quadratic polynomial functions were adopted. Another set of random points in the design space were chosen to check these models. Surrogate models were continuously updated with additional sample points until the accuracy requirement was satisfied. Then, these RSM models were used to replace the numerical ones in the following optimization process.

During optimization, the Multi-Island Genetic Algorithm (MIGA; Wang *et al.* 2015) was employed. Genetic algorithms (GA) are widely used due to their advantage to treat complex non-linear optimizations. MIGA, a further development of GA, divides each population of individuals into several sub-populations called islands, and traditional genetic operations are performed on each island separately. Several individuals are then selected from each island and migrated to different ones periodically. The migration operation maintains the diversity of probable solutions and prevents the premature phenomena.

OPTIMIZATION RESULTS

During this optimization study, peak heat flux at the stagnation point under the hypersonic condition was regarded as the objective function, while the lift-to-drag ratio at landing condition was treated as the constraint. Coefficients of control points in Eq. 3 were taken as design variables. The optimization problem is described as:

$$\left.\begin{array}{l} \min q_{ws} \\ \text{s.t. } K \geq K_0 \end{array}\right\} \tag{7}$$

where: K and K_0 refer to the lift-to-drag ratio of the optimal and baseline airfoil, respectively.

According to the description in the sections "Space Shuttle Re-Entry Description" and "Numerical Model", 2 typical flight conditions were considered (Table 1).

Table 1. Flight conditions used for optimization.

Flight condition	Altitude (km)	Pressure (Pa)	Temperature (K)	Velocity (m/s)	α (°)
Hypersonic	74	4	210	6,700	40
Landing	0	101,325	288.2	100	15

Isight was used to integrate MATLAB® code, CFD-GEOM as well as CFD-FASTRAN to conduct DOE and the optimization process. First, 150 sample points were selected using LHS to conduct CFD analyses for both hypersonic and landing conditions, and the design space is: a_1, a_5, a_6, a_{10} are among [−0.01, 0.01], while $a_2 \sim a_4$ and $a_7 \sim a_9$ are among [−0.02, 0.02]. Then, RSM surrogate models were constructed for both objective and constraint functions. Another 20 random points were used to evaluate the accuracy of the surrogate model. Details are shown in Table 2, where RMSE stands for root mean square error. It is shown that the approximations for heat flux and lift-to-drag ratio are of high quality.

Table 2. Evaluation of the surrogate model.

Parameter	RMSE
q_w	0.03587
K	0.03534

Afterwards, the surrogate model was used to replace the previous CFD models to carry out the optimization process. The critical parameters of MIGA are: the sub-population size is 10, the number of islands is 10, the number of generations is 30, the rate of crossover and mutation are 0.9 and 0.01, respectively, the rate of migration is 0.1, and the migration interval is 5.

The optimal results are shown in Table 3, where C_d represents the drag coefficient. The characteristics of both baseline and

Table 3. Optimization results.

Parameters		Baseline	Optimal		Increment (%)	Forecast error (%)
			RSM	CFD		
Hypersonic condition	q (W/m²)	444,011	411,252	410,236	-7.61	0.25
	T (K)	1,768.7	1,734.8	1,734.0	−1.96	0.05
Landing condition	C_l	1.45276	1.46372	1.45905	0.43	0.05
	C_d	0.04244	0.04219	0.04252	0.19	−0.78
	K	34.23091	34.69353	34.31444	0.24	1.10

optimal airfoils are presented. Compared with the baseline airfoil results, the optimal one has a less severe aerodynamic heating environment at the stagnation point under hypersonic condition and maintains the lift-to-drag ratio at the same level when landing. Specifically, the peak heat flux is reduced by about 7.61%.

The normalized leading-edge profiles of the baseline and optimal airfoils are illustrated in Fig. 8, where c represents the airfoil chord length. The optimal airfoil is flatter around the stagnation point. Specifically, the radius of curvature at the stagnation point is 0.758 m for the optimal case, while it is 0.690 m for the baseline one. The result is consistent with that of Eq. 1, *i.e.*, the heat flux at the stagnation point is inversely proportional to the square root of the nose radius of the leading edge.

Figure 9 shows the heat flux distribution of the upper and lower surfaces of both airfoils. The maximum heat flux is reduced for the optimal case.

The optimized variables were also input to perform the CFD analysis. As shown in Table 3, the relative error between the RSM and CFD results is very small, which indicates that the surrogate model has a fairly good accuracy. The heat flux contours of the baseline and optimal airfoils are shown in Fig. 10, where the stagnation point positions are also marked.

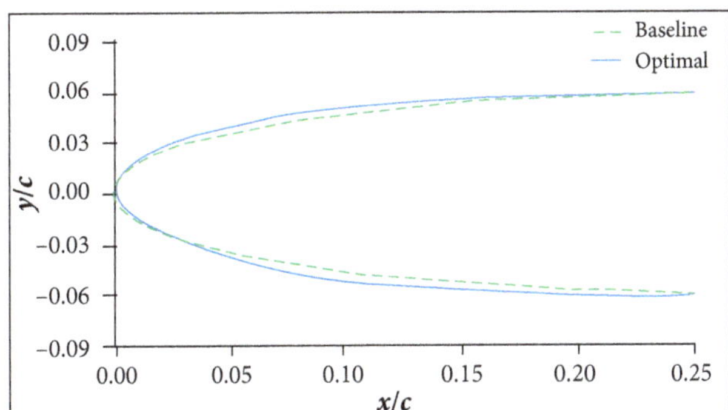

Figure 8. Shape of baseline and optimal airfoils.

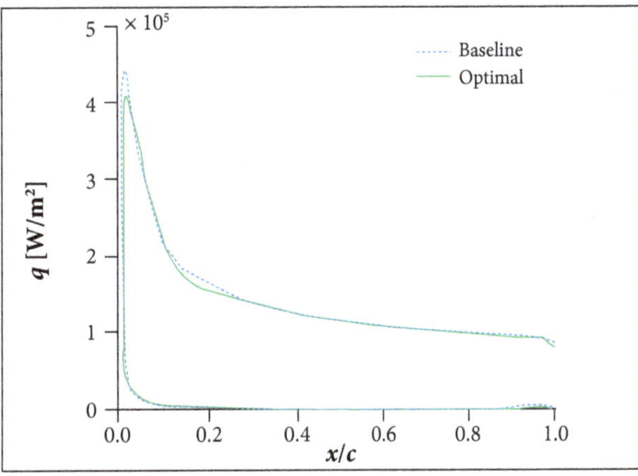

Figure 9. Heat flux distribution of baseline and optimal airfoils.

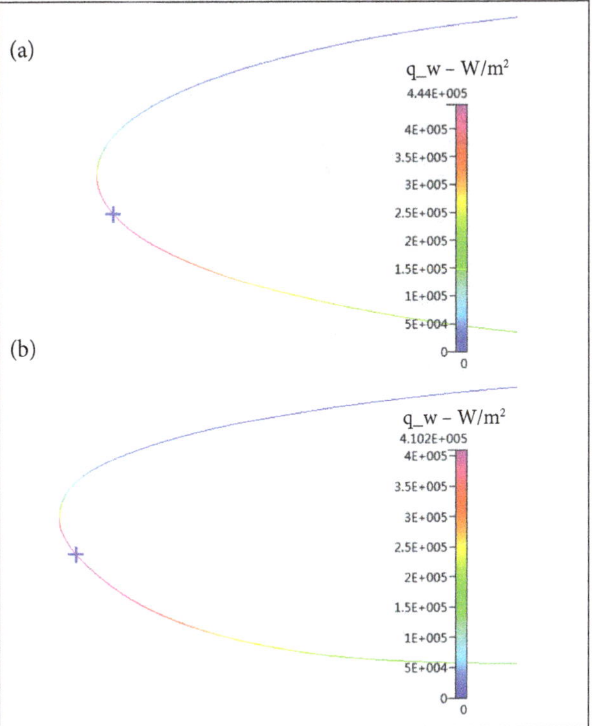

Figure 10. Heat flux contours of airfoil leading edge. (a) Baseline; (b) Optimal.

CONCLUSION

An aerothermodynamic optimization procedure considering the landing aerodynamic performance has been developed for NACA 0012 airfoil. In the optimization study, a modified Hicks-Henne type function is first adopted to parametrically describe the airfoil leading edge. CFD models are then established and further validated to simulate the hypersonic and landing problem. An optimization approach composed of DOE, RSM and MIGA is used to obtain the optimal airfoil. It is found that the surrogate model results agree well with the CFD ones. The optimal airfoil has a lower peak heat flux at the stagnation point compared with the baseline one. Meanwhile, the lift-to-drag ratio at landing condition is nearly the same as that of the baseline airfoil.

ACKNOWLEDGEMENTS

This study was supported by the Funding of Jiangsu Innovation Program for Graduate Education (Grant No. CXLX13_163), the Fundamental Research Funds for the Central Universities (Grant No. NZ2016101) and a project funded by the Priority Academic Program Development of Jiangsu Higher Education Institutions (PAPD).

AUTHOR'S CONTRIBUTION

Conceptualization, Wang Z and Kretov A; Methodology, Zhou C and Zhi J; Investigation, Zhou C and Zhi J; Writing – Original Draft, Zhou C, Zhi J and Kretov A; Writing – Review & Editing, Zhou C and Wang Z; Funding Acquisition, Wang Z; Resources, Wang Z; Supervision, Wang Z and Kretov A.

REFERENCES

Allen HJ, Eggers Jr AJ (1958) A study of the motion and aerodynamic heating of ballistic missiles entering the Earth's atmosphere at high supersonic speeds. NACA-TR-1381. Washington: NACA.

Bian Y, Zhong J (1986) Heat transfer of high temperature boundary layer. Beijing: Science Press. In Chinese.

Buckley HP, Zingg DW (2013) Approach to aerodynamic design through numerical optimization. AIAA J 51(8):1972-1981. doi: 10.2514/1.J052268

Dassault Systèmes Simulia Corp. (2012) Isight 5.7 User's Guide. Providence: Dassault Systèmes Simulia Corp.

Hicks RM, Henne PA (1978) Wing design by numerical optimization. J Aircraft 15(7):407-412. doi: 10.2514/3.58379

Koziel S, Leifsson LT (2014) Multi-objective airfoil design using variable-fidelity CFD simulations and response surface surrogates. AIAA 2014-0289. Proceedings of the 10th AIAA Multidisciplinary Design Optimization Conference; National Harbor: USA.

Ladson CL (1988) Effects of independent variation of Mach and Reynolds numbers on the low-speed aerodynamic characteristics of the NACA 0012 airfoil section. NASA-TM-4074. Washington: NASA.

Launius RD, Jenkins DR (2012) Coming home: reentry and recovery from space. Washington: National Aeronautics and Space Administration.

Li P, Zhang B, Chen Y (2012) An effective transonic airfoil optimization method using Response Surface Model (RSM). Journal of Northwestern Polytechnical University 3:395-401. In Chinese.

Liem RP, Mader CA, Martins JRRA (2015) Surrogate models and mixtures of experts in aerodynamic performance prediction for aircraft mission analysis. Aerosp Sci Technol 43:126-151. doi: 10.1016/j.ast.2015.02.019

Lu Z (2009) Aerodynamics. Beijing: Beihang University Press. In Chinese.

McKay MD, Beckman RJ, Conover WJ (1979) A comparison of three methods for selecting values of input variables in the analysis of output from a computer code. Technometrics 21(2):239-245. doi: 10.2307/1268522

Park C, Joh C, Kim Y (2009) Multidisciplinary design optimization of a structurally nonlinear aircraft wing via parametric modeling. Int J Precis Eng Man 10(2):87-96. doi: 10.1007/s12541-009-0032-1

Powers BG (1986) Space Shuttle longitudinal landing flying qualities. J Guid Contr Dynam 9(5):566-572. doi: 10.2514/3.20147

Rochelle WC, Roberts BB, D'Attorre L, Bilyk MA (1973) Shuttle orbiter re-entry flowfields at high angle of attack. J Spacecraft Rockets 10(12):783-789. doi: 10.2514/3.61969

Sellers JJ (2004) Understanding space: an introduction to astronautics. New York: McGraw-Hill.

Sun R, Chen G, Zhou C, Zhou LW, Jiang JH (2013) Multidisciplinary design optimization of adaptive wing leading edge. Sci China Technol Sci 56(7):1790-1797. doi: 10.1007/s11431-013-5250-1

Tetzman DG (2010) Simulation and optimization of spacecraft re-entry trajectories (Master's thesis). Minneapolis: University of Minnesota.

Wang YZ, Li F, Zhang X, (2015) Composite wind turbine blade aerodynamic and structural integrated design optimization based on RBF Meta-Model. Materials Science Forum 813:10-18. doi: 10.4028/www.scientific.net/MSF.813.10

Xia C, Chen W (2015) Gradient-based aerothermodynamic optimization of a hypersonic wing profile. Procedia Engineering 126:189-193. doi: 10.1016/j.proeng.2015.11.214

Yang B, Xu Q, He L, Zhao LH, Gu CG, Ren P (2015) A novel global optimization algorithm and its application to airfoil optimization. Journal of Turbomachinery 137(4):041011. doi: 10.1115/1.4028712

Zhou C, Wang Z, Zhi J (2014) Aerodynamic optimization design of adaptive airfoil leading edge based on Isight. Journal of Shanghai Jiaotong University 48(8):1122-1126. In Chinese.

Permissions

All chapters in this book were first published in JATM, by Departamento de Ciência e Tecnologia Aeroespacial; hereby published with permission under the Creative Commons Attribution License or equivalent. Every chapter published in this book has been scrutinized by our experts. Their significance has been extensively debated. The topics covered herein carry significant findings which will fuel the growth of the discipline. They may even be implemented as practical applications or may be referred to as a beginning point for another development.

The contributors of this book come from diverse backgrounds, making this book a truly international effort. This book will bring forth new frontiers with its revolutionizing research information and detailed analysis of the nascent developments around the world.

We would like to thank all the contributing authors for lending their expertise to make the book truly unique. They have played a crucial role in the development of this book. Without their invaluable contributions this book wouldn't have been possible. They have made vital efforts to compile up to date information on the varied aspects of this subject to make this book a valuable addition to the collection of many professionals and students.

This book was conceptualized with the vision of imparting up-to-date information and advanced data in this field. To ensure the same, a matchless editorial board was set up. Every individual on the board went through rigorous rounds of assessment to prove their worth. After which they invested a large part of their time researching and compiling the most relevant data for our readers.

The editorial board has been involved in producing this book since its inception. They have spent rigorous hours researching and exploring the diverse topics which have resulted in the successful publishing of this book. They have passed on their knowledge of decades through this book. To expedite this challenging task, the publisher supported the team at every step. A small team of assistant editors was also appointed to further simplify the editing procedure and attain best results for the readers.

Apart from the editorial board, the designing team has also invested a significant amount of their time in understanding the subject and creating the most relevant covers. They scrutinized every image to scout for the most suitable representation of the subject and create an appropriate cover for the book.

The publishing team has been an ardent support to the editorial, designing and production team. Their endless efforts to recruit the best for this project, has resulted in the accomplishment of this book. They are a veteran in the field of academics and their pool of knowledge is as vast as their experience in printing. Their expertise and guidance has proved useful at every step. Their uncompromising quality standards have made this book an exceptional effort. Their encouragement from time to time has been an inspiration for everyone.

The publisher and the editorial board hope that this book will prove to be a valuable piece of knowledge for researchers, students, practitioners and scholars across the globe.

List of Contributors

Roberto da Cunha Follador
Departamento de Ciência e Tecnologia Aeroespacial - Instituto de Estudos Avançados - São José dos Campos/SP - Brazil
Departamento de Ciência e Tecnologia Aeroespacial - Instituto Tecnológico de Aeronáutica - Divisão de Engenharia Mecânica - São José dos Campos/SP - Brazil

Luís Gonzaga Trabasso
Departamento de Ciência e Tecnologia Aeroespacial - Instituto Tecnológico de Aeronáutica - Divisão de Engenharia Mecânica - São José dos Campos/SP - Brazil

Zhen Rong
Zhejiang University - School of Aeronautics and Astronautics - Hangzhou/Zhejiang - China
Ministry of Education - Beihang University - Key Laboratory of Fluid
Mechanics - Beijing/China

Xueying Deng, Baofeng Ma and Bing Wang
Ministry of Education - Beihang University - Key Laboratory of Fluid Mechanics - Beijing/China

Arnaldo Forgas Júnior and Jorge Otubo
Departamento de Ciência e Tecnologia Aeroespacial - Instituto Tecnológico de Aeronáutica - Divisão de Engenharia Aeronáutica e Mecânica - São José dos Campos/SP - Brazil

Rodrigo Magnabosco
Fundação Educacional Inaciana - São Bernardo do Campo/SP - Brazil

Seyedeh Nasrin Hosseini and Seyed Mohammad Hossein Karimian
Amirkabir University of Technology - Aerospace Engineering Department - Tehran/Tehran - Iran

Puran Singh
Amity School of Engineering & Technology - Mechanical & Automation Engineering Department - FEM Laboratory - New Delhi/Delhi - India

Debashis Pramanik and Ran Vijay Singh
Manav Rachna International University - Faculty of Engineering Technology - Mechanical Engineering Department - Faridabad/Haryana - India

Henrique Gazzetta Junior
Empresa Brasileira de Aeronáutica - São José dos Campos/SP - Brazil

Cleverson Bringhenti, João Roberto Barbosa and Jesuíno Takachi Tomita
Departamento de Ciência e Tecnologia Aeroespacial - Instituto Tecnológico de
Aeronáutica - Divisão de Engenharia Aeronáutica e Mecânica - São José dos Campos/SP - Brazil

Michail Zhelamskij
Polytechnical University - Department of Measurement and Technologies - Saint Petersburg - Russia

Ronald Izidoro Reis and Luiz Claudio Pardini
Departamento de Ciência e Tecnologia Aeroespacial - Instituto de Aeronáutica e Espaço - Divisão de Materiais - São José dos Campos/SP - Brazil

Wilson Kiyoshi Shimote
Departamento de Ciência e Tecnologia Aeroespacial - Instituto de Aeronáutica e Espaço - Divisão de Propulsão Espacial - São José dos Campos/SP - Brazil

Carlos Breviglieri
Departamento de Ciência e Tecnologia Aeroespacial - Instituto Tecnológico de Aeronáutica - Divisão de Ciências da Computação - São José dos Campos/SP - Brazil.

João Luiz F Azevedo
Departamento de Ciência e Tecnologia Aeroespacial - Instituto de Aeronáutica e Espaço - Divisão de Aerodinâmica - São José dos Campos/SP - Brazil

Alan Fernando Ney Boss
Departamento de Ciência e Tecnologia Aeroespacial - Instituto Tecnológico de Aeronáutica - Ciências e Tecnologias Espaciais - São José dos Campos/SP - Brazil

Antonio Carlos da Cunha Migliano
Departamento de Ciência e Tecnologia Aeroespacial - Instituto Tecnológico de Aeronáutica - Ciências e Tecnologias Espaciais - São José dos Campos/SP - Brazil
Departamento de Ciência e Tecnologia Aeroespacial - Instituto de Estudos Avançados - Divisão de Física Aplicada - São José dos Campos/SP - Brazil

Ingrid Wilke
Rensselear Polytechnic Institute - Department of Physics, Applied Physics and Astronomy - Terahertz and Ultrafast Spectroscopy Laboratory - Troy/NY - USA

Mohsen Dehghani Mohammad-abadi and Seyed Hamid Jalali-Naini
Tarbiat Modares University - Faculty of Mechanical Engineering - Aerospace Group - Tehran/Tehran - Iran

Melis De Bruyn Neto, Koshun Iha and José Atílio Fritz Fidel Rocco
Departamento de Ciência e Tecnologia Aeroespacial - Instituto Tecnológico de Aeronáutica - Divisão de Ciências Fundamentais - São José dos Campos/SP - Brazil

Rita de Cássia Mendonça Sales
Departamento de Ciência e Tecnologia Aeroespacial - Instituto Tecnológico de Aeronáutica - Divisão de Ciências Fundamentais - São José dos Campos/SP - Brazil
Centro Estadual de Educação Tecnológica Paula Souza - Faculdade de Tecnologia de São José dos Campos - Prof. Jessen Vidal - São José dos Campos/SP - Brazil

Hossein Bonyan Khamseh
Universidade Federal de Minas Gerais - Escola de Engenharia Elétrica - Departamento de Engenharia Elétrica - Belo Horizonte/MG - Brazil

Anwar Ul Haque, Waqar Asrar and Erwin Suleiman
International Islamic University Malaysia - Department of Mechanical Engineering - Experimental and Computational Thermofluid Mechanic Research Group - Kuala Lumpur - Malaysia

Asharf Ali Omar
University of Tripoli - Department of Aeronautical Engineering - Tripoli - Libya

Hamid Reza Talesh Bahrami and Hamid Saffari
Iran University of Science and Technology - School of Mechanical Engineering - Liquefied Natural Gas Research Laboratory - Tehran/Tehran - Iran

Jeffrey Newcamp, Wim Verhagen, Heiko Udluft and Richard Curran
Delft University of Technology - Faculty of Aerospace Engineering - Air Transport and Operations Section - Delft/South Holland - Netherlands

Huadong Sun, Jianqiao Yu and Siyu Zhang
Beijing Institute of Technology - School of Aerospace Engineering - Beijing - China

Chen Zhou, Zhijin Wang, Jiaoyang Zhi and Anatolii Kretov
Nanjing University of Aeronautics and Astronautics - College of Aerospace Engineering - Minister Key Discipline Laboratory of Advanced Design Technology of Aircraft -Nanjing/Jiangsu - China

Index

www.ingramcontent.com/pod-product-compliance
Lightning Source LLC
Chambersburg PA
CBHW082015190326
41458CB00010B/3199

* 9 7 8 1 6 3 2 3 8 6 6 3 2 *